Molecular Neurobiology of Alzheimer Disease and Related Disorders

To Megumi Takeda (September 12, 1957 – February 4, 2002)

Molecular Neurobiology of Alzheimer Disease and Related Disorders

Editors

Masatoshi Takeda *Osaka*
Toshihisa Tanaka *Osaka*
Ramón Cacabelos *Coruña*

140 figures, 72 in color, and 18 tables, 2004

KARGER

Basel · Freiburg · Paris · London · New York ·
Bangalore · Bangkok · Singapore · Tokyo · Sydney

Prof. Masatoshi Takeda
Department of Psychiatry and
Behavioral Proteomics
Osaka University
Graduate School of Medicine
Osaka, Japan

Dr. Toshihisa Tanaka
Department of Psychiatry and
Behavioral Proteomics
Osaka University
Graduate School of Medicine
Osaka, Japan

Prof. Ramón Cacabelos
EuroEspes Biomedical Research Center
Institute for CNS Disorders
Bergondo, Coruña

Library of Congress Cataloging-in-Publication Data

Molecular neurobiology of Alzheimer disease and related disorders / editors, Masatoshi Takeda, Toshihisa Tanaka, Ramón Cacabelos
p. ; cm.
Includes bibliographical references and index.
ISBN 3–8055–7603–X (hard cover)
1. Alzheimer's disease–Molecular aspects. 2. Molecular neurobiology. I. Takeda, Masatoshi, 1949-. II. Tanaka, Toshihisa. III. Cacabelos, Ramón.
[DNLM: 1. Alzheimer Disease–metabolism. 2. Alzheimer Disease–physiopathology. 3. Alzheimer Disease–genetics. 4. Neurobiology. WT 155 M7183 2003]
RC523.M663 2003
616.8′3107–dc22

2003055886

Bibliographic Indices. This publication is listed in bibliographic services, including Current Contents® and Index Medicus.

Drug Dosage. The authors and the publisher have exerted every effort to ensure that drug selection and dosage set forth in this text are in accord with current recommendations and practice at the time of publication. However, in view of ongoing research, changes in government regulations, and the constant flow of information relating to drug therapy and drug reactions, the reader is urged to check the package insert for each drug for any change in indications and dosage and for added warnings and precautions. This is particularly important when the recommended agent is a new and/or infrequently employed drug.

www.karger.com
Printed in Switzerland on acid-free paper by Reinhardt Druck, Basel
ISBN 3–8055–7603–X

Contents

Foreword

The dawn of psychogeriatrics in Japan was celebrated with the symposium entitled 'Psychiatry for the Elderly' in the frame of the annual meeting of the Japanese Society of Psychiatry and Neurology in 1954, on which occasion Professor Ziro Kaneko (Osaka University), Professor Tadashi Inose (Yokohama City University), and Professor Naotake Shinfuku (Tottori University) delivered their lectures on the psychological process of aging, neuropathology of aging and psychopathology of aging, respectively. The proceedings of the symposium entitled *The Psychiatric Aspects of Senility* (Igagu-shoin, Tokyo, 1956) were published as a monograph in Japanese; this was an epoch-making achievement in Japanese psychiatry because the interest in psychogeriatrics had been so sparse until then.

In the 1960s, dementia in Japanese elderly people was mainly regarded to be cerebrovascular dementia. Most Alzheimer's disease patients were unrecognized and there were only a few case reports of early-onset Alzheimer's disease. In those days, basic research in Alzheimer's disease was confined to neuropathology or histochemistry. Electron microscopy, however, revealed the unique structure of paired helical filaments in Alzheimer brains, which triggered biochemical research aimed at elucidating the mechanism of paired helical filament formation. My colleagues at the Department of Neuropsychiatry, Osaka University and I found that soluble proteins were insolubilized in Alzheimer brains, which was reported at the International Meeting of Neuropathology in Budapest in 1974. This report, which attracted considerable interest and stimulated neurochemical research on the dementia brain in several leading institutes, implied that neurochemical or biochemical research could be successfully applied to elucidate the pathogenesis of Alzheimer's disease. I am proud of this contribution of the Department of Neuropsychiatry, Osaka University which I chaired at that time and I am happy to observe the strong trend of psychogeriatric research launched by Professor J. Kaneko, as mentioned above, and pursued under the leadership of the present chairman, Professor M. Takeda.

Most of us would agree with the recognition that research activity in the Department of Neuropsychiatry, Osaka University, has played an important role in Alzheimer research and the Department achieved a solid reputation as one of the leading research institutes in psychogeriatrics.

The success of the 21st Annual Meeting of the Japanese Dementia Study Society and the International Symposium on Neurobiology of Alzheimer's Disease and Related Disorders in October 2002 appears to be additional evidence for this. The International Symposium, especially, had an impact on Alzheimer research in this country, gathering many scientists from major research institutes in Japan and abroad to exchange their research findings. I would say the program of the symposium was well suited to stimulate young researchers in this field.

This monograph contains selected papers presented at the symposium and, just like the monograph The Psychiatric Aspects of Senility published half a century ago, will certainly contribute to promote scientific research in this field. It is my pleasure to write a foreword to this book and I would like to congratulate this collaborative achievement of Professors Masatoshi Takeda, Toshihisa Tanaka, and Ramón Cacabelos, who dedicated their time to this monograph, cultivating the long tradition of research of the Department of Neuropsychiatry, Osaka University.

Prof. emeritus of Osaka University Tsuyoshi Nishimura, Osaka

Preface

The average life expectancy of human beings had remained essentially unchanged since ancient times until the end of the 18th century, and it was only in those two centuries that our life expectancy increased. Since then, from an aging society (the elderly exceeding 7%), our society has become an aged society (the 65-year-olds and over exceeding 14% of the total population). In Japan this has happened in 2000; thus Japan has transited from an aging to an aged society in only 24 years, which is the most rapid transition in the world – almost four times faster than many European countries. The last national census of Japan reported that the average life expectancy of the Japanese was 85.23 years for females and 78.32 years for males in the year 2002. The Japanese now enjoy the longest average life expectancy, whereas in 1947 it was only 53.96 years for females and 50.06 years for males. Due to this rapid extension of life expectancy, Japanese society is now facing strains and problems related to its high proportion of elderly people (17%), and its very high percentage (7%) of very old people (above 75 years old).

In many European countries, the increase in the elderly population has already brought about some changes and modifications in the social life system, but there are still many things to be implemented to build a new society in which people can lead mutually cooperative lives regardless of their biological age. In the 21st century, the elderly population will increase all over the world because developing countries are showing a more rapid increase in the elderly population at the present time. By the year 2025, 70% of the elderly will live in developing countries, and by the year 2050, 80% of the elderly will be found in present developing countries. These facts indicate that the rapid increase in the elderly population is a global social problem to be solved by taking advantage of information and experience from all countries. In a sense, Japan is the top runner in terms of society aging, and the Japanese experience may serve as an example to younger societies in other countries.

Alzheimer's disease is the most malignant disease in aged societies. It affects 6–10% of the elderly population, causing impairment in cognitive functions and significant disability in daily living for more than 10 years. In Japan it affects 750,000 individuals, and by the year 2035, this number will have increased to 1.5 million.

Neurofibrillary tangles, amyloid deposits and neuronal loss are the three hallmarks of Alzheimer's disease. Neurofibrillary tangles and amyloid plaques are insoluble depositions with unique structural characteristics, abundantly observed in Alzheimer brains and to some extent in normal aged brains. Due to the insolubility of these unique structures in Alzheimer brain tissues, they were difficult to study by usual biochemical methods in the past. In 1980s, owing to the use of a solubilization method with formic acid or perchloric acid, the neurobiological study of Alzheimer's disease made significant progress. The major neurobiological findings include partial identification of the amino acid sequence of amyloid precursor protein (APP) (1984), identification of amyloid precursor protein gene on chromosome 21 (1987), detection of mutations in APP with familial Alzheimer's disease (1991), identification of apolipoprotein E4 as a significant risk (1993), discovery of presenilin-1 and presenilin-2 (1994). Some neurobiological research outcomes have been applied in the clinical treatment of patients with Alzheimer's disease. Acetylcholine esterase inhibitors are now widely used to treat Alzheimer patients. Tau and beta-amyloid protein levels can be useful as biological diagnostic markers of Alzheimer's disease.

Active research is going on, aiming to elucidate the pathogenesis of Alzheimer's disease. Major topics of neurobiological study of Alzheimer's disease include the unraveling of the molecular mechanisms of neurofibrillary tangle formation in neuronal and glial cells; the molecular processing of amyloid precursor protein in intracellular organelles and in extracellular space, and the molecular mechanism of neuronal loss. In this book, these major topics are covered by leading scientists in the field of neurobiology of Alzheimer's disease.

Alzheimer's disease attracted researchers from diverse academic fields, including clinical, basic and social sciences. It is essential to promote the understanding of this formidable disease and to share new findings among researchers in the field.

Clinical and basic research of Alzheimer's disease has been the main interest of the Department of Psychiatry and Behavioral Proteomics,

Osaka University Graduate School of Medicine, since the time of Professor Jiro Kaneko and Professor Tsuyoshi Nishimura, and it has been our great pleasure to compile this book as a mile stone of the activity in our Department.

In October 2002, the Department of Psychiatry and Behavioral Proteomics, Osaka University Graduate School of Medicine organized the 21st Annual Meeting of the Dementia Study Academy of Japan, in which more than 400 researchers in this field got together to discuss their research progress in clinical and basic fields of dementia study. The three-day meeting program included two official symposia entitled 'Neurobiology of Amyloid and Presenilins' and 'Neurodegeneration Mechanism with Tau, Syneclein and Neurofilaments', two satellite symposia entitled 'Early Diagnosis of Alzheimer's Disease' and 'Treatment of Alzheimer's Disease', four luncheon seminars of 'Strategy for BPSD', 'Neuroimaging of Dementia', 'Normal Pressure Hydrocephalus', and 'Treatment of Vascular Dementia', in addition to 85 general presentations.

In conjunction with the 21st Annual Meeting of the Dementia Study Academy of Japan, we organized an International Symposium on the 'Molecular Neurobiology of Alzheimer Disease and Related Disorders'. The articles in this book were selected from papers presented at this two-day International Symposium, which was very successful, with the participation of eight leading scientists from the USA, Canada and Europe. They are: Dr. Greg M. Cole (University of California), Dr. Khalid Iqbal (New York State Institute for Basic Research), Dr. Peter St. George-Hyslop (University of Toronto), Dr. Konrad Beyreuther (University of Heidelberg), Dr. Ramon Cacabelos (EuroEspes Biomedical Research Center), Dr. Harold Hampel (University of Munich), Dr. Inge Grundke-Iqbal (New York State Institute for Basic Research), and Dr. Roger M. Nitsch (University of Zurich).

We were very happy to host leading scientists from all over the world and are thankful to the speakers, and especially to the authors of the articles of this book, which, we believe, will be useful not only to basic scientists but also to clinicians interested in Alzheimer's disease and related disorders.

Masatoshi Takeda, MD, PhD
Toshihisa Tanaka, MD, PhD
Ramón Cacabelos, MD, PhD

October 5, 2002 First day at Hotel Osaka Sun Palace

October 5, 2002
(First row) from left to right
Ramón Cacabelos, Khalid Iqbal, Roger Nitsch, Inge Grundke-Iqbal, Masatoshi Takeda, Konrad Bayreuther, Greg Cole, Peter St. George-Hyslop, Harald Hampel
(Second row)
Takashi Kudo, Tetsuaki Arai, Katsuya Urakami, Katsuhiko Yanagisawa, Nobuo Yanagusawa, Takeshi Tabira, Hiroshi Mori, Tomohiro Miyasaka, Yoshitaka Tatebayashi, Toshihisa Tanaka
(Third row)
Kouzin Kamino, Masaki Nishimura, Taisuke Tomita, Taiichi Katayama, Kazunori Imaizumi, Hideo Kimura , Shun Shimohama, Hisahi Tanii, Akihiko Nunomura, Masayasu Okochi

October 6, 2002 Second day at Hotel Osaka Sun Palace

October 6, 2002

(First row) from left to right

Ramón Cacabelos, Harald Hampel, Khalid Iqbal, Inge Grundke-Iqbal, Konrad Bayreuther, Yasuo Ihara, Masatoshi Takeda, Roger Nitsch, Greg Cole

(Second row)

Takashi Kudo, Toshihisa Tanaka, Katsuhiko Yanagisawa, Akihiko Takashima, Takeshi Ishihara, Takeshi Tabira, Akihiko Nunomura, Tetsuaki Arai

(Third row)

Yoshitaka Tatebayashi, Taiichi Katayama, Hiroshi Mori, Masaki Nishimura, Tomohiro Miyasaka,Taizo Taniguchi, Kazunori Imaizumi, Masayasu Okochi

Takeda M, Tanaka T, Cacabelos R (eds): Molecular Neurobiology of Alzheimer Disease and Related Disorders. Basel, Karger, 2004, pp 1–16

Methods of Regulating Alzheimer Pathogenesis: Diet, Oxidative Damage and Inflammation

Greg M. Cole, Takashi Morihara, Giselle P. Lim, Frederic Calon, Bruce Teter, Fusheng Yang, Sally A. Frautschy

VA Medical Center, GRECC, Greater Los Angeles Healthcare System, Sepulveda, Calif., USA

The Amyloid Cascade in Alzheimer's Disease – Inflammation, Oxidative Damage and Synapse Loss

While the most obvious lesions diagnostic of Alzheimer's disease (AD) are extracellular β-amyloid plaques containing large deposits of aggregated, fibrillar amyloid-β protein (Aβ) and neurofibrillary tangles of filamentous 'hyperphosphorylated' τ-protein, synapse loss appears to be a better and more proximal correlate of cognitive deficits. AD brain has significant but very selective neuron loss with a 20–40% loss in cortical synaptophysin or presynaptic terminals that progresses throughout the disease and correlates with the clinical decline. Postsynaptic defects have been less well studied, even though the loss of the postsynaptic markers neurogranin and drebrin has been reported to be a much larger 70–80% [1].

While amyloid deposits are widely distributed in AD, neuron loss is much more limited and better correlated with neurofibrillary tangles in selected cortical and hippocampal layers; this has led to a longstanding debate on the significance of tangles versus amyloid. The view that tangles and tau pathology are the critical events has received new support from the discovery of tau mutations in a subset of frontal temporal dementia (FTD) cases [2]. However, data from transgenics expressing normal human tau genes have not resolved the

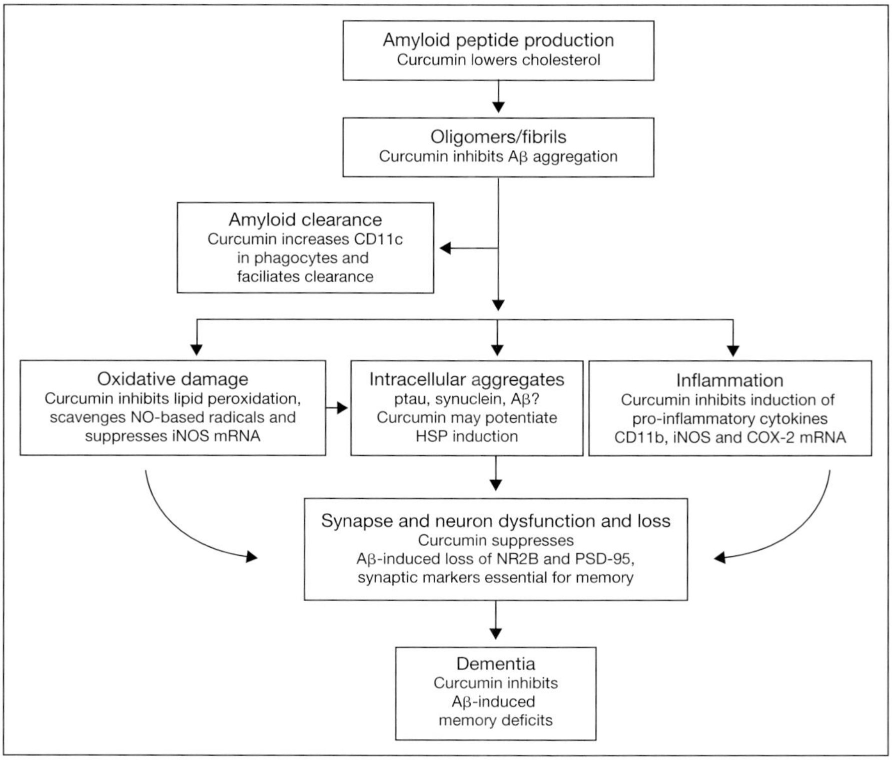

Fig. 1. The amyloid cascade and curcumin intervention.

controversy since they lack both neuron loss and tangles, even when co-expressed with amyloid-β precursor protein (APP) mutations. Instead, in animal models, only the mutant human tau (FTD mutations) is linked to tangle formation [3, 4], and it is still unresolved whether these mice have neurodegeneration equivalent to AD brain.

Because Aβ deposits occur early in AD and Down's syndrome and increased Aβ42 amino acid peptide production is a direct consequence of numerous mutations known to cause AD, amyloid peptides are reasonably hypothesized to play a causal role in the disease [5]. Additional support for this hypothesis has come from evidence that amyloid peptides under aggregating conditions can be acutely toxic to cultured neurons from rodents or humans [6]. In essence, the amyloid hypothesis suggests that elevated Aβ42 leads to excessive accumulation of amyloid fibrils or oligomer intermediates that in turn cause neurotoxicity, both directly as well as via effects on inflammation, oxidative damage and secondary accumulation of intraneuronal phosphotau and synuclein (fig. 1).

A challenge for the amyloid cascade hypothesis has been that transgenic mice that develop abundant amyloid pathology, including neuritic plaques, fail to show tangles or significant neuron loss [7, 8]. At least some of the APP transgenic mice have ~20–30% focal synaptophysin loss, but this has not been clearly related to memory deficits which typically appear early in the pathogenesis [9–11].

APP transgenic mice clearly model some amyloid cascade events, particularly amyloid deposition and neuritic plaque formation. Importantly, they have a microglial/inflammatory response [12], elevated oxidative damage [13, 14], clear but limited synaptotoxicity [7, 8, 11] and cognitive deficits related to the Aβ [15–18]. They are therefore a useful model for testing the impact of potentially disease-modifying environmental risk factors and drugs directed at the amyloid cascade.

Protective Factors Reducing the Risk of Alzheimer's Disease in Epidemiological Studies

The identification of autosomal dominant genetic risk factors has been critical in producing reasonable causal pathways and a hope for new drugs for treatment. The majority of AD cases are not early onset and autosomal dominant, but are late onset, probably related, in part, to incompletely penetrant genes, for example, the relatively potent risk conferred by APOE4. That is, these weak genetic risk factors, do not invariably cause AD by themselves but require the presence of other risk factors. The most certain of these is aging, which appears to result in limited AD-like pathology even in nondemented humans, dogs, some primates and various other mammals, but not in rodents. While aging is one of the most important risk factors, aging alone is not sufficient, nor is amyloid pathology. AD likely arises from the interaction between genetic and environmental risk factors, which may offer immediately available opportunities to reduce disease risk. For minimum risk of toxicity and maximum public health impact, prevention should focus on evaluation of inexpensive agents with a long history of use.

Nonsteroidal Anti-Inflammatory Drugs

Perhaps the best established protective factor for AD is chronic use of nonsteroidal anti-inflammatory drugs (NSAIDs) [19, 20], which is consistent with AD-associated brain inflammation. Amyloid peptide aggregates interact with multiple microglial receptors and directly cause activation with attendant increases in toxic cytokines, superoxide and nitric oxide production [21]. Aβ can also directly activate complement pathways [22, 23]. In addition to Aβ, neuronal injury can activate a secondary inflammatory response. Based on extensive evidence for an inflammatory response in AD brain involving microglial and complement activation and cytokine cascades [20, 24], epidemiological methods were employed to search for evidence of protection afforded by NSAID use that would support an important causal role for inflammation. Reduced risk was found not only in arthritis sufferers taking chronic high NSAID doses, but also in community-based studies where relatively safe and lower, 'analgesic' doses of over-the-counter NSAIDs like ibuprofen and naproxen were frequently used [20, 24, 25]. In a study of twins and sib pairs, NSAID usage, most commonly ibuprofen, was associated with both reduced risk and delayed onset sufficient to account for reduced risk [26].

Antioxidants

Increased oxidative damage to proteins, DNA, RNA and lipids occurs in AD compared to control brains [27] and this damage occurs early [28]. Aβ aggregates cause oxidative damage to neuronal cultures by elevating hydrogen peroxide production [29], binding metals [30], forming peptide radicals [31] or inducing microglial activation [32]. Antioxidants can protect against Aβ toxicity, suggesting they might slow pathogenesis [33]. In fact, diets rich in antioxidants [34] and vitamin E in particular [35, 36] appear to reduce AD risk. Unlike NSAIDs, which appear to affect only disease risk, antioxidants may also have an impact on progression [36]. However, a clinical trial found that vitamin E had only a modest (but significant) effect in reducing AD progression [37].

Statins

Cholesterol has been associated with increased amyloid in the brain [38] and increased Aβ production [39]. Epidemiological data found increased dietary fat and cholesterol associated with increased AD risk [40], but follow-up research on the Rotterdam cohort has not confirmed this result [41]. Consistent with a causal role for cholesterol, several epidemiological studies associated the use of cholesterol-lowering statins with large reductions in AD risk [42–44]. Although outside the scope of this review, dietary cholesterol increases and statins reduce amyloid accumulation in APP transgenic mice [45, 46].

Dietary Fish and n–3 Fatty Acids

Polyunsaturated fatty acids are important targets for oxidative damage because they generate lipoxyl radicals as products resulting in autocatalytic lipid peroxidation with additional aldehyde products that attack macromolecules. While proteins and nucleic acids are typically considered the ultimate targets, two of the most peroxidizable fatty acids, arachidonic acid [C20:4 (n–6), AA] and docosahexaenoic acid [C22:6 (n–3), DHA], play important functional roles. AA is the major substrate for enzymatic oxidation by the cyclooxygenase and lipoxygenase pathways. DHA is concentrated in synapses and a potential ligand for the PPAR and retinoid receptor (RXR, LXR) transcription factors. DHA also modulates membrane fluidity, membrane enzymes, G protein and channel activities [47]. Precursors in the n–6 series beginning with linoleic acid are used to synthesize AA and precursors in the n–3 series beginning with linolenic acid are used to synthesize DHA. Because of the precursor pathways and their regulation, the absolute dietary levels of AA or DHA are less important than the ratio of n–6/n–3 fatty acids in regulating inflammatory and cardiovascular pathways. A dietary intake n–6/n–3 fatty acid ratio of about 4:1 has been considered fairly optimal. WHO recommends between 3:1 and 4:1 while the Japanese Society for Lipid Nutrition recommends 2:1 [48]. Elevated n–6/n–3 dietary fat increases AD risk [49, 50]. Japanese who move to Brazil and eat more meat and less fish have higher overall dementia and AD rates [51]. High levels of fish consumption have been associated with reduced risk of age-related cognitive decline [52] and dementia, including AD [40].

Low DHA has also been associated with AD risk in the US. Those in the bottom 50% of serum DHA levels in the Framingham study had increased AD risk of 67%, and in those that also had E4, risk of low scores on the Mini Mental State Examination rose 400% [53]. Similarly, low dietary intake and low blood levels of DHA appear to increase risk for AD [54]. However, other studies have been negative. In our view, it is not only dietary intake deficiency, but focal oxidation leading to local deficiency that is likely to be important in AD. Global measures of nonenzymatic oxidation of both AA and DHA to isoprostanes are clearly increased in AD [55–59].

In conclusion, there is both an epidemiological literature and a rationale for AD risk reduction with NSAIDs, antioxidants, statins and n–3 fatty acids. Evidence that any of these agents can suppress or delay AD pathology in rodent models indicates they have a very good chance of working in humans where they have already been associated with reduced AD risk.

Testing Epidemiological Risk Factors in Animal Models

Amyloid Cascade Interventions – Nonsteroidal Anti-Inflammatory Drugs

Because the strongest epidemiological support for a single protective NSAID has been for ibuprofen, our first choice was to test the effect of ibuprofen in the Tg2576 HuAPPsw line where plaque formation begins at ~10 months and robust plaque-associated microgliosis was evident by 16 months of age [60].

Chronic treatment (from 10 to 16 months old) with 375 ppm ibuprofen *reduced* plaque burden and levels of sodium dodecyl sulfate-insoluble, formic acid extracted, total Aβ measured with ELISA by 40–50%. Interleukin-1β and GFAP protein levels were significantly reduced. Ibuprofen also reduced microglia-stained area per plaque, a marker of dystrophic neurites (ubiquitin), and caspase activation per plaque [61], suggesting that it had an impact not only on amyloid, but on the peri-plaque response to amyloid as well. Other studies have also found that ibuprofen or other NSAIDs can limit amyloid accumulation [62], raising the issue of the mechanism of amyloid reduction. Ibuprofen can prevent or reverse age-related cognitive deficits (water maze) in APPsw mice [Hsiao-Ashe et al., unpubl. obs.]. These studies demonstrate a potential for ibuprofen as an effective intervention.

Ibuprofen and Amyloid Reduction

In our initial efforts to determine a mechanism for the impact of ibuprofen on amyloid we ruled out effects on APP levels [61]. However, it is important to note that because the transgene is driven by the prion promoter, several possible NSAID effects on APP expression will be absent in the transgenic model but might be significant in AD patients. Amyloid may be cleared by microglial phagocytosis via scavenger receptors and ibuprofen could stimulate this by increasing CD36 receptors via PPAR-γ. Alternatively, amyloid-activated complement pathways, leading to C3b/iC3b opsonization could also enhance clearance through CD11 receptors [63]. However, ibuprofen had no effect on the level of expression of CD36, C1q or CD11 [Morihara et al., unpubl. obs.]. In summary, we found no evidence to support an effect of ibuprofen on amyloid clearance.

Evidence was recently discovered, indicating a novel COX-independent mechanism of action for selected NSAIDs, including ibuprofen, via reducing Aβ42 production [64]. We confirmed in vitro that Aβ42 production in HEK293 cells expressing APPsw was selectively reduced by ibuprofen, and was also reduced by profen *R*-enantiomers (*R*-ibuprofen and *R*-flurbiprofen) with weak COX inhibitory activity [65]. These results are entirely consistent with data from Koo and Golde [64], suggesting that Aβ42 reduction does not require COX inhibition. *R*-flurbiprofen or related compounds with little or no COX inhibitor activity have the potential to be used at high enough doses to chronically limit Aβ42 production in the absence of significant side-effects. Whether ibuprofen itself can be used for this purpose is not yet clear. The in vitro doses required for this are readily attained in plasma, but may be difficult to achieve in brain, suggesting that the Aβ42-lowering activity of ibuprofen will be restricted by dose-limiting toxicity. Nevertheless, in vivo evidence for selective Aβ42 reduction by NSAIDs [64] argues that an effect on gamma secretase activity may slow amyloid accumulation. It is also possible that amyloid suppression in vivo may involve other, more traditional NSAID targets related to COX inhibition and control of the inflammatory response in glia.

One possibility we have examined is suppression of the astrocyte-derived, pro-amyloidogenic proteins α_1-antichymotrypsin (ACT) and apolipoprotein E (ApoE). Both of these factors bind Aβ in AD brain and regulate amyloid formation in vitro and in vivo [66, 67]. We have observed significant reductions in mRNA for ApoE in ibuprofen-treated transgene-negative animals [Teter et al., unpubl. obs.] and in a murine homologue for ACT in ibuprofen-treated transgene-positive animals [Morihara et al., submitted]. These actions may require only relatively low, COX-inhibitory doses, and may be shared with naproxen and other NSAIDs that do not reduce Aβ42 production, but which appear to reduce AD risk.

Amyloid Cascade Interventions – Antioxidants

Because high dose vitamin E treatment had a modest impact on AD progression, we sought to test a potentially more potent antioxidant intervention and set up a screen of potential treatments using a rat CNS Aβ-infusion model [68, 69]. We chose to test curcumin, a well-studied and purified compound from the turmeric spice that is a potent antioxidant and scavenger of OH, O_2^- and NO radicals. Curcumin inhibits brain lipid peroxidation (associated with τ-aggregation) [70] 5–10 times better than α tocopherol (vitamin E) and is more effective in scavenging NO-based radicals [71] associated with α-synuclein pathology [72]. In addition to curcumin itself, whose CNS bioavailability is limited by glucoronidation, metabolites including tetrahydrocurcumin, ferulic acid and vanillin are also potent antioxidants that likely contribute to in vivo antioxidant activity [73]. Unlike vitamin E, chronic administration of the major tetrahydrocurcumin metabolite from 13 months of age can significantly extend both mean and maximum life span in male C57Bl/6 mice [74]. Curcumin is a novel anti-inflammatory that controls inflammation by inhibiting AP-1- and NFκB-driven expression of cytokines [75], iNOS [76] and Cox-2 [77]. It is tolerated well at chronic high doses and was screened by the US National Toxicology Program prior to making the short list of compounds under consideration for cancer chemoprevention by the US National Cancer Institute [78]. Its safety at active doses is indicated by the long history of use as a turmeric extract for multiple indications in traditional Indian (Ayurvedic) and Chinese medicine for thousands of years, notably for promoting wound healing and control of inflammation. Like aspirin, which was also discovered as a traditional anti-inflammatory medical extract, curcumin has more than one beneficial effect. These data suggested curcumin and related species (curcuminoids) and metabolites might afford greater protection than vitamin E by controlling inflammation and oxidative damage and promoting CNS lesion 'healing'.

Because of the strong preclinical safety and broad-spectrum efficacy data and identified structure, we chose to first test curcumin in a rat Aβ infusion model [69]. This model is also being used to test the efficacy of steroids [79], vitamin E and other antioxidants and antioxidant cocktails [Frautschy et al., unpubl. obs.]. Results indicated that dietary curcumin at 2,000 ppm was a potent anti-Aβ compound in vivo. More detailed follow-up studies in the Aβ infusion model showed 500 ppm dietary curcumin reduced lipid peroxidation (F2-isoprostanes), reduced Aβ deposition by 80% and prevented post-synaptic marker (NR2B, PSD-95) loss and Aβ-induced cognitive deficits in acquisition in the Morris water maze [69]. While there was a reduction in plaque-independent microglia, consistent with an anti-inflammatory activity, we also saw an increase in the microglial response to the diffuse plaques. We then sought to confirm and extend these results in the Tg2576 APPsw mouse, using the same 10–16 months of age treatment protocol used for the ibuprofen study. In this model, 160 ppm dietary curcumin reduced oxidized protein (measured as carbonyls) by 50–70%, interleukin-1β by 57% and Aβ burden and Aβ levels (by ELISA) by 43–50% [13]. Similar to the results in the rat infusion model, microgliosis was also inhibited by 33% in neuron layers, but microgliosis was stimulated by 250% adjacent to plaques.

Amyloid Cascade Interventions – Mechanism of Curcumin Inhibition of β-Amyloidosis

The reduction in Aβ accumulation by curcumin observed in both mouse and rat models might be due to reductions in Aβ production or aggregation, or to an increase in clearance. Curcumin did not reduce total Aβ or Aβ42 production in HEK293 cells in vitro, but because curcumin can lower plasma and tissue cholesterol [80, 81], it may be able to indirectly lower Aβ production in vivo. Nevertheless, the observation that the drug reduced exogenously infused Aβ accumulation argues that postproduction effects are likely important. Curcumin itself can directly bind to plaques in vitro and in vivo and directly inhibit Aβ aggregation in vitro with an IC_{50} below that required for inhibition of lipid peroxidation in vitro [Yang et al., unpubl. obs.]. This suggests direct targeting of Aβ aggregation.

Curcumin also reduced total Aβ (by ELISA) in unfixed cryostat human AD slices incubated in vitro with murine microglia, but had no effect in the absence of microglia. Immunolabeling for Aβ suggested curcumin stimulated phagocytosis of Aβ by the murine microglia. Similarly, we used confocal double-labeling to observe apparent microglial phagocytosis of plaques in 16-month-old Tg2576 mice fed curcumin (160 ppm), suggesting that this effect occurred in vivo. Because confocal double-labeling cannot resolve the intracellular location of microglial amyloid, careful ultrastructural studies are being conducted to confirm phagocytosis of amyloid. However, other evidence suggests that curcumin promotes a microglial phagocytic phenotype.

Microglia display a complex array of phenotypic stages characterized both morphologically and by stage-specific marker expression. Kloss et al. [82, 83] have analyzed and staged microglial activation by changes in the pattern of integrins, using the facial nerve transection model in mice to induce microglia activation without allowing injury-associated monocyte invasion.

They define 5 stages:

- Stage 0 'resting': αMβ2 (CD11b, complement receptor 3, highly ramified)
- Stage 1 'alert': αMβ2 and its ligand, ICAM-1 (hypertrophied with reduced ramification)
- Stage 2 'homing and adhesion': α5β1, α6β1 (limited MHCI, B7.2, reduced CD11b, ICAM-1)
- Stage 3a 'phagocytosis': upregulation of stage 2 markers, CD11b and appearance of CD11c (αXβ2)
- Stage 3b 'bystander microglia': ramified, not phagocytic, with very high α4β1 integrin + most stage 3a markers (MHCI, B7.2, ICAM-1) but without CD11c/D18 (αXβ2).

'Bystander activation' can be induced by diffusible molecules from glial 'nodules' at sites of injury and probably by glia at plaques. Most microglia in APP transgenics are not phagocytic when analyzed at the ultrastructural level [84] and appear to fit the description of being arrested at the 'bystander microglia' stage. Consistent with that view, we find that the majority of periplaque microglia show little or no CD11c [Yang, unpubl. obs.]. Because CD11b is on resting microglia, but upregulated in stage 3b, it is a marker that should be detectable in the resting state, but also a useful index of increasing activation, while CD11c is a phagocyte-specific marker. We measured the expression of these markers, using real-time RT-PCR. With this approach, CD11c but not CD11b mRNA was induced in the cortex of APP transgenics relative to transgene-negative animals at 16 months of age while both were induced by 22 months. Curcumin (160 ppm) significantly reduced CD11b but increased CD11c mRNA in the cortex [Morihara et al., unpubl. obs.]. Plaque-associated, CD11c-labeled microglia that had a nonramified phagocytic morphology were markedly increased in the curcumin-treated group. Collectively, these data provide support for mechanisms of curcumin action, involving both direct inhibition of Aβ aggregation and stimulation of the periplaque microglial phagocytic phenotype, leading to amyloid clearance, possibly via C3b/iC3b opsonized Aβ aggregates binding upregulated CD11c.

Neurodegeneration in AD vs. APPsw Mice

Like many neurodegenerative disorders, AD has relatively selective neuron loss in uniquely vulnerable populations including those in the hippocampal CA1, the entorhinal cortex layer II, and other tangle-vulnerable cortical layers and subcortical nuclei. There is also a ~20–50% loss in synaptophysin or presynaptic terminals in vulnerable regions like association cortex which correlates with clinical decline. While region-dependent dendritic decline is less well studied, the loss of the postsynaptic markers neurogranin and drebrin has been reported to be more profound (70–80% loss) [1, 85]. In contrast, despite extensive amyloid pathology, human APPsw transgenic mice and even bigenic mice coexpressing mutant presenilin have not shown comparable levels of region-dependent neuron and presynaptic marker loss. The most common explanation has been the lack of tangle formation. In addition to rare neuron loss, some transgenic models have ~20–30% focal synaptophysin loss in parts of the hippocampus, but this modest loss has not been clearly related to cognitive deficits. Instead, cognitive deficits in the mouse models either precede most of the pathology or correlate with the 'maintained' synaptophysin that is associated with extensive sprouting [10].

In the Tg2576 APPsw mouse, presynaptic marker loss (using Western blots), can only be seen in the oldest mice (24–30 months of age) when there is a ~30% drop in synaptophysin [Cole et al., unpubl. obs.]. Consistent with postsynaptic loss in AD, old (22 month old) APPsw mice show loss of postsynaptic markers, the most robust of which is a 60% loss of the dendritic spine actin-binding protein, drebrin, that also shows severe loss in AD.

Fish Oil and n–3 Fatty Acids

As discussed above, DHA is enriched in synapses and DHA is reduced in diets associated with AD risk. DHA levels in AD brain are down, at least in part, because oxidized DHA is increased in AD. In addition to the multiple useful effects of DHA in the brain [47], DHA can protect against apoptotic neurodegeneration in a neuronal cell line in vitro [86]. A major part of this protective effect may be due to control of the PI-3 to Akt kinase pathway that phosphorylates the proapoptotic Bcl-xl/Bcl-2 death promoter proteins and thus inhibits caspase activation [87].

Based on these observations, we hypothesized that the standard rodent chow may be neuroprotective because it is enriched in n–3 fatty acids (soy oil and fish meal) and the ratio of n–6/n–3 is ~4:1, optimized for rodent development and health. Breeder chow has slightly increased fat and an n–6/n–3 ratio of about 7:1. On this basis we removed the fish and soy sources of n–3 fatty acids and added n–6 fatty acid (safflower oil rich) to create an extreme 'bad American diet' or BAD diet depleted of n–3 (n–6/n–3 ratio of about 85:1).

This type of extreme BAD diet with n–6 linoleic acid from safflower, corn or coconut oil has been previously used to deplete DHA from the brain. Because of tissue reserves, it is typically necessary to use these extreme diets from weaning or even in the maternal diet and across generations in order to deplete DHA in the CNS [47]. However, based on the hypothesis that increased CNS lipid peroxidation in the transgene-positive mice would accelerate DHA oxidation and depletion, we began our diet study using aging adult mice. Transgene-positive and transgene-negative APPsw mice aged to 17 months on breeder chow were used. From 17 to 22.5 months of age, groups of transgene-negative and transgene-positive APPsw mice were placed on 3 different diets: Standard breeder chow (n–6/n–3, ~7:1), BAD diet (n–6/n–3, ~85:1), BAD diet + DHA (n–6/n–3, ~7:1).

After sacrifice, we examined Aβ levels (by ELISA) in detergent-insoluble fractions, carbonyls as a measure of oxidative damage and synaptic markers (by Western blots) in the cortex. Aβ levels were increased by the BAD diet relative to standard breeder chow and they were very significantly reduced by BAD+DHA diet relative to BAD diet. Ongoing studies suggest that this is most likely attributable to regulation of APP processing to Aβ by increasing membrane fluidity, since DHA fluidizes membranes, the opposite of cholesterol which reduces membrane fluidity and promotes the generation of Aβ from APP [88]. DHA also reduced oxidative damage indexed by carbonyls, an effect that is probably secondary to reducing the pro-oxidant Aβ, but may also reflect regulation of antioxidant enzymes [89]. The most surprising result was that BAD diet caused a dramatic loss of postsynaptic markers, like drebrin and PSD-95, in the membrane fraction. The losses were much more pronounced in the APPsw transgene-positive animals, reaching 95% loss for drebrin. The postsynaptic marker reductions were almost completely reversed by BAD+DHA diets. BAD diet increased oxidative damage and caspase activation (detected by antibody to caspase-cleaved actin on Western blots). These changes were also prevented by DHA treatment of transgene-positive animals. Immuno-ultrastructural analysis revealed caspase-cleaved actin in the postsynaptic density that is increased in the APPsw transgene-positive mice [Triller and Rostaing, Ecole Normale Supérieure, Paris France, unpubl. obs.]. Collectively, these results suggest that dietary DHA can protect against some aspects of Aβ-dependent neurodegeneration in transgene-positive animals.

Conclusions

In AD pathogenesis, DHA oxidation and depletion may be important events in a cycle of Aβ-induced lipid peroxidation, synaptic membrane alterations leading to increased Aβ production, caspase activation and selective postsynaptic marker loss, notably drebrin. Because DHA is the most oxidizable target for damage and is enriched in synapses, it may be especially vulnerable to the synaptic Aβ accumulation in AD. Dietary essential fatty acids (n–6/n–3), particularly DHA, modulate synaptic marker loss in APP transgenics and may play a role in controlling synapse loss in AD patients. DHA deficiency caused by oxidation and/or diet appears to induce selective postsynaptic caspase activation (synaptosis) accompanied by and correlating with postsynaptic marker loss.

While the effects of BAD diet and DHA depletion are exaggerated in APPsw transgene-positive animals, limited caspase activation and postsynaptic marker loss are also apparent in transgene-negative animals. This is consistent with our hypothesis that the standard rodent diets with optimal essential fatty acid levels are neuroprotective and may blunt the expression of neurodegeneration in transgenic mouse models of human neurodegenerative diseases. Further, supplementation with antioxidants and NSAIDs should limit synaptic oxidative damage to AA and DHA and may slow AD progression. Many of the factors that reduce the risk for AD, reduce amyloid accumulation in AD models and they may synergize. Curcumin may be more effective than other single approaches because it is not only an antioxidant/NSAID, but also an amyloid-binding compound capable of inhibiting aggregation and stimulating phagocytic clearance of amyloid. Whatever the choice of agents or design of cocktails, ultimately, targeting multiple rather than single pathways in the amyloid cascade should prove most effective.

References

1 Harigaya Y, Shoji M, Shirao T, Hirai S: Disappearance of actin-binding protein, drebrin, from hippocampal synapses in Alzheimer's disease. J Neurosci Res 1996;43:87–92.

2 Bird TD, Nochlin D, Poorkay P, Cherrier M, Kaye J, Payami H, Peskind E, Lampe TH, Nemens E, Boyer PJ, Schellenberg GD: A clinical pathological comparison of three families with frontotemporal dementia and identical mutations in the tau gene (P301L). Brain 1999;122:741–756.

3 Götz J, Chen F, van Dorpe J, Nitsch RM: Formation of neurofibrillary tangles in P301L tau transgenic mice induced by Aβ42 fibrils. Science 2001;293:1491–1495.

4 Lewis J, Dickson DW, Lin W-L, Chisholm L, Corral A, Jones G, Yen S-H, Sahara N, Skipper L, Yager D, Eckman C, Hardy J, Hutton J, McGowan E: Enhanced neurofibrillary degeneration in transgenic mice expressing mutant Tau and APP. Science 2001;293:1487–1491.

5 Selkoe DJ: Alzheimer's disease: Genotypes, phenotypes, and treatments. Science 1997;275:630–631.

6 Yankner BA: Mechanisms of neuronal degeneration in Alzheimer's disease. Neuron 1996;16:921–932.

7 Irizarry MC, McNamara M, Fedorchak K, Hsiao K, Hyman BT: APP_{sw} Transgenic mice develop age-related Aβ deposits and neuropil abnormalities, but no neuronal loss in CA1. J Neuropathol Exp Neurol 1997;56:965–973.

8 Irizarry MC, Soriano F, McNamara M, Page KJ, Schenk D, Games D, Hyman BT: Aβ deposition is associated with neuropil changes, but not with overt neuronal loss in the human amyloid precursor protein V717F (PDAPP) transgenic mouse. J Neurosci 1997;17:7053–7059.

9 Sommer B, Sturchler-Pierrat C, Abramowski D, Wiederhold Kh, Calhoun M, Jucker M, Kelly P, Staufenbiel M: Transgenic approaches to model Alzheimer's disease. Rev Neurosci 2000;11:47–51.
10 King DL, Arendash GW: Maintained synaptophysin immunoreactivity in Tg2576 transgenic mice during aging: Correlations with cognitive impairment. Brain Res 2002;926:58–68.
11 Mucke L, Masliah E, Yu GQ, Mallory M, Rockenstein EM, Tatsuno G, Hu K, Kholodenko D, Johnson-Wood K, McConlogue L: High-level neuronal expression of abeta 1–42 in wild-type human amyloid protein precursor transgenic mice: Synaptotoxicity without plaque formation. J Neurosci 2000;20:4050–4058.
12 Frautschy SA, Yang F, Irizarry M, Hyman B, Saido TC, Hsiao K, Cole GM: Microglial response to amyloid plaques in APPsw transgenic mice. Am J Pathol 1998;152:307–317.
13 Lim GP, Chu T, Yang F, Beech W, Frautschy SA, Cole GM: The curry spice curcumin reduces oxidative damage and amyloid pathology in an Alzheimer transgenic mouse. J Neurosci 2001;21: 8370–8377.
14 Pratico D, Uryu K, Leight S, Trojanoswki JQ, Lee VM: Increased lipid peroxidation precedes amyloid plaque formation in an animal model of Alzheimer amyloidosis. J Neurosci 2001;21: 4183–4187.
15 Hsiao K, Chapman P, Nilsen S, Eckman C, Harigaya Y, Younkin S, Yang F, Cole G: Correlative memory deficits, Aβ elevation and amyloid plaques in transgenic mice. Science 1996;274:99–102.
16 Morgan D, Diamond DM, Gottschall PE, Ugen KE, Dickey C, Hardy J, Duff K, Jantzen P, DiCarlo G, Wilcock D, Connor K, Hatcher J, Hope C, Gordon M, Arendash GW: Aβ peptide vaccination prevents memory loss in an animal model of Alzheimer's disease. Nature 2000;408:982–985.
17 Chen G, Chen KS, Knox J, INglis J, Bernard A, Martin SJ, Justice A, McConlogue L, Games D, Freedman SB, Morris RGM: A learning deficit related to age and β-amyloid plaques in a mouse model of Alzheimer's disease. Nature 2000;408:975–979.
18 Westerman MA, Cooper-Blacketer D, Mariash A, Kotilinek L, Kawarabayashi T, Younkin LH, Carlson GA, Younkin SG, Ashe KH: The relationship between abeta and memory in the Tg2576 mouse model of Alzheimer's disease. J Neurosci 2002;22:1858–1867.
19 McGeer PL, Schulzer M, McGeer EG: Arthritis and anti-inflammatory agents as possible protective factors for Alzheimer's disease: A review of 17 epidemiologic studies. Neurology 1996;47: 425–432.
20 in 't Veld BA, Ruitenberg A, Hofman A, Launer LJ, van Duijn CM, Stijnen T, Breteler MM, Stricker BH: Nonsteroidal antiinflammatory drugs and the risk of Alzheimer's disease. N Engl J Med 2001;345:1515–1521.
21 Akiyama H, Barger S, Barnum S, Bradt B, Bauer J, Cole GM, Cooper NE, Eikelenboom P, Emmerling M, Fiebich BL, Finch CE, Frautschy S, Griffin WS, Hampel H, Hull M, Landreth G, Lue L, Mrak R, Mackenzie IR, Mcgeer PL, O'Banion MK, Pachter J, Pasinetti G, Plata-Salaman C, Rogers J, Rydel R, Shen Y, Streit W, Strohmeyer R, Tooyoma I, Van Muiswinkel FL, Veerhuis R, Walker D, Webster S, Wegrzyniak B, Wenk G, Wyss-Coray T: Inflammation and Alzheimer's disease. Neurobiol Aging 2000;21:383–421.
22 Rogers J, Cooper NR, Webster S, Schultz J, McGeer PL, Styren SD, Civin WH, Brachova L, Bradt B, Ward P, Lieberburg I: Complement activation by β-amyloid in Alzheimer disease. Proc Natl Acad Sci USA 1992;89:10016–10020.
23 Webster S, Bradt B, Rogers J, Cooper N: Aggregation state-dependent activation of the classical complement pathway by the amyloid β peptide. J Neurochem 1997;69:388–398.
24 Stewart WF, Kawas C, Corrada M, Meter EJ: Risk of Alzheimer's disease and duration of NSAID use. Neurology 1997;48:626–632.
25 Zandi PP, Anthony JC, Hayden KM, Mehta K, Mayer L, Breitner JC: Reduced incidence of AD with NSAID but not H_2 receptor antagonists: The Cache County Study. Neurology 2002;59: 880–886.
26 Breitner JCS, Welsh KA, Helms MJ, Gaskell PC, Gau BA, Roses AD, Pericak-Vance MA, Saunders AM: Delayed onset of Alzheimer's disease with nonsteroidal anti-inflammatory and histamine H_2 blocking drugs. Neurobiol Aging 1995;16:523–530.
27 Perry G, Smith MA: A central role for oxidative damage in the pathogenesis and therapeutics of Alzheimer's disease. Alzheimer's Res 1997;2:319–324.

28 Nunomura A, Perry G, Aliev G, Hirai K, Takeda A, Balraj EK, Jones PK, Ghanbari H, Wataya T, Shimohama S, Chiba S, Atwood CS, Petersen RB, Smith MA: Oxidative damage is the earliest event in Alzheimer disease. J Neuropathol Exp Neurol 2001;60:759–767.
29 Behl C, Davis JB, Lesley R, Schubert D: Hydrogen peroxide mediates amyloid β-protein toxicity. Cell 1994;77:817–827.
30 Opazo C, Huang X, Cherny RA, Moir RD, Roher AE, White AR, Cappai R, Masters CL, Tanzi RE, Inestrosa NC, Bush AI: Metalloenzyme-like activity of Alzheimer's disease beta-amyloid. Cu-dependent catalytic conversion of dopamine, cholesterol, and biological reducing agents to neurotoxic H(2)O(2). J Biol Chem 2002;277:40302–40308.
31 Butterfield DA, Kanski J: Methionine residue 35 is critical for the oxidative stress and neurotoxic properties of Alzheimer's amyloid beta-peptide 1–42. Peptides 2002;23:1299–1309.
32 Meda L, Cassatella MA, Szendrei GI, Otvos L, Baron P, Villalba M, Ferrari D, Rossi F: Activation of microglial cells by beta-amyloid protein and interferon-gamma. Nature 1995;374:647–650.
33 Behl C, Davis J, Cole GM, Schubert D: Vitamin E protects nerve cells from amyloid β-protein toxicity. Biochem Biophys Res Commun 1992;186:944–950.
34 Grundman M: Vitamin E and Alzheimer disease: The basis for additional clinical trials. Am J Clin Nutr 2000;71 (suppl):630S–636S.
35 Morris MC, Evans DA, Bienias JL, Tangney CC, Bennett DA, Aggarwal N, Wilson RS, Wilson RS, Scherr PA: Dietary intake of antioxidant nutrients and the risk of incident Alzheimer disease in a biracial community study. JAMA 2002;287:33230–33237.
36 Engelhart MJ, Geerlings MI, Ruitenberg A, van Suiten JC, Hofman A, Witteman JCM, Breteler MMB: Dietary intake of antioxidants and risk of Alzheimer disease. JAMA 2002; 287:3223–3229.
37 Sano M, Ernesto C, Thomas RG, Klauber MR, Schafer K, Grundman M, Woodbury P, Growdon J, Cotman CW, Pfeiffer E, et al: A controlled trial of selegiline, alpha-tocopherol, or both as treatment for Alzheimer's disease. The Alzheimer's Disease Cooperative Study. N Engl J Med 1997; 336:1216–1222.
38 Sparks DL, Scheff SW, Hunsaker JC, Liu H, Landers T, Gross DR: Induction of Alzheimer-like β-amyloid immunoreactivity in the brains of rabbits with dietary cholesterol. Exp Neurol 1994;126:88–94.
39 Simons M, Keller P, Dichgans J, Schulz JB: Cholesterol and Alzheimer's disease: Is there a link? Neurology 2001;57:1089–1093.
40 Kalmijn S, Launer LJ, Ott A, Witteman JC, Hofman A, Breteler MM: Dietary fat intake and the risk of incident dementia in the Rotterdam study. Ann Neurol 1997;42:776–782.
41 Engelhart MJ, Geerlings MI, Ruitenberg A, Van Swieten JC, Hofman A, Witteman JC, Breteler MM: Diet and risk of dementia: Does fat matter?: The Rotterdam Study. Neurology 2002;59:1915–1921.
42 Wolozin B, Kellman W, Ruosseau P, Celesia GG, Siegel G: Decreased prevalence of Alzheimer disease associated with 3-hydroxy-3-methylglutaryl coenzyme A reductase inhibitors. Arch Neurol 2000;57:1439–1443.
43 Blauw GJ, Shepherd J, Murphy MB: Dementia and statins. Prosper study group. Lancet 2001;357:881.
44 Jick H, Zornberg GL, Jick SS, Seshadri S, Drachman DA: Statins and the risk of dementia. Lancet 2000;356:1627–1631.
45 Refolo LM, Pappolla MA, Malester B, LaFrancois J, Bryant-Thomas T, Wang R, Tint GS, Sambamurti K, Duff K: Hypercholesterolemia accelerates the Alzheimer's amyloid pathology in a transgenic mouse model. Neurobiol Dis 2000;7:321–331.
46 Refolo LM, Pappola MA, LaFrancois J, Malester B, Schmidt SD, Thomas-Bryant T, Tint GS, Wang R, Mercken M, Petanceska SS, Duff KE: A cholesterol-lowering drug reduces β-amyloid pathology in a transgenic mouse model of Alzheimer's disease. Neurobiol Dis 2001;8:890–899.
47 Salem N Jr, Litman B, Kim HY, Gawrisch K: Mechanisms of action of docosahexaenoic acid in the nervous system. Lipids 2001;36:945–959.
48 Horrocks LA, Yeo YK: Health benefits of docosahexaenoic acid (DHA). Pharmacol Res 1999;40:211–225.
49 Otsuka M: Analysis of dietary factors in Alzheimer's disease: Clinical use of nutritional intervention for prevention and treatment of dementia (in Japanese). Nippon Ronen Igakkai Zasshi 2000;37:970–973.

50 Otsuka M, Yamaguchi K, Ueki A: Similarities and differences between Alzheimer's disease and vascular dementia from the viewpoint of nutrition. Ann N Y Acad Sci 2002;977:155–161.
51 Yamada T, Kadekaru H, Matsumoto S, Inada H, Tanabe M, Moriguchi EH, Moriguchi Y, Ishikawa P, Ishikawa AG, Taira K, Yamori Y: Prevalence of dementia in the older Japanese-Brazilian population. Psychiatry Clin Neurosci 2002;56:71–75.
52 Kalmijn S, Feskens EJM, Launer LJ, Kromhout D: Polyunsaturated fatty acids, antioxidants and cognitive function in very old men. Am J Epidemiol 1997;145:33–41.
53 Kyle DJ, Schaefer E, Patton G, Beiser A: Low serum docosahexaenoic acid is a significant risk factor for Alzheimer's dementia. Lipids 1999;34 Suppl:S245.
54 Conquer JA, Tierney MC, Zecevic J, Bettger WJ, Fisher RH: Fatty acid analysis of blood plasma of patients with Alzheimer's disease, other types of dementia, and cognitive impairment. Lipids 2000;35:1305–1312.
55 Praticò D, Lee VM-Y, Trojanowski JQ, Rokach J, Fitzgerald GA: Increased F2-isoprostanes in Alzheimer's disease: Evidence for enhanced lipid peroxidation in vivo. FASEB J 1998;12: 1777–1783.
56 Montine TJ, Markesbery WR, Roberts RJI, Morrow JD: Cerebrospinal fluid F2-isoprostane levels are increased in Alzheimer's disease. Ann Neurol 1998;44:410–413.
57 Roberts LJ, Montine TJ, Markesbery WR, Tapper AR, Hardy P, Chemtob S, Dettbarn WD, Morrow JD: Formation of isoprostane-like compounds (neuroprostane) in vivo from docosahexaenoic acid. J Biol Chem 1998;273:13605–13612.
58 Nourooz-Zadeh J, Liu EHC, Yhlen B, Anggard EE, Halliwell B: F4-Isoprostanes as specific marker of docosahexaenoic acid peroxidation in Alzheimer's disease. J Neurochem 1999;72: 734–740.
59 Reich EE, Markesbery WR, Roberts LJ 2nd, Swift LL, Morrow JD, Montine TJ: Brain regional quantification of F-ring and D-/E-ring isoprostanes and neuroprostanes in Alzheimer's disease. Am J Pathol 2001;158:293–297.
60 Lim GP, Yang F, Chu T, Chen P, Beech W, Teter B, Tran T, Ubeda O, Ashe KH, Frautschy SA, Cole GM: Ibuprofen suppresses plaque pathology and inflammation in a mouse model for Alzheimer's Disease. J Neurosci 2000;20:5709–5714.
61 Lim GP, Yang F, Chu T, Gahtan E, Ubeda O, Beech W, Overmier JB, Hsiao Ashe K, Frautschy SA, Cole GM: Ibuprofen effects on Alzheimer pathology and open field activity in APPsw transgenic mice. Neurobiol Aging 2001;22:683–691.
62 Jantzen PT, Connor KE, DiCarlo G, Wenk GL, Wallace JL, Rojiani AM, Coppola D, Morgan D, Gordon MN: Microglial activation and beta-amyloid deposit reduction caused by a nitric oxide-releasing nonsteroidal anti-inflammatory drug in amyloid precursor protein plus presenilin-1 transgenic mice. J Neurosci 2002;22:2246–2254.
63 Wyss-Coray T, Mucke L: Inflammation in neurodegenerative disease–a double-edged sword. Neuron 2002;35:419–432.
64 Weggen S, Eriksen JL, Das P, Sagi SA, Wang R, Pietrzik CU, Findlay KA, Smith TE, Murphy MP, Bulter T, Kang DE, Marquez-Sterling N, Golde TE, Koo EH: A subset of NSAIDs lower amyloidogenic Abeta42 independently of cyclooxygenase activity. Nature 2001;414:212–216.
65 Morihara T, Chu T, Ubeda O, Beech W, Cole GM: Selective inhibition of Aβ42 production by NSAID R-enantiomers. J Neurochem 2002;83:1–4.
66 Mucke L, Yu GQ, McConlogue L, Rockenstein EM, Abraham CR, Masliah E: Astroglial expression of human alpha(1)-antichymotrypsin enhances Alzheimer-like pathology in amyloid protein precursor transgenic mice. Am J Pathol 2000;157:2003–2010.
67 Nilsson LN, Bales KR, DiCarlo G, Gordon MN, Morgan D, Paul SM, Potter H: Alpha-1-antichymotrypsin promotes beta-sheet amyloid plaque deposition in a transgenic mouse model of Alzheimer's disease. J Neurosci 2001;21:1444–1451.
68 Frautschy SA, Yang F, Calderón L, Cole GM: Rodent models of Alzheimer's disease: Rat Aβ infusion approaches to amyloid deposits. Neurobiol Aging 1996;17:311–321.
69 Frautschy SA, Hu W, Miller SA, Kim P, Harris-White ME, Cole GM: Phenolic anti-inflammatory antioxidant reversal of Aβ-induced cognitive deficits and neuropathology. Neurobiol Aging 2001;22:991–1003.
70 Guttmann RP, Erickson AC, Johnson GV: Tau self-association: stabilization with a chemical cross-linker and modulation by phosphorylation and oxidation state. J Neurochem 1995;64:1209–1215.

71 Christen S, Woodall AA, Shigenaga MK, Southwell-Keely PT, Duncan MW, Ames BN: Gamma-tocopherol traps mutagenic electrophiles such as NO_x and complements alpha-tocopherol: Physiological implications. Proc Natl Acad Sci USA 1997;94:3217–3222.
72 Giasson BI, Duda JE, Murray IV, Chen Q, Souza JM, Hurtig HI, Ischiropoulos H, Trojanowski JQ, Lee VM: Oxidative damage linked to neurodegeneration by selective alpha-synuclein nitration in synucleinopathy lesions. Science 2000;290:985–989.
73 Okada K, Wangpoengtrakul C, Tanaka T, Toyokuni S, Uchida K, Osawa T: Curcumin and especially tetrahydrocurcumin ameliorate oxidative stress-induced renal injury in mice. J Nutr 2001;131:2090–2095.
74 Kitani K, Osawa T: Tetrahydrocurcumin prolongs survival curves of male C57Bl mice. Age Annual Meeting, San Diego 2002.
75 Abe Y, Hashimoto S, Horie T: Curcumin inhibition of inflammatory cytokine production by human peripheral blood monocytes and alveolar macrophages. Pharmacol Res 1999;39:41–47.
76 Spitz DR, Oberley LW: An assay for superoxide dismutase activity in mammalian tissue homogenates. Anal Biochem 1989;179:8–18.
77 Goel A, Boland CR, Chauhan DP: Specific inhibition of cyclooxygenase-2 (COX-2) expression by dietary curcumin in HT-29 human colon cancer cells. Cancer Lett 2001;172:111–118.
78 Kelloff GJ, Crowell JA, Hawk ET, Steele VE, Lubet RA, Boone CW, Covey JM, Doody LA, Omenn GS, Greenwald P, Hong WK, Parkinson DR, Bagheri D, Baxter GT, Blunden M, Doeltz MK, Eisenhauer KM, Johnson K, Knapp GG, Longfellow DG, Malone WF, Nayfield SG, Seifried HE, Swall LM, Sigman CC: Strategy and planning for chemopreventive drug development: Clinical development plan: Curcumin. J Cell Biochem Suppl 1996;26:72–85.
79 Harris-White ME, Simmons M, Nash D, Miller SA, Chu T, Teter B, Cole GM, Frautschy SA: Estrogen (E_2) and glucocorticoid (Gc) effects on microglia and Aβ clearance in vitro and in vivo. Neurochem Int 2001;39:435–448.
80 Hussain MS, Chandrasekhara N: Effect on curcumin on cholesterol gall-stone induction in mice. Indian J Med Res 1992;96:288–291.
81 Suni KB, Kuttan R: Effect of oral curcumin administration on serum peroxides and cholesterol levels in human volunteers. Indian J Physiol Pharmacol 1992;36:273–275.
82 Kloss CU, Bohatschek M, Kreutzberg GW, Raivich G: Effect of lipopolysaccharide on the morphology and integrin immunoreactivity of ramified microglia in the mouse brain and in cell culture. Exp Neurol 2001;168:32–46.
83 Kloss CU, Werner A, Klein MA, Shen J, Menuz K, Probst JC, Kreutzberg GW, Raivich G: Integrin family of cell adhesion molecules in the injured brain: Regulation and cellular localization in the normal and regenerating mouse facial motor nucleus. J Comp Neurol 1999;411:162–178.
84 Stalder M, Phinney A, Probst A, Sommer B, Staufenbiel M, Jucker M: Association of microglia with amyloid plaques in brains of APP23 transgenic mice. Am J Pathol 1999;154:1673–1684.
85 Shim KS, Lubec G: Drebrin, a dendritic spine protein, is manifold decreased in brains of patients with Alzheimer's disease and Down syndrome. Neurosci Lett 2002;324:209–212.
86 Kim HY, Akbar M, Lau A, Edsall L: Inhibition of neuronal apoptosis by docosahexaenoic acid (22:6n–3). Role of phosphatidylserine in antiapoptotic effect. J Biol Chem 2000;275: 35215–35223.
87 Akbar M, Kim HY: Protective effects of docosahexaenoic acid in staurosporine-induced apoptosis: Involvement of phosphatidylinositol-3 kinase pathway. J Neurochem 2002;82:655–665.
88 Simons K, Ehehalt R: Cholesterol, lipid rafts, and disease. J Clin Invest 2002;110:597–603.
89 Hossain MS, Hashimoto M, Gamoh S, Masumura S: Antioxidative effects of docosahexaenoic acid in the cerebrum versus cerebellum and brainstem of aged hypercholesterolemic rats. J Neurochem 1999;72:1133–1138.

Dr. Gregory M. Cole
Research 151 – VA Medical Center – Greater Los Angeles Healthcare System
16111 Plummer Street, Sepulveda, CA 91343 (USA)
Tel. +1 818 891 7711, ext 9949, Fax +1 818 895 5835, E-Mail gmcole@ucla.edu

Takeda M, Tanaka T, Cacabelos R (eds): Molecular Neurobiology of Alzheimer Disease and Related Disorders. Basel, Karger, 2004, pp 17–30

The RNA-Binding Protein Causes Aberrant Splicing of Presenilin-2 Pre-mRNA in Sporadic Alzheimer's Disease

Taiichi Katayama[a,b], *Takayuki Manabe*[a,b], *Kazunori Imaizumi*[c], *Naoya Sato*[a], *Junichi Hitomi*[a,b], *Takashi Kudo*[d], *Takeshi Yanagita*[a], *Shinsuke Matsuzaki*[a,b], *Akila Mayeda*[e], *Masaya Tohyama*[a,b]

[a] Department of Anatomy and Neuroscience, Graduate School of Medicine, Osaka University, Suita,
[b] CREST, Japan Science and Technology Corporation, Osaka,
[c] Division of Structural Cell Biology, Nara Institute of Science and Technology (NAIST), Takayama,
[d] Department of Clinical Neuroscience, Psychiatry, Graduate School of Medicine, Osaka University, Suita, Japan and
[e] Department of Biochemistry and Molecular Biology, University of Miami School of Medicine, Miami, Fla., USA

Alzheimer's disease (AD) is a neurodegenerative disorder, clinically characterized by progressive loss of memories and other cognitive abilities. Pathologically, severe neuronal loss, glial proliferation, extracellular deposition of senile plaque composed of amyloid-β protein (Aβ) and intraneuronal neurofibrillary tangles are found in AD brains [1]. However, direct relationships between these morphological changes and the molecular mechanisms of AD onset have not been established. Most cases of familial AD are caused by mutations in three different genes [2, 3]: the amyloid precursor protein (APP) gene located on chromosome 21, the presenilin-1 (PS) gene found on chromosome 14, the presenilin-2 (PS2) gene located on chromosome 1. However, despite extensive research, little is known about the causative mechanisms of sporadic AD, which accounts for over 90% of AD cases. Because the pathological observations of both familial and sporadic AD brains are thought to be identical or quite similar, genes mutated in familial AD are considered to be logical candidates for

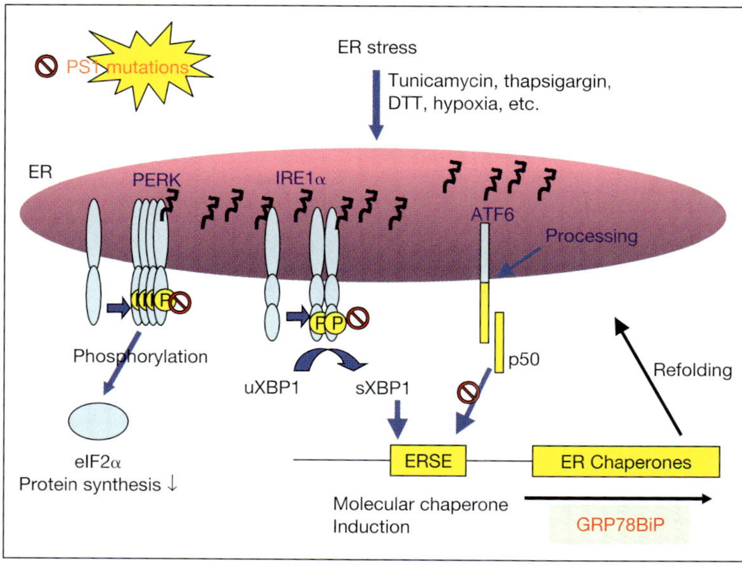

Fig. 1. PS1 mutations downregulate the signaling pathway of the unfolded protein response. This scheme shows mechanisms of ER stress response. Activation of PERK results in phosphorylation of eIF2α, and leads to inhibition of translation initiation. Autophosphorylation and dimerization of IRE1α causes activation of endonuclease domains that have the potential to cleave uXBP1, and generate an activated form of XBP1 (sXBP1). ATF6 is cleaved at or close to the cytosolic face of the membrane in response to ER stress. The N-terminal cytoplasmic domain (p50ATF6), which contains the DNA-binding, dimerization and transactivation domains, is translocated into the nucleus and activates the transcription of ER molecular chaperone genes (such as GRP78/BiP) containing the ER stress response element (ERSE), which is thought to be a regulatory element of the promoter regions conserved in ER molecular chaperone genes in mammalian cells. We reported that down-regulation of BiP induction by FAD-linked PS1 mutant is due to attenuated signaling of the UPR through decreased levels of phosphorylated IRE1 and inhibition of activation of ATF6 under ER stress conditions. Moreover, PS1 mutants also inhibited the phosphorylation of PERK. Therefore, it is possible that mutant PS1 perturbs the functions of each ER stress transducer and inhibits its downstream signal.

further investigation of the etiology of both ADs. Therefore, we have analyzed the function of PS1 mutants, since mutations of PS1 are responsible for many cases of familial AD, and have recently clearly shown that PS1 mutations down-regulate the signaling pathway of the unfolded protein response [4–6]. That is, we found out that a PS1 mutation causes endoplasmic reticulum (ER) dysfunction (fig. 1).

Alternative splicing of primary mRNA transcripts is a potent strategy for the regulation of gene expression in eukaryotes [7–9]. Variation in the selection

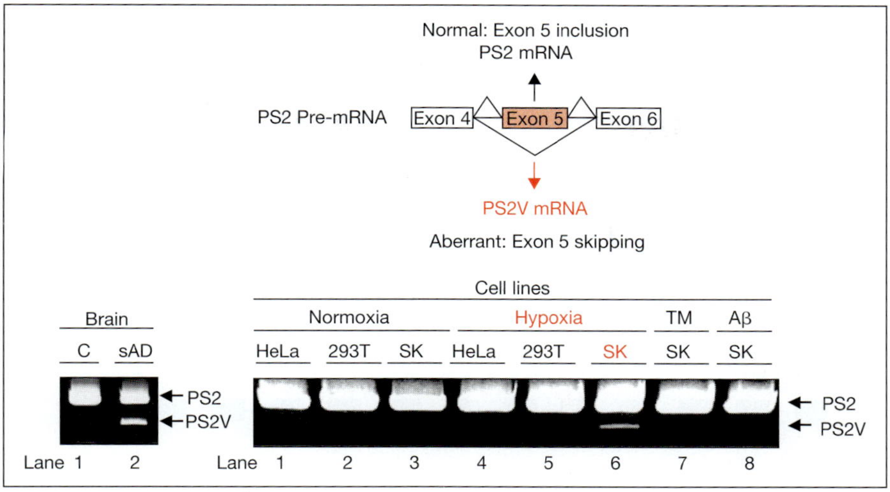

Fig. 2. Detection of the normal PS2 transcript and the aberrant isoform PS2V in the brain of a patient with sporadic AD and various cell lines (left panel). Total RNA was extracted from a representative brain from a patient with sporadic AD (sAD) or an age-matched control brain (C). RT-PCR-amplified products were separated on a polyacrylamide gel and visualized by ethidium bromide staining. Arrows indicate the positions of the normal PS2 transcript and the aberrant PS2V transcript (lacking exon 5) (right panel). Total RNA was extracted from various cell lines subjected to different stresses and RT-PCR was performed to detect the corresponding PS2 and PS2V transcripts as in left panel. The identity of PS2V was verified by cDNA sequencing. From Manabe et al. [16].

of the alternative exon results in the production of different protein isoforms from the same gene in response to tissue-specific physiologically or developmentally regulated states. The alternative spliced isoforms may have a distinct function, but occasionally they lack proper functions altogether. In fact, isoforms with aberrant functions have been reported to be associated with certain neurodegenerative disorders, such as frontotemporal disease and parkinsonism [10], spinal muscular atrophy [11], amyotrophic lateral sclerosis.

Given these observations, we alternatively examined spliced products of such specific genes in sporadic AD brain tissues using RT-PCR. We had already observed a shorter RT-PCR product in sporadic AD brains, and the product was identified as the transcript lacking exon 5 of PS2 (PS2V), which was preferentially expressed in sporadic AD brains compared with those of age-matched disease controls [12] (fig. 2).

We have not detected any other splicing variant of other genes using various primer sets of these genes. Only PS2 could be detected.

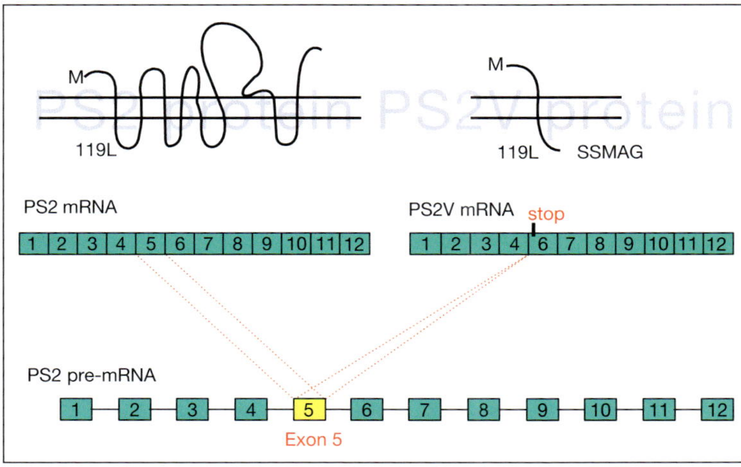

Fig. 3. Schematic representation of the splice variant detected in AD brains based on DNA sequence analysis. PS2 exons were numbered as described before [24]. Note that the loss of exon 5 caused the protein to be out of frame in exon 6, which contains residues from the first methionine to No. 119 leucine (L) and an additional 5 amino acid residues (SSMAG) at its carboxy terminus. From Sato et al. [12].

Cell Types and Stress Conditions for the Production of PS2V

We determined the optimal cell types and stress conditions for producing the aberrant splice variant of the PS2 gene transcript.

We found that under hypoxic conditions the SK-N-SH neuroblastoma cell line also produced the shorter PS2V isoform besides the PS2 product. In contrast, only the full-length PS2 product was detected in both HeLa and HEK-293T cells under hypoxic conditions. It is interesting that the loss of exon 5 caused the protein to be out of frame in exon 6 which contains residues from the first methionine to No. 119 leucine and an additional 5 amino acid residues (SSMAG) at its carboxy terminus (fig. 3).

PS2V Was Detected in Sporadic Alzheimer's Disease Brains and Caused Increases in the Production of Aβ

To confirm that PS2V was located in sporadic AD brains, we carried out a immunohistochemical analysis using the anti-SSMAG antibodies. PS2V immunoreactivity was detected in sporadic AD brain hippocampus. Furthermore, PS2V-immunoreactive cells were observed in the CA1 region of the hippocampus [13]. The cells expressing PS2V proteins showed a strongly apoptotic morphology [13, 14]. In addition, PS2V immunoreactivity was observed in all specimens from sporadic AD brains.

Further investigation revealed that the PS2V protein causes ER dysfunction, similar to familial AD-linked PS1 mutants, as well as significant increases in the production of both Aβ(1–40) and Aβ(1–42) [13]. Since the PS2V isoform was detected in SK-N-SH cells only under hypoxic conditions, we assumed that a PS2 pre-mRNA-binding factor, which leads to the skipping of exon 5 of PS2, is present in nuclear extracts from hypoxic SK-N-SH cells.

HMGA1a Is a PS2 Pre-mRNA-Binding Factor

To detect the PS2 pre-mRNA binding factor, we performed a pre-mRNA-binding assay by ultraviolet cross-linking using various labeled probes as indicated (fig. 4).

When using probe No. 0, the binding activity was clearly demonstrated in nuclear extract from hypoxic SK-N-SH cells but not normoxic cells.

Furthermore, this binding factor bound to the 3′ end of exon 5 specifically under hypoxic conditions (probe No. 5). Here, the protein was isolated through several steps of purification. We obtained a partial amino acid sequence and compared it with known proteins in the databases. This search resulted in a perfect match with the human high mobility group protein A1a (HMGA1a) [15]. This protein is known as DNA-binding protein, as previously reported. However, our results provide the first experimental evidence that HMGA1a binds to RNA in a specific manner [16].

HMGA1a Binds to a Specific Sequence of PS2 Pre-mRNA

To demonstrate whether HMGA1a is sufficient for this specific signal, immunodepletion of HMGA1a from the nuclear extracts of hypoxic SK-N-SH cells was performed. Immunoblotting analysis revealed depletion of HMGA1a, and the depleted extracts showed significant loss of binding to the No. 5 probe. Taken together, we conclude that HMGA1a is the bona fide protein factor responsible for specific binding to probe No. 5. Then we checked whether HMGA1a binding is sequence specific or not.

There are two repeated homologous sequences in the 3′ terminus of PS2 exon 5 (indicated as red bundle; fig. 4). To examine whether HMGA1a actually recognizes these tandem sequences, we prepared six more short RNA probes for the UV cross-linking assays as indicated. In contrast to the strong binding activity to probe No. 5, we could not detect any significant binding to probe No. 6,

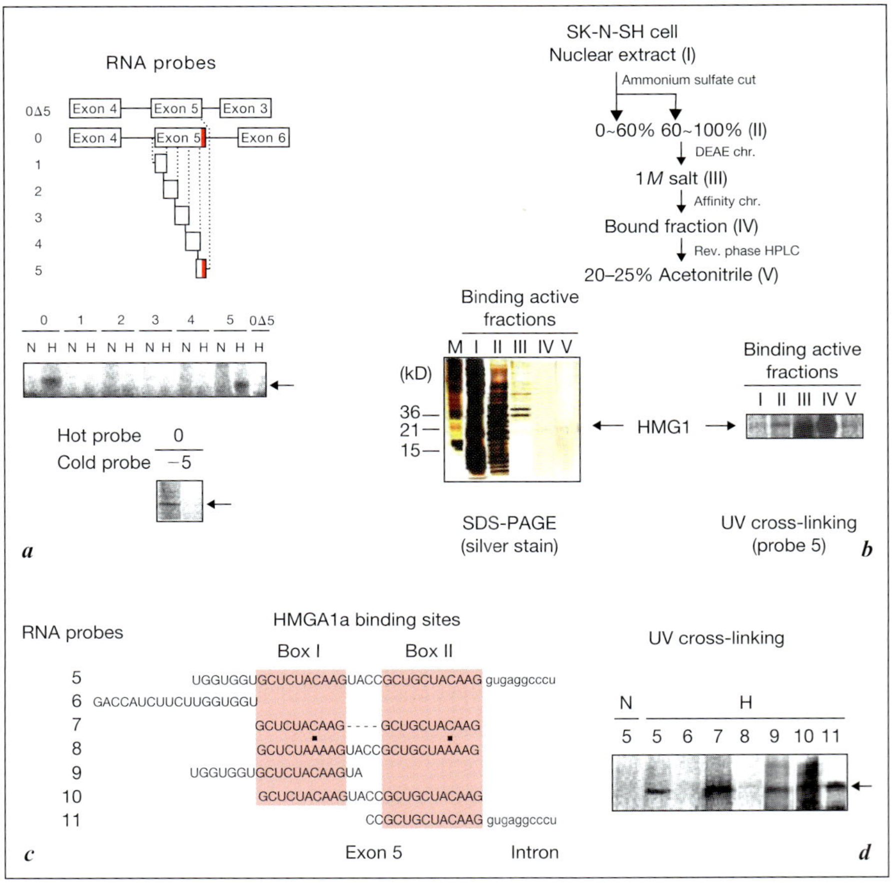

Fig. 4. UV cross-linking assays of PS2 exon 5 binding protein. ***a*** Schematic representations of the seven RNA probes used. Boxes and solid lines represent exons and introns, respectively. The red shading indicates the detected binding site. All probes were uniformly ^{35}S-labeled by in vitro transcription as described in a previous paper [16]. Nuclear extracts from SK-N-SH cells in normoxia (N) or hypoxia (H) were analyzed in UV cross-linking assays with each of the ^{35}S-labeled probes and subjected to SDS-PAGE (10–20% gradient gel). Arrow indicates the position of HMGA1a protein (~18 kD). ***a*** (lower panel) Nuclear extracts from SK-N-SH cells in hypoxia (H) were preincubated with a 100-fold excess of the indicated nonradioactive (Cold) probe No. 5. The preincubated reactions were used for the UV cross-linking assay with the ^{35}S-labeled probe (Hot), and subjected to SDS-PAGE (15% gel). Arrow indicates the position of HMGA1a protein. ***b*** Purification and characterization of human HMGA1a Upper panel: purification profile of the specific binding activity to No. 5 probe from nuclear extracts of hypoxia-induced SK-N-SH cells. Individual fractions are designated with Roman numerals. Lower left panel: protein analysis of each fraction (25 μl/fraction) on 15% SDS-PAGE followed by silver staining. M: Molecular weight (kD) markers. Lower right: UV cross-linking assays (with No. 5 probe) to evaluate binding

which is upstream of the tandem sequences and importantly, the binding is fully abolished when a point mutation (C to A) was introduced into each of the tandem sequences (No. 8).

These results demonstrate that HMGA1a binding to PS2 pre-mRNA is sequence specific.

As mentioned above, we had two questions: 'Is the expression of HMGA1a induced by hypoxic stimuli?' and 'Does HMGA1a colocalize with an authentic splicing factor under hypoxic conditions?'

The Expression of HMGA1a Is Induced by Hypoxic Stimuli

To clarify these questions, we tested its expression at both the message and the protein product levels. As indicated in figure 5, Northern blotting analyses and immunoblotting analyses showed that HMGA1a messenger and protein levels are gradually increased in SK-N-SH cells expressing PS2V under hypoxic stress.

However, in hypoxia-exposed HEK 293T cells (fig. 5a, b, middle panel) or tunicamycin-treated SK-N-SH cells (fig. 5a, b, right panel), where PS2V was not induced, no significant increase in HMGA1a mRNA levels was observed. Immunoblotting analyses using anti-HMGA1 antibody detected higher levels of HMGA1a protein in nuclear extracts of SK-N-SH cells under hypoxia compared to those cultured under normoxia. HMGA1a protein was barely detected in HEK 293T cells exposed to hypoxia (fig. 5a, b, middle panel), and very low levels in tunicamycin-treated SK-N-SH cells (fig. 5a, b, right panel). We confirm that there was no difference in induction of HSP70, as a positive control, in any cell line (data not shown).

HMGA1a Colocalizes with an Authentic Splicing Factor under Hypoxic Conditions

To determine the subcellular localization of HMGA1a/A1b proteins, we exposed SK-N-SH cell cultures to hypoxic conditions for 21 h and detected the

Fig. 4. (continued)
activity of each purified fraction (15 μl/fraction). ***c*** (left panel) Sequences of the RNA probes (No. 5–11). Exons and introns are shown in uppercase and lowercase, respectively. The homologous tandem sequences are shaded in pink boxes. ***d*** (right panel): UV cross-linking assay with the seven RNA probes shown in left panel. Arrow indicates the position of HMGA1a protein.

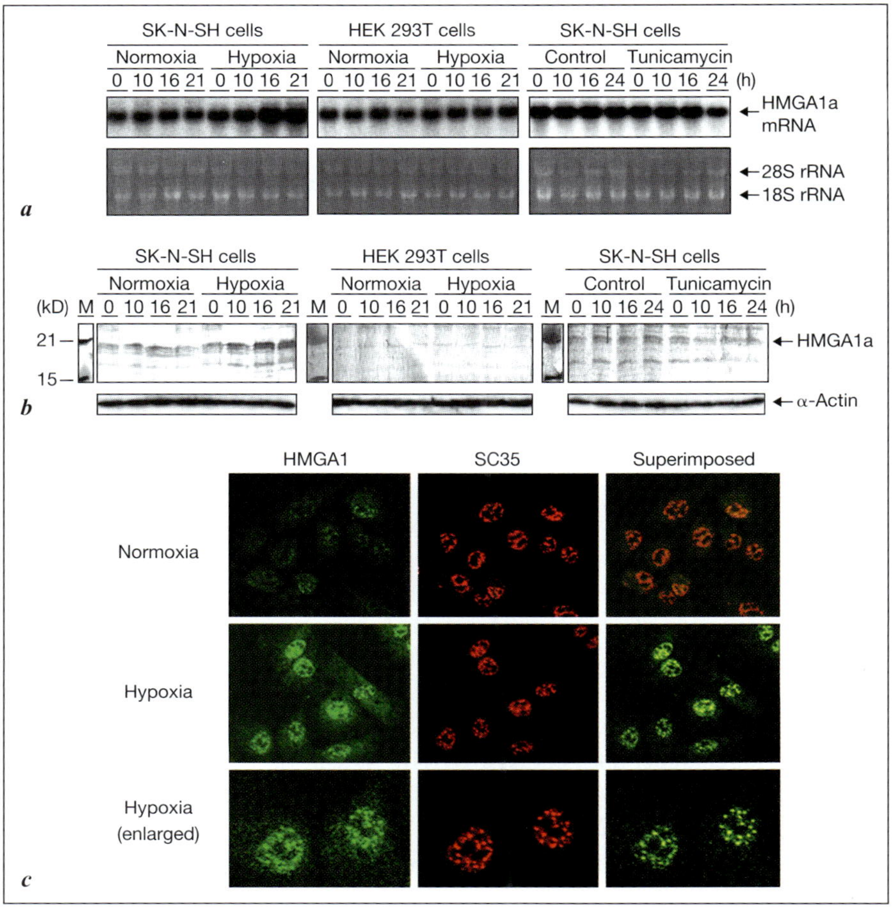

Fig. 5. Effects of various stresses on the expression of HMGA1a mRNA and protein in cultured cell lines. ***a*** Each cell line was exposed to normoxia or hypoxia and then harvested at the indicated time (left and middle panels). Tunicamycin-treated SK-N-SH cells were recovered at the indicated times after the treatment (right panel). Total RNAs were separated by formaldehyde-formamide-agarose gel electrophoresis and subjected to Northern blotting assays using a ^{32}P-labeled HMGA1a cDNA probe. Detection of rRNAs by denaturing PAGE is shown (lower panel) as an internal control. Arrows show the positions of the indicated RNAs. ***b*** Nuclear fractions from normoxia or hypoxia were separated by SDS-PAGE and subjected to immunoblotting assays using an anti-HMGA1 antibody. Expression levels of α-actin were used as an internal control (lower panel). Arrow indicates the positions of the HMGA1a (upper panel) and α-actin (lower panel). ***c*** Effects of hypoxia on the subcellular localization of endogenous HMGA1 protein in SK-N-SH cells. SK-N-SH cells were exposed to normoxia or hypoxia for 21 h. Cells were double-immunostained with anti-SC35 antibody and anti-HMGA1, followed by staining with Cy3- and FITC-conjugated secondary antibodies, respectively, and analyzed by immunofluorescence microscopy. Endogenous HMGA1 (green) and SC35 (red) images were superimposed and the yellow-green color indicates colocalization of the two proteins in nuclear speckles.

presence of HMGA1a protein by immunofluorecence microscopy with anti-HMGA1 antibody. To indicate the presence and location of essential splicing factors, an antibody against SR protein SC35, which shows typical nuclear speckle localization, was used as well [17]. In cells exposed to normoxic conditions, HMGA1a localized mainly to nuclei together with weak and diffuse immunoreactivity in the cytoplasm, and it did not colocalize with SC35, which localized in the typical nuclear speckles (fig. 5c, 'Normoxia'). However, in hypoxia-exposed SK-N-SH cells, more potent immunoreactivity was observed as nuclear speckles, which colocalized with SC35, and there was a decrease in cytoplasmic distribution compared to that observed in normoxic-conditioned cells (fig. 5c, 'Hypoxia'). We conclude that HMGA1a is reorganized in the nuclear speckles in SK-N-SH cells under hypoxic stimulation close to where splicing of PS2 pre-mRNA takes place [18].

These results suggest that HMGA1a protein can change its subcellular localization for its function. Previously, it was reported that HMGA1a/A1b proteins are mainly localized to the heterochromatin mass in actively growing 3T3 fibroblasts, whereas in quiescent cells they are more diffusely distributed [19]. Therefore, an important part of the alternative splicing function of HMGA1a in SK-N-SH cells may depend on its timely induction and dynamic relocalization in nuclei after hypoxic stimulation.

PS2V Is Produced by the Expression of HMGA1a

In SK-N-SH cells, HMGA1a was expressed under hypoxic conditions, conditions which led to the production of PS2V. We tested whether PS2V would also be produced by the overexpression of HMGA1a without hypoxia. For this purpose we transiently overexpressed HMGA1a and checked PS2V.

Markedly higher levels of HMGA1a protein expression were observed in nuclear extracts from HMGA1a-transfected SK-N-SH and HEK 293T cells compared to mock-transfected cells (fig. 6a). The product corresponding to the full-length PS2 mRNA could be detected by RT-PCR assays of total RNA from mock-transfected SK-N-SH cells, (fig. 6b, left panel). In contrast, when using total RNA from HMGA1a-transfected cells, RT-PCR yielded a shorter product in addition to the full-length product. The amount of the shorter product detected was proportional to the amount of transfected-HMGA1a cDNA (fig. 6, left panel). Due to overexpression of HMGA1a, PS2V was also detected in nonneuronal HEK 293T cells (fig. 6a, right panel), even though PS2V could not be induced in this cell line by hypoxia (fig. 2). These results suggest that the production of PS2V is solely controlled through the induction of HMGA1a regardless of cell type. This observation is consistent with the fact that the

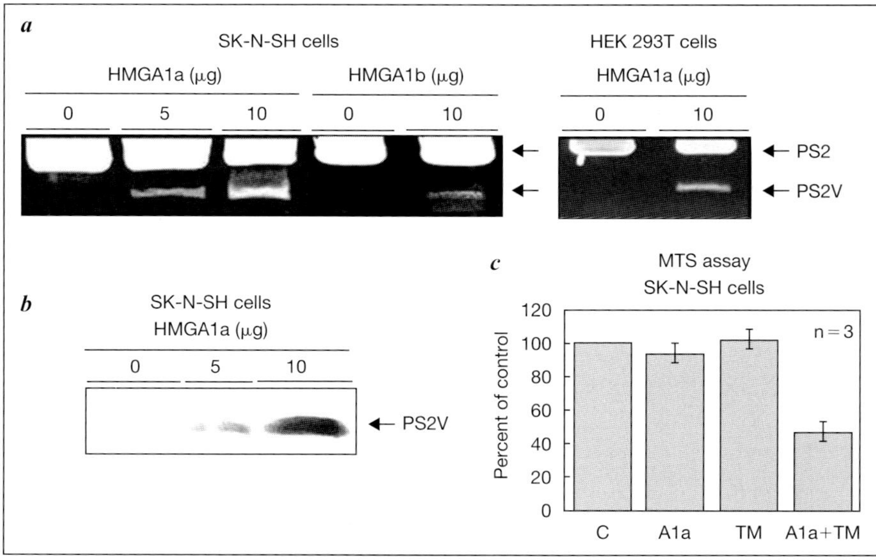

Fig. 6. Effects of overexpression of HMGA1 on PS2 pre-mRNA splicing in cultured cells. ***a*** Generation of aberrant PS2V mRNA was detected by RT-PCR. The indicated cell lines were transiently transfected with HMGA1a and cells were harvested 24 h after transfection and total RNA from harvested cells was analyzed by RT-PCR with specific primers. Amplified products were separated on a PAGE and visualized with ethidium bromide. Arrows indicate the positions of the normal PS2 mRNA (upper arrow) and aberrant PS2V isoform (lower arrow). ***b*** Generation of PS2V protein detected by immunoblotting analysis. Transient transfection was performed as in ***a*** and PS2V protein in nuclear extracts from the transfected cells were immunoprecipitated by using anti-PS2N antibody (against PS2 and PS2V), followed by immunoblotting assay using PS2V specific anti-SSMAG antibody (SSMAG is a PS2V-specific C-terminal peptide). ***c*** SK-N-SH cells were transfected with 5 μg of HMGA1a, followed by treatment with 1 μg/ml tunicamycin (TM) for 10 h, and then cell death assay was performed at 24 h after transfections, subsequently to the MTS assay. From Manabe et al. [16].

induction of HMGA1a protein by hypoxia occurs only in neuronal SK-N-SH cell lines. Finally, we confirmed that the PS2V protein product was induced in nuclear extracts from HMGA1a-transfected SK-N-SH cells in a dose-dependent manner (fig. 6b), furthermore, cells overexpressing HMGA1a were more sensitive to ER stress (fig. 6c), as observed in the cells expressing PS2V [12].

Interference of U1 snRNP Binding to 5′ Splice Site by HMGA1a

We examined the interaction between HMGA1a and U1-70K, one of the integral protein components of U1 snRNP [20–22]. By immunoprecipitation with either U1-70K antibody or anti-HMGA1 antibody, we found that HMGA1a has a strong affinity for U1-70K protein in hypoxia-induced SK-N-SH cells, but not in normoxia. This immunoprecipitation was due to direct protein interaction between U1-70K and HMGA1a, and we clarified that HMGA1a binds to integral component U1-70K of U1 snRNP, but not to free U1-70K. Furthermore, the production of either HMGA1a-induced or hypoxia-induced PS2V was clearly repressed by cotransfection of U1-70K, but not by cotransfection of another U1 snRNP protein, U1-A. We demonstrated that the HMGA1a binding site of PS2 pre-mRNA exists in the upstream of the 5′ splice of the exon 5. Therefore, it is plausible that the strong HMGA1a affinity for U1-70K tethers U1 snRNP to the upstream of the 5′ splice site and interferes with proper binding of U1 snRNP to the 5′ splice site. Overexpressed U1-70K, which is trapped by HMGA1a, could release U1 snRNP and allows U1 snRNP proper binding to the 5′ splice site (fig. 7).

Levels of HMGA1a Protein in Brains of Patients with Sporadic Alzheimer's Disease

Finally, we examined the levels of HMGA1a proteins in the brain tissue of patients with sporadic AD. Immunoblotting with HMGA1 antibody showed that the levels of HMGA1a were increased in total lysates and nuclear fractions from the hippocampus of patients with sporadic AD compared with those of age-matched controls. The same results were shown in immunohistochemistry.

In this report, we demonstrated that HMGA1a is a mediator of aberrant PS2 pre-mRNA splicing, or production of deleterious PS2V protein. We also observed increased steady-state levels of HMGA1a in brains of patients with sporadic AD. In our experiments, HMGA1a was induced by hypoxia only in a neuronal cell line. It was reported that microinfarction is closely correlated with a history of AD [23]. Furthermore, at the acute stage of cerebral infarction in the area of the middle cerebral artery, PS2V was detected in the region of the penumbra around the ischemic core [unpublished data, pers. commun.]. Therefore, sustained hypoxia caused by cerebral microinfarction could be one of the necessary, though not sufficient, triggers to induce HMGA1a-mediated production of the aberrant PS2V isoform in patients with sporadic AD.

Conclusion

We found that the hypoxia-induced HMGA1 protein markedly accumulated in the nuclear speckles, consistent with its function as a splicing modulator.

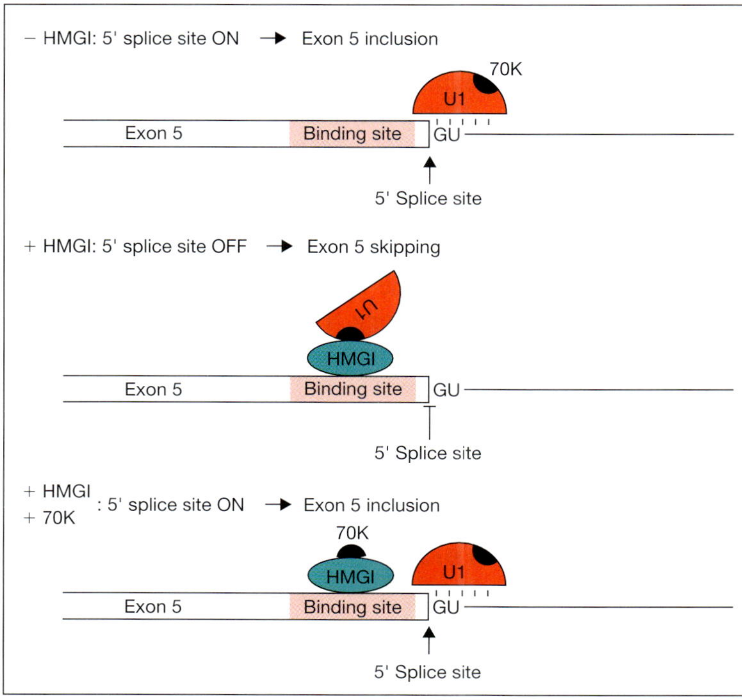

Fig. 7. Interaction of HMGA1a and U1-70K proteins and the effect of U1-70K overexpression on PS2 pre-mRNA splicing in vivo. The model for the mechanism of HMGA1a-induced aberrant exon 5 skipping. Hypoxia-induced HMGA1a protein binds to the conserved HMGA1a binding site (upstream of the 5′ splice of the middle exon 5). U1 snRNP is tethered by binding between HMGA1a and U1-70K component, which interferes with proper binding of U1 snRNP to the 5′ splice site. The disabled 5′ splice site causes exon 5 skipping between upstream exon 4 and downstream exon 6 (PS2V production). Overexpression of free U1-70K, which binds to HMGA1a, allows proper U1 snRNP allocation at the 5′ splice site, and exon 5 inclusion is restored (no PS2V production).

It was previously reported that HMGA1 proteins are mainly localized to heterochromatin masses in actively growing 3T3 fibroblasts, whereas in quiescent cells, they are more diffusely distributed [19]. Therefore, an important part of the splicing modulation function of HMGA1a in SK-N-SH cells may depend on its timely induction, triggered by hypoxic stimulation, which leads to its functional localization in nuclei. Considering all these features, HMGA1a appears to be a novel substrate-specific splicing modulator rather than a general alternative splicing factor.

In summary, we propose a novel mechanism for the development of sporadic AD cases involving induced aberrant splicing of a particular gene in the absence of any mutations. We demonstrated that HMGA1a is a key factor that binds to PS2 pre-mRNA and causes its aberrant splicing that may be an initial trigger leading to neuronal cell death in sporadic AD. Our mechanistic insight is likely to have a considerable impact on the therapy and prevention of sporadic AD.

Acknowledgements

The authors gratefully acknowledge APRO Science Co., Ltd., for their assistance in peptide sequencing, and Dr. K. Ogita for his advice on protein purification techniques. We are grateful to the Netherlands Brain Bank for the use of brain samples. We thank Drs. C.T. Moraes and F. Huijing for critical reading of the manuscript. A.M. was supported by the Institutional Research Grant (IRG-98–277–04) from A.C.S. and the Florida Biomedical Research Program Grant (BM031) from F.D.H.

References

1 Selkoe DJ: Normal and abnormal biology of the beta-amyloid precursor protein. Annu Rev Neurosci 1994;17:489–517.
2 Goate A, et al: Segregation of a missense mutation in the amyloid precursor protein gene with familial Alzheimer's disease. Nature 1991;349:704–706.
3 Rogaev EI, et al: Familial Alzheimer's disease in kindreds with missense mutations in a gene on chromosome 1 related to the Alzheimer's disease type 3 gene. Nature 1995;376:775–778.
4 Katayama T, et al: Presenilin-1 mutations downregulate the signalling pathway of the unfolded-protein response. Nat Cell Biol 1999;1:475–485.
5 Imaizumi K, Katayama T, Tohyama M: Presenilin and the UPR (letter). Nat Cell Biol 2001;3:E104.
6 Katayama T, et al: Disturbed activation of ER stress transducers by FAD-linked PS1 mutations J Biol Chem 2001;276:43446–43454.
7 Wang Y-C, Selvakumar M, Helfman DM: Alternative pre-mRNA splicing; in Krainer AR (ed): Eukaryotic mRNA Processing. Oxford, IRL Press, 1997, pp 242–279.
8 Cooper TA, Mattox W: The regulation of splice-site selection, and its role in human disease. Am J Hum Genet 1997;61:259–266.
9 Lopez AJ: Alternative splicing of pre-mRNA: developmental consequences and mechanisms of regulation. Annu Rev Genet 1998;32:279–305.
10 Hutton M, et al: Association of missense and 5′-splice-site mutations in tau with the inherited dementia FTDP-17. Nature 1998;393:702–705.
11 Lefebvre S, et al: Identification and characterization of a spinal muscular atrophy-determining gene. Cell 1995;80:155–165.
12 Sato N, et al: A novel presenilin-2 splice variant in human Alzheimer's disease brain tissue. J Neurochem 1999;72:2498–2505.
13 Sato N, et al: Increased production of β-amyloid and vulnerability to endoplasmic reticulum stress by an aberrant spliced form of presenilin-2. J Biol Chem 2001;276:2108–2114.
14 Manabe T, et al: The cytosolic inclusion bodies that consist of splice variants that lack exon 5 of the presenilin-2 gene differ obviously from Hirano bodies observed in the brain from sporadic cases of Alzheimer's disease patients. Neurosci Lett 2002;328:198–200.
15 Reeves R, Nissen MS: The A.T-DNA-binding domain of mammalian high mobility group I chromosomal proteins. A novel peptide motif for recognizing DNA structure. J Biol Chem 1990;265: 8573–8582.

16 Manabe T, et al: Induced HMGA1a expression causes aberrant splicing of Presenilin-2 pre-mRNA in sporadic Alzheimer's disease. Cell Death Diff 2003;10:698–708.
17 Spector DL, Fu X-D, Maniatis T: Associations between distinct pre-mRNA splicing components and the cell nucleus. EMBO J 1991;10:3467–3481.
18 Misteli T, Spector DL: The cellular organization of gene expression. Curr Opin Cell Biol 1998;10:323–331.
19 Martelli AM, et al: Intranuclear distribution of HMGI/Y proteins. An immunocytochemical study. J Histochem Cytochem 1998;46:863–864.
20 Krämer A: The structure and function of proteins involved in mammalian pre-mRNA splicing. Annu Rev Biochem 1996;65:367–409.
21 Will CL, Lührmann R: snRNP structure and function; in Krainer AR (ed). Eukaryotic mRNA Processing. Oxford, IRL Press, 1997, pp 130–173.
22 Burge CB, Tuschl T, Sharp PA: Splicing of precusors to mRNAs by the spliceosomes; in Gesteland RF, Cech TR, Atkins JF (eds): The RNA World, ed 2, Cold Spring Harbor, Cold Spring Harbor Laboratory Press, 1999, pp 525–560.
23 Kalaria RN: Small vessel disease and Alzheimer's dementia: Pathological considerations. Cerebrovasc Dis 2002;13(suppl 2):48–52.
24 Prihar G, et al: Structure and alternative splicing of the presenilin-2 gene. Neuroreport 1996;7:1680–1684.

Taiichi Katayama
Department of Anatomy and Neuroscience
Graduate School of Medicine, Osaka University
Suita, Osaka 565–0871 (Japan)
Tel. +81 6 6879 3221, Fax +81 6 6879 3229, E-Mail katayama@anat2.med.osaka-u.ac.jp

Takeda M, Tanaka T, Cacabelos R (eds): Molecular Neurobiology of Alzheimer Disease and Related Disorders. Basel, Karger, 2004, pp 31–41

Alzheimer's γ-Secretase Mechanism Produces Amyloid-β-Protein Like Peptides Simultaneously with Release of Intracellular Signaling Fragments

Masayasu Okochi, Akio Fukumori, Yumi Satoh, Nuripa Aidaralieva, Hisashi Tanii, Kojin Kamino, Toshihisa Tanaka, Takashi Kudo, Masatoshi Takeda

Department of Post-Genomics and Diseases, Division of Psychiatry and Behavioral Proteomics, Osaka University Graduate School of Medicine, Osaka, Japan

The amyloid hypothesis posits that the process by which secreted soluble amyloid β-protein (Aβ) turns into its aggregated insoluble form is essential for the development of Alzheimer's disease (AD). This has been the leading hypotheses to explain the pathogenesis of AD [1]. Aβ, originally identified biochemically and always present as insoluble amyloid fibrils in senile plaques of AD brains, were found to be released physiologically from cells in the form of soluble peptides [2, 3]. The amyloid hypothesis was a logical solution of this contradiction. Senile plaques and neurofibrillary tangles (NFT) are pathological structures characteristic of AD. However, although senile plaques are AD specific, NFT occur as broader and more general lesions in neurodegenerative diseases [4]. This indicates that senile plaques are related to the AD-specific pathological process, whereas NFT are more closely related to general processes of neurodegeneration.

Aβ is generated by sequential cleavages of the β-amyloid precursor protein (βAPP) [5]. Causative mutations for familial AD have been identified in *presenilins (PSs)* and *βAPP* genes [6]. It has been proposed that PS, as functions of these genes, are proteolytic enzymes (γ-secretases) and βAPP is thought to be one of their substrates [7], which is consistent with the fact that proteolytic cleavage (γ-cleavage) is directly responsible for Aβ generation [5].

These findings emphasize the importance of the Aβ peptide for understanding the pathological process of AD.

Except for unusual conditions, all pathological AD mutants of PS and βAPP affect the precision of the γ-cleavage site of βAPP [1], that is, the mutants cause a partial shift of the γ-cleavage site in the direction of the C-terminal with 2–3 amino acids [1]. As a result, the generative ratio of Aβ species ending 42 (Aβ numbering) in relation to that of the major Aβ species, ending 40 is upregulated [8]. Because (1) fibrillization of Aβ42 is much faster than that of Aβ40, and (2) Aβ42 is the major accumulating Aβ species in AD, the relative upregulation of Aβ42 in familial AD plays a central role in the insolubilization and accumulation of Aβ in the brain [1]. Aβ42 deposition in SP is also an invariant phenotype of sporadic AD, and is observed in the majority of AD cases. However, because of the highly aggregative nature of Aβ42, it has been very difficult to determine whether the precision of the γ-cleavage is affected, and thus whether Aβ42 generation is upregulated in sporadic AD brains. Nevertheless, Aβ42 peptide could be not only, as seen earlier, a substance which regulates the AD pathological process, but theoretically also one of the most effective biomarkers for AD. However, again because of its extremely aggregative nature, the Aβ42 level in CSF or peripherally of AD patients usually decreases and does not reflect its generation [9], which makes it difficult to use this level as a prediagnostic marker of AD.

We have recently found that a group of peptides may be secreted by the same mechanism as that for γ-secretase of βAPP [10, 11]. We also found that the precision of this cleavage is affected by familial AD-associated PS1 mutations similar to the pathological endoproteolysis of βAPP [10]. Therefore, by measuring these Aβ-like peptides instead of Aβ, it may be possible to determine whether γ-cleavage of βAPP is affected in sporadic AD brain. Further, it is theoretically possible that this might lead to the use of peptide levels as a prediagnostic marker for AD.

Intramembranous Endoproteolysis Is Essential for the Novel Signaling Paradigm

It is well known that signal transduction plays an important role in neural functions. In its classical form, the signal transduction paradigm is understood to mean that ligand binding to cell surface receptors induces activation of intracellular kinases or ion influx into cytosol, which functions as a second messenger. It is true that this simple paradigm has helped to explain a number of signaling events. Recently, however, a novel signaling mechanism has been proposed, in which membrane-anchored cell surface receptors themselves

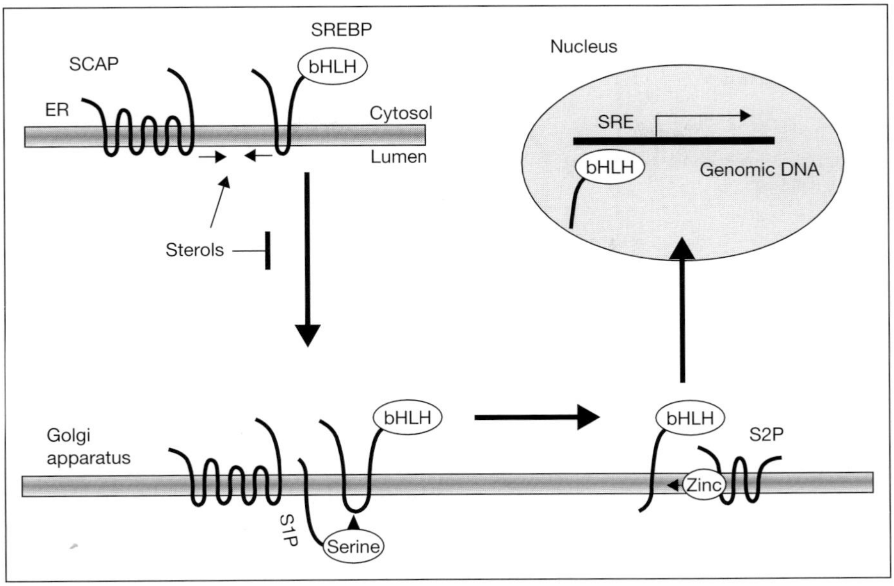

Fig. 1. Endoproteolysis of SREBPs: Concept of RIP. Serine of S1P indicates a proteolytic active center and zinc of S2P metal ion binding to the active center.

undergo sequential endoproteolysis upon ligand binding, and their intracellular domains directly translocate to the nucleus and function as transcription modifiers [12, 13].

The biochemical characteristic of such a signaling mechanism is the importance of intramembranous endoproteolysis which releases the cytosolic domain of the receptor from membranes [12, 13]. That is, fragments which translocate to the nucleus and modify transcription are immediately generated by a special form of intramembranous endoproteolysis. This cleavage is known as regulated intramembranous proteolysis (RIP) [13]. RIP is an as yet largely unknown endoproteolysis which can hydrolyse a peptide bond in a highly hydrophobic environment. RIP was first described in connection with the sequential endoproteolysis of sterol regulatory element-binding protein (SREBP) (fig. 1) [13], a membrane-bound transcription factor which regulates cholesterol homeostasis. SREBP cleavage-activating protein (SCAP), a sensor for intracellular sterols, recognizes a reduction in sterols and transports SREBPs from the endoplasmic reticulum (ER) to the Golgi membrane. The transported SREBPs are then sequentially cleaved by two Golgi-resident membrane proteases, site-1 protease (S1P) and site-2 protease (S2P), which release from the membrane the basic helix-loop-helix-leucine zipper (bHLH-Zip)

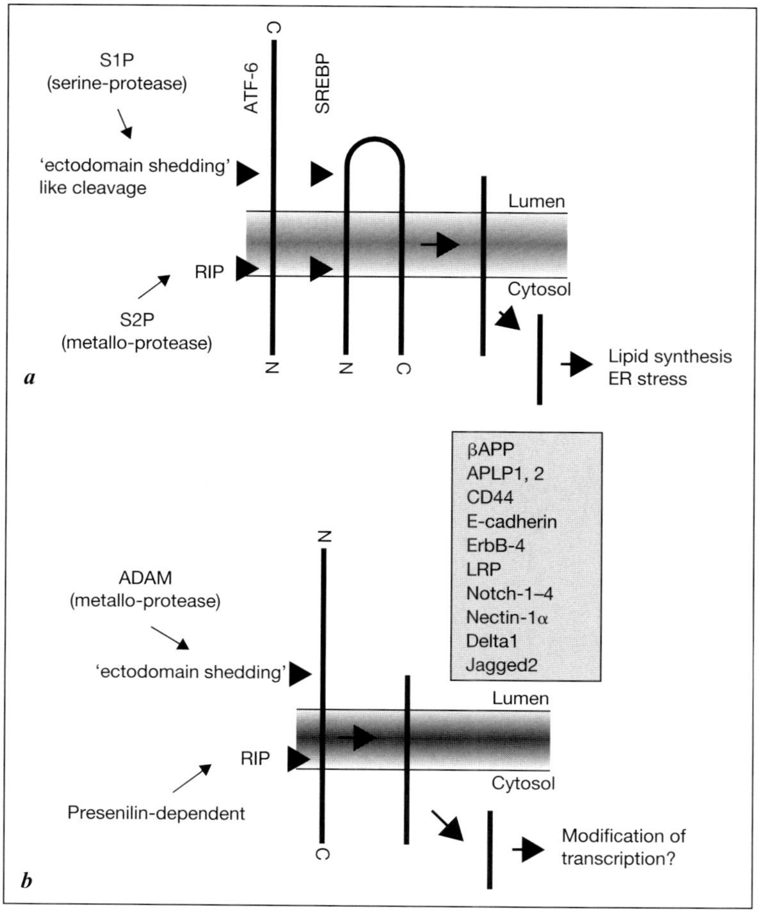

Fig. 2. ***a*** Sequential cleavages of ATF-6 and SREBPs share common features. ***b*** Common sequential cleavage mechanism (RIP) when substrates share the type-1 topology in their transmembrane domain.

domain as NTF. The functional bHLH-Zip domain then translocates to the nucleus and binds to the sterol regulatory element (SRE), which resides in the enhancer or promoter region of the target genes. When the intracellular cholesterol level increases, generation of the SCAP/SREBP complex is eliminated, which then inhibits release of the bHLH-Zip domain from the membrane and is followed by a decrease in the transcription of all target genes.

Recently, a type II membrane-anchored transcriptional factor ATF6, which is activated in ER stress response, has been shown to be a substrate for the sequential cleavages by S1P and S2P (fig. 2a) [14]. Striking similarities in

Table 1. List of putative polytopic TM I-Clips

protease	class	TM topology	substrates
S2P family	Metallo	type-2	SREBP, ATF6
Rhomboid family	Serine	type-1	TGF-α
Presenilins	Aspartic	type-1	APP, Notch, CD44, Erb-B4, etc.
SPP family	Aspartic	type-2	Signal peptide remnants

All proteases identified so far are soluble or single TM proteins, whereas all candidates for I-Clips are putative polytopic TM proteins. Moreover, I-Clips generally contain their proteolytic active centers in the hydrophobic sequence. These emphasize unusual characteristics of I-Clips. Amino acid sequences around active centers of I-Clips are reportedly not similar to conventional proteases but, in some cases, almost identical between I-Clips, which indicates that some unknown common mechanism might underlie this mysterious proteolysis.

endoproteolysis of ATF6 and SREBP can easily be found. In both cases, 'ectodomain shedding' by S1P triggers intramembranous endoproteolysis by S2P, which in turn generates NTF that translocate to the nucleus [15]. Induction of GRP78, an ER chaperone, is eliminated in cells lacking S2P [14]. Interestingly, when both S1P and S2P are involved in RIP, the transmembrane domain of the substrates seems to share type II topology.

On the other hand, when a disintegrin and metallo-protease (ADAM)- and PS-dependent γ-secretase mechanism is involved in RIP, the transmembrane domains of the substrate receptors appear to have a type I topology (fig. 2b). In addition to Notch receptors [16, 17], βAPP [18], ErbB-4 [19], E-cadherin [20], LRP [21], CD44 [11, 22], nectin-1α [23], Delta1 [31], and Jagged2 [31] have so far been identified as substrates for this mechanism. Although still controversial, these proteins are basically thought to undergo 'extracellular shedding' which is a prerequisite for consecutive PS-dependent proteolysis. Intramembrane cleaving proteases (I-Clips) are summarized in table 1.

PS comprise eight potent transmembrane proteins with both an N- and a C-terminus in cytosol [24] and occurring in high molecular weight complexes (~500 kD) [25]. PS produce γ-secretase activity, which generates both the C-terminus of Aβ and the N-terminus of the βAPP intracellular cytoplasmic domain (AICD) [26]. Genomic knock-out of PS1 or PS1/2 causes Notch phenotype in vivo, which shows that the major function of PS is to mediate Notch signaling [27]. Notch signaling was found to be a common signaling mechanism for metazoans which plays an essential role in neural differentiation from ectoderm [12]. Recently, however, this signaling has been found to play

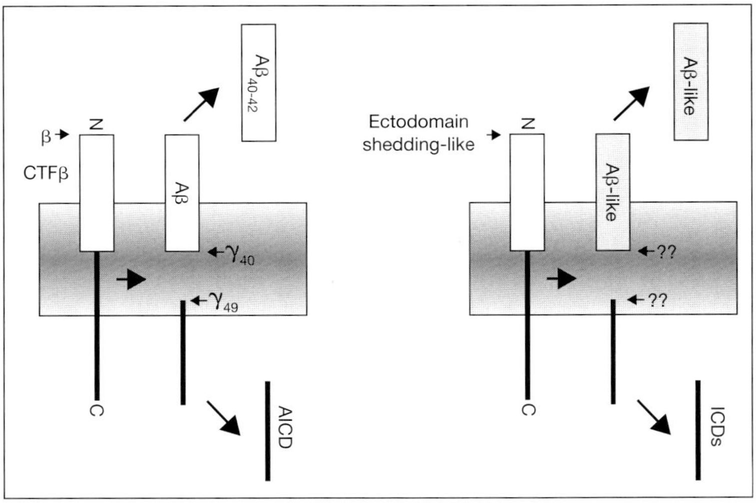

Fig. 3. Aβ-like peptides are secreted through signal transduction mediated by PS-dependent RIP.

various roles not only during development but also in adulthood. Notch signaling is realized only when Notch ligands (DSL family proteins) expressed in signaling cells bind to Notch receptors expressed in the signal-receiving cells. Upon binding to ligands, Notch receptors undergo sequential endoproteolysis, which results in the release of the cytosolic C-terminal fragment, NICD (Notch intracellular cytoplasmic domain), which is believed to directly translocate to the nucleus and regulate transcription of target genes [12].

Notch-1-β and CD44-β Peptides, Aβ-Like Fragments, Are Physiologically Secreted

We have analyzed in detail the PS-dependent intramembranous proteolysis of Notch-1 [10] and CD44 [11] and found that, as a result of the endoproteolysis, the Aβ-like Notch (Notch-1 Aβ-like peptide: Nβ) or CD44 (CD44 Aβ-like peptide: CD44β) fragment was extracellularly secreted as NTF [10, 11] (fig. 3). This indicates that at least several peptides that contain a transmembrane domain-like Aβ are secreted in vivo (fig. 4a). We suggest that secretion of peptides containing the transmembrane domain may be a phenomenon common to all substrates for PS-dependent endoproteolysis. Interestingly, the C-termini of these secreted peptides do not directly correspond to the N-termini of cytosolic C-terminal fragments (CTFs) functioning as signaling molecules,

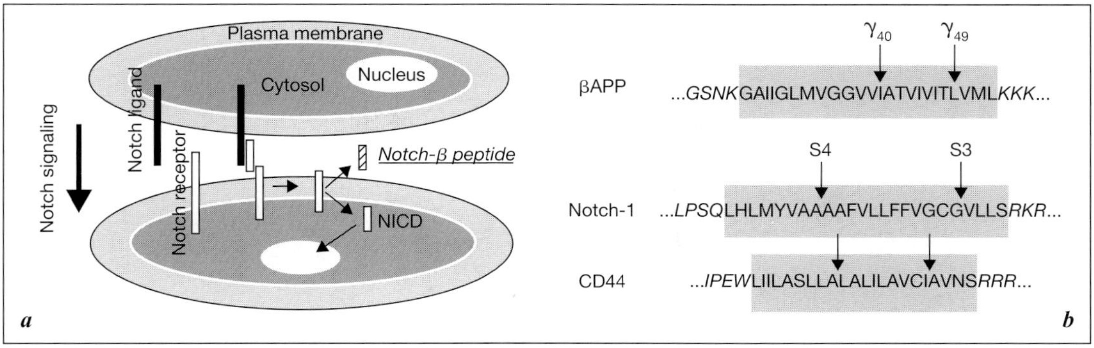

Fig. 4. ***a*** Notch signaling accompanies secretion of Notch-β peptide. By sequential endoproteolysis of Notch-1 shown in figures 2b, 3, an Aβ-like peptide, Nβ, is released. ***b*** PS-dependent intramembranous proteolysis, which we termed the 'dual cleavage' mechanism. Arrows indicate proteolytic cleavage sites. Small transmembrane peptides between 2 cleavage sites (arrows) have not yet been identified.

but are formed by distinct proteolysis upstream of N-termini of CTFs (fig. 4b). Thus, intramembranous endoproteolysis, which liberates an Aβ-like peptide, essentially consists of a distinct dual endoproteolysis, which we have termed 'dual cleavage' mechanism (fig. 4b) [10, 11]. These findings seem to indicate that 'dual cleavage' is necessary to degrade and liberate transmembrane peptides from membrane.

An important finding is that, similar to the pathological cut of βAPP, the precision of the γ-cleavage-generating C-terminus of Nβ is affected by familial-AD-associated PS1/2 mutations (fig. 5) [10]. These mutations were found to cause a partial shift in the cleavage site that generates increased levels of Nβ species whose C-termini are elongated by 2–4 amino acids [10]. This means that the level of secretion of Nβ1733-35 compared to that of Nβ1731, the most abundant Nβ species, is upregulated in the mutant-expressing cells [10]. We therefore suggest that secretion of Aβ-like peptides such as Nβ share the same γ-secretase mechanism as that of Aβ (see also fig. 2b, 3, 4b, 5).

Level of an Elongated Aβ-Like Peptide as a Substitute for Aβ May Reflect AD-Associated Pathological Impairment of γ-Secretase

Although the findings are only preliminary, Nβ, like Aβ, did not seem to aggregate, fibrillate nor accumulate in AD brains [Okochi and Arai,

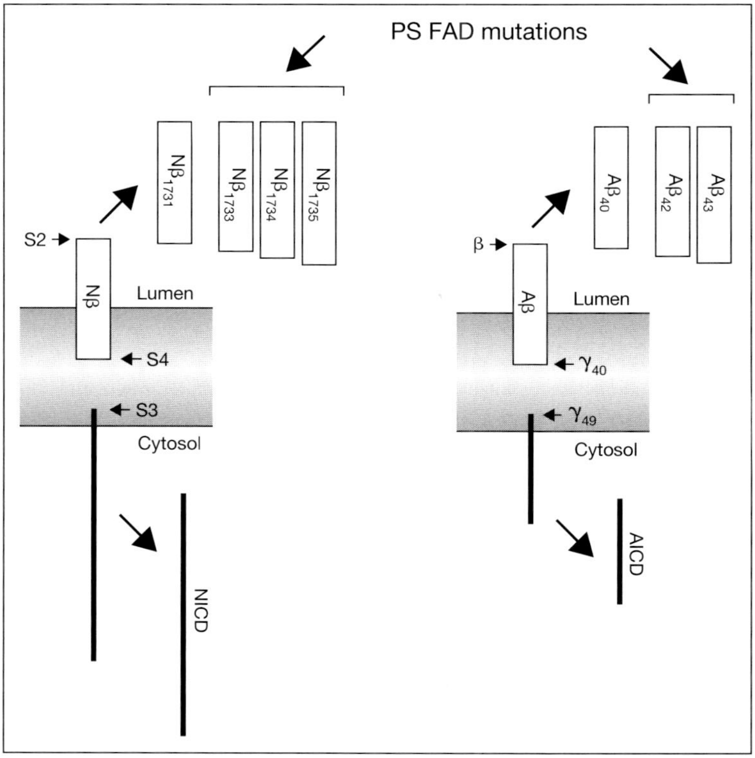

Fig. 5. Familial-AD-associated PS mutations affect γ-cleavage and elongate Nβ. These mutants showed a very similar effect on the precision of the γ-cleavage for the two distinct substrates. The magnitude of the effect, as analyzed so far, was not dependent on the substrates but on mutations of PS. In other words, PS mutants, while dramatically upregulating Aβ42 generation, simultaneously increase the level of elongated Nβ. This seems to indicate that mutations affect direct interaction between PS and their substrates.

unpubl. obs.]. Therefore, by measuring Nβ or the level of Nβ1733-35 relative to that of Nβ1731 in CSF or peripheral, it may be possible to determine the level of γ-secretase activity or Aβ42/40 generation ratio in patients with sporadic AD (see also fig. 5). Aβ deposition leading to AD may gradually take place over a number of years. It is likely that, in the process, γ-secretase activity is upregulated [28] or the precision of γ-cleavage in the brain is affected [1]. Therefore, by measuring the level of Aβ-like peptide or the relative level of elongated species in healthy individuals, it may be possible to diagnose those who are likely to develop AD before they show symptoms.

It Is Worthwhile Studying Whether Secreted Nβ Peptide Level Is Upregulated in Cells of Human Malignancies

Various kinds of examinations have demonstrated that expression levels of Notch receptors including Notch-1 are strongly upregulated in tumor cells, which indicates that Notch signaling is promoted in human malignancies [29]. Very recent evidence indicates a novel mode of cross-talk between the epidermal growth factor receptor/Ras/mitogen-activated protein kinase cascade and the Notch pathway [30]. Moreover, there are indications that oncogenic mutants of Ras observed in 25–50% of human cancer perform an oncogenic function through Notch signaling [29]. One might therefore argue that Notch signaling, a local cell signaling which suppresses cell differentiation and promotes proliferation, may be involved in tumor genesis itself. However, since no one knows how local cell signaling can be monitored, no attempts have been made so far, to measure the signaling level or to evaluate the level in relation to diagnosis or therapy for human tumor in vivo. We have discovered that, for each Notch-signaling fragment produced, a kind of 'peptide evidence' of the signaling event is definitely secreted extracellularly (see also fig. 4a). By measuring this Nβ peptide level, therefore, the level of Notch signaling may be assessed.

Acknowledgement

We wish to thank Dr. Tetsuaki Arai of the Department of Neuropathology, Tokyo Institute of Psychiatry, for pathological characterization of anti-Nβ antibodies and our colleagues at the Laboratory for Biochemistry, especially Dr. Naohiko Matsumoto, for critically reading the manuscript. This research project has been conducted in close collaboration with Prof. Christian Haass and Dr. Harald Steiner of the Department of Biochemistry, Laboratory for Alzheimer's Disease Research, Ludwig Maximilian University, Munich. This work was supported by grants from the Ministry of Education, Science, Culture and Sports (14017060 and 14770499 to M.O.) and the Ministry of Health and Welfare (14121601 to M.T. and M.O. and 13080101 to M.T.).

References

1 Selkoe DJ: Alzheimer's disease: Genes, proteins, and therapy (review). Physiol Rev 2001;741–766.
2 Haass C, Schlossmacher MG, Hung AY, Vigo-pelfrey C, Mellon A, Ostaszewski BL, Lieberburg I, Koo EH, Schenk D, Teplow DB, et al: Amyloid beta-peptide is produced by cultured cells during normal metabolism. Nature 1992;359:322–325.
3 Shoji M, Golde TE, Ghiso J, Cheung TT, Estus S, Shaffer LM, Cai XD, McKay DM, Tintner R, Frangione B, et al: Production of the Alzheimer amyloid beta protein by normal proteolytic processing. Science 1992;258:126–129.
4 Hardy J: Genetic dissection of primary neurodegenerative diseases (review). Biochem Soc Symp 2001;67:51–57.

5 Haass C, Selkoe DJ: Cellular processing of beta-amyloid precursor protein and the genesis of amyloid beta-peptide (review). Cell 1993;75:1039–1042.

6 St George-Hyslop PH: Molecular genetics of Alzheimer's disease review. Biol Psychiatry 2000; 47:183–199.

7 Wolfe MS, Xia W, Ostaszewski BL, Diehl TS, Kimberly WT, Selkoe DJ: Two transmembrane aspartates in presenilin-1 required for presenilin endoproteolysis and gamma-secretase activity. Nature 1999;398:513–517.

8 Steiner H, Haass C: Intramembrane proteolysis by presenilins (review). Nat Rev Mol Cell Biol 2000;1:217–224.

9 Teunissen CE, de Vente J, Steinbusch HW, De Bruijn C: Biochemical markers related to Alzheimer's dementia in serum and cerebrospinal fluid. Neurobiol Aging 2002;23:485–508.

10 Okochi M, Steiner H, Fukumori A, Tanii H, Tomita T, Tanaka T, Iwatsubo T, Kudo T, Takeda M, Haass C: Presenilins mediate a dual intramembranous gamma-secretase cleavage of Notch-1. EMBO J 2002;21:5408–5416.

11 Lammich S, Okochi M, Takeda M, Kaether C, Capell A, Zimmer AK, Edbauer D, Walter J, Steiner H, Haass C: Presenilin-dependent intramembrane proteolysis of CD44 leads to the liberation of its intracellular domain and the secretion of an Abeta-like peptide. J Biol Chem 2002;277:44754–44759.

12 Mumm JS, Kopan R: Notch signaling: From the outside in (review). Dev Biol 2000;228:151–165.

13 Brown MS, Ye J, Rawson RB, Goldstein JL: Regulated intramembrane proteolysis: A control mechanism conserved from bacteria to humans (review). Cell 2000;100:391–398.

14 Ye J, Rawson RB, Komuro R, Chen X, Dave UP, Prywes R, Brown MS, Goldstein JL: ER stress induces cleavage of membrane-bound ATF6 by the same proteases that process SREBPs. Mol Cell 2000;6:1355–1364.

15 Yoshida H, Matsui T, Yamamoto A, Okada T, Mori K: XBP1 mRNA is induced by ATF6 and spliced by IRE1 in response to ER stress to produce a highly active transcription factor. Cell 2001; 107:881–891.

16 Mizutani T, Taniguchi Y, Aoki T, Hashimoto N, Honjo T: Conservation of the biochemical mechanisms of signal transduction among mammalian Notch family members. Proc Natl Acad Sci USA 2001;98:9026–9031.

17 Saxena MT, Schroeter EH, Mumm JS, Kopan R: Murine notch homologs (N1-4) undergo presenilin-dependent proteolysis. J Biol Chem 2001;276:40268–40273.

18 Naruse S, Thinakaran G, Luo JJ, Kusiak JW, Tomita T, Iwatsubo T, Qian X, Ginty DD, Price DL, Borchelt DR, Wong PC, Sisodia SS: Effects of PS1 deficiency on membrane protein trafficking in neurons. Neuron 1998;21:1213–1221.

19 Lee HJ, Jung KM, Huang YZ, Bennett LB, Lee JS, Mei L, Kim TW: Presenilin-dependent gamma-secretase-like intramembrane cleavage of ErbB4. J Biol Chem 2002;277:6318–6323.

20 Marambaud P, Shioi J, Serban G, Georgakopoulos A, Sarner S, Nagy V, Baki L, Wen P, Efthimiopoulos S, Shao Z, Wisniewski T, Robakis NK: A presenilin-1/gamma-secretase cleavage releases the E-cadherin intracellular domain and regulates disassembly of adherens junctions. EMBO J 2002;21:1948–1956.

21 May P, Reddy YK, Herz J: Proteolytic processing of low density lipoprotein receptor-related protein mediates regulated release of its intracellular domain. J Biol Chem 2002;277:18736–18743.

22 Okamoto I, Kawano Y, Murakami D, Sasayama T, Araki N, Miki T, Wong AJ, Saya H: Proteolytic release of CD44 intracellular domain and its role in the CD44 signaling pathway. J Cell Biol 2001;155:755–762.

23 Kim DY, Ingano LA, Kovacs DM: Nectin-1alpha, an immunoglobulin-like receptor involved in the formation of synapses, is a substrate for presenilin/gamma-secretase-like cleavage. J Biol Chem 2002;277:49976–49981.

24 Li X, Greenwald I: Membrane topology of the *C. elegans* SEL-12 presenilin. Neuron 1996;17: 1015–1021.

25 Edbauer D, Winkler E, Haass C, Steiner H: Presenilin and nicastrin regulate each other and determine amyloid beta-peptide production via complex formation. Proc Natl Acad Sci USA 2002;99: 8666–8671.

26 Haass C, Steiner H: Alzheimer disease gamma-secretase: A complex story of GxGD-type presenilin proteases. Trends Cell Biol 2002;12:556–562.

27 Wong PC, Zheng H, Chen H, Becher MW, Sirinathsinghji DJ, Trumbauer ME, Chen HY, Price DL, Van der Ploeg LH, Sisodia SS: Presenilin 1 is required for Notch1 and Dll1 expression in the paraxial mesoderm. Nature 1997;387:288–292.
28 Ikeuchi T, Dolios G, Kim SH, Wang R, Sisodia SS: Familial Alzheimer disease-linked presenilin 1 variants enhance production of both Abeta 1–40 and Abeta 1–42 peptides that are only partially sensitive to a potent aspartyl protease transition-state inhibitor of 'gamma-secretase'. J Biol Chem 2003;278:7010–7018.
29 Weijzen S, Rizzo P, Braid M, Vaishnav R, Jonkheer SM, Zlobin A, Osborne BA, Gottipati S, Aster JC, Hahn WC, Rudolf M, Siziopikou K, Kast WM, Miele L: Activation of Notch-1 signaling maintains the neoplastic phenotype in human Ras-transformed cells. Nat Med 2002;8:979–986.
30 Shaye DD, Greenwald I: Endocytosis-mediated downregulation of LIN-12/Notch upon Ras activation in *Caenorhabditis elegans*. Nature 2002;420:686–690.
31 Ikeuchi T, Sisodia SS: The Notch ligands, Delta1 and Jagged2, are substrates for presenilin-dependent "gamma-secretase" cleavage. J Biol Chem 2003;278:7751–7754.

Masayasu Okochi
Department of Post-Genomics and Diseases
Division of Psychiatry and Behavioral Proteomics
Osaka University Graduate School of Medicine, 565–0871 Osaka (Japan)
Tel. +81 6 6879 3053, Fax +81 6 6879 3059, E-Mail mokochi@psy.med.osaka-u.ac.jp

Takeda M, Tanaka T, Cacabelos R (eds): Molecular Neurobiology of Alzheimer Disease and Related Disorders. Basel, Karger, 2004, pp 42–51

Pivotal Role of Neurofibrillary Degeneration in Alzheimer Disease and Therapeutic Targets

Khalid Iqbal, Alejandra del C. Alonso, Ezzat El-Akkad, Cheng-Xin Gong, Niloufar Haque, Sabiha Khatoon, Hitoshi Tanimukai, Ichiro Tsujio, Inge Grundke-Iqbal

New York State Institute for Basic Research in Developmental Disabilities, Staten Island, N.Y., USA

Alzheimer disease (AD), which is the single major cause of dementia, is a multifactorial disease. In less than 5% of the cases the disease cosegregates with certain mutations in β-amyloid precursor protein (β-APP), presenilin-1 or presenilin-2 (for review, see ref. [1]). Over 95% of the AD cases are not associated with any known mutations, and the nature of the etiological agent(s), which is most probably some metabolic and or environmental factor, is not as yet understood [2]. Thus, based on multi-etiology, AD offers a large number of therapeutic targets. However, any such therapy is likely to be effective on only a subset of AD, some of which, like AD caused by β-APP mutations, might be only a small fraction of the total AD cases. It is, therefore, very important to identify a therapeutic target(s) that can be effective in most, if not all, cases of AD. Inhibition of neurofibrillary degeneration offers such a target.

Independent of the etiology, whether genetic or nongenetic, neurofibrillary degeneration and β-amyloidosis are the two histopathological hallmark lesions of AD. Whereas neurofibrillary degeneration is associated with the clinical expression of AD, i.e. dementia, β-amyloidosis alone in the absence of neurofibrillary degeneration does not produce the disease. Hereditary cerebral hemorrhage with amyloidosis of the Dutch type (HCHWA-D) and sporadic cerebral amyloid angiopathy (CAA) are characterized by extensive β-amyloidosis without any dementia [3–5]. Some of the normal aged individuals have as much β-amyloid plaques in the brain as typical cases of AD [6–9]. Furthermore, there are several other neurodegenerative conditions, such as the Guam parkinsonism-dementia

Table 1. Tauopathies/diseases characterized by the accumulation of abnormally hyperphosphorylated tau

- Alzheimer disease
- Down syndrome, adult cases
- Guam parkinsonism-dementia complex
- Dementia pugilistica
- Pick disease
- Dementia with argyrophilic grains
- Frontotemporal dementia
- Corticobasal degeneration
- Pallidopontonigral degeneration
- Progressive supranuclear palsy
- Gerstmann-Sträussler-Scheinker disease with tangles

complex, dementia pugilistica, frontotemporal dementia with parkinsonism linked to chromosome 17 (FTDP-17) and progressive supranuclear palsy, which are characterized by neurofibrillary degeneration of the Alzheimer type and dementia but in the absence of β-amyloidosis. Furthermore, the discovery of mutations in the tau gene and their cosegregation with the disease in the inherited FTDP-17 has confirmed that abnormalities in tau protein as a primary event can lead to neurodegeneration and dementia [10–12]. Neither β-amyloidosis nor neurofibrillary degeneration is unique to AD but disorders with the latter lesion called tauopathies are associated with dementia (table 1). The neurofibrillary degeneration of the Alzheimer type is seen only sparsely in aged animals and in experimentally induced conditions. Thus, all these findings taken together suggest that neurofibrillary degeneration plays a pivotal role in the pathogenesis of AD and related tauopathies and that inhibition of this lesion is a promising therapeutic target for these diseases. Identification of specific therapeutic pharmacological targets requires understanding of the molecular mechanism by which this lesion might cause neurodegeneration.

Neurofibrillary Degeneration

Microtubule-associated protein (MAP) tau, which promotes the assembly and maintains the structure of microtubules in a normal mature neuron is a family of six isoforms that differ from one another in having three or four microtubule-binding repeats (R) of 31–32 amino acids each, and two, one or none amino terminal inserts (N) of 29 amino acids each [13, 14]. Tau is abnormally hyperphosphorylated in AD and in this form is the major protein subunit of paired helical filaments (PHF) [15–19]. In AD brains the levels of tau, but not the mRNA for this protein [20], are 4- to 8-fold increased as compared to age-matched control brains and this increase is in the form of the abnormally hyperphosphorylated tau [21, 22]. The abnormally hyperphosphorylated tau is found in AD brains in two subcellular pools, i.e., (1) as polymerized into neurofibrillary tangles of PHF mixed with straight filaments (SF); and (2) in a nonfibrillized form in the cytosol [17, 23, 24]. The tau polymerized in neurofibrillary tangles is apparently inert and behaves like normal tau in promoting microtubule assembly only on enzymatic dephosphorylation in vitro when released from PHF/tangles [25, 26]. Whereas the cytosolic abnormally hyperphosphorylated tau (AD P-tau) which can be as much as ~40% of the total abnormal tau in AD brains [24] does not interact with tubulin/microtubules but instead sequesters normal tau, MAP1 and MAP2, causing inhibition and disassembly of microtubules in vitro [27–29]. Furthermore, the association between AD P-tau and normal tau is not saturable and results in the formation of tangles of ~2.1-mm filaments in vitro [28]; the association between AD P-tau and MAP1 or MAP2 is weaker than that between the AD P-tau and normal tau and does not result in the formation of filaments [29].

Fig. 1. Etiology and molecular mechanism of neurofibrillary degeneration and its relationship to β-amyloidosis. Mutations in β-APP, presenilin-1 and presenilin-2 account for less than 5% of all cases of AD. The remaining 95% of AD cases appear to represent the sporadic form of the disease. Although β-amyloidosis and neurofibrillary degeneration are the two histopathological hallmarks of AD, no temporal relationship between the two lesions has been established to date and neurofibrillary degeneration but not β-amyloidosis appears to be required for the clinical expression of the disease, i.e. dementia. While β-amyloidosis can occur without associated dementia such as in normal brain aging, hereditary cerebral hemorrhage with amyloidosis of the Dutch type (HCHWA-D) and sporadic cerebral amyloid angiopathy (CAA), neurofibrillary degeneration is always associated with dementia. Besides AD, neurofibrillary changes, which are made up of the abnormally hyperphosphorylated τ, are seen in a number of related disorders, called tauopathies (table 1). The cosegregation of certain mutations in the tau gene with the disease in inherited cases of frontotemporal dementia with Parkinsonism linked to chromosome 17 (FTDP-17), a tauopathy demonstrates that neurofibrillary degeneration can occur without β-amyloidosis, and can be a primary cause of neurodegeneration and associated dementia. In AD, more than one etiological factor initiates one or more specific signal transduction cascades, which result in a protein phosphorylation/dephosphorylation imbalance and abnormal hyperphosphorylation of tau. The abnormally hyperphosphorylated tau sequesters normal tau, MAP1 and MAP2 and leads to neurodegeneration. This toxic behavior of the abnormal tau is attenuated by its polymerization into PHF/neurofibrillary tangles, which slowly and gradually, i.e. over a period of several years too, probably as a space-occupying lesion, contributes to the neuronal degeneration. Glycosylation appears to stabilize the neurofibrillary changes. Ubiquitination probably represents an unsuccessful attempt of the affected neurons to degrade the tangles.

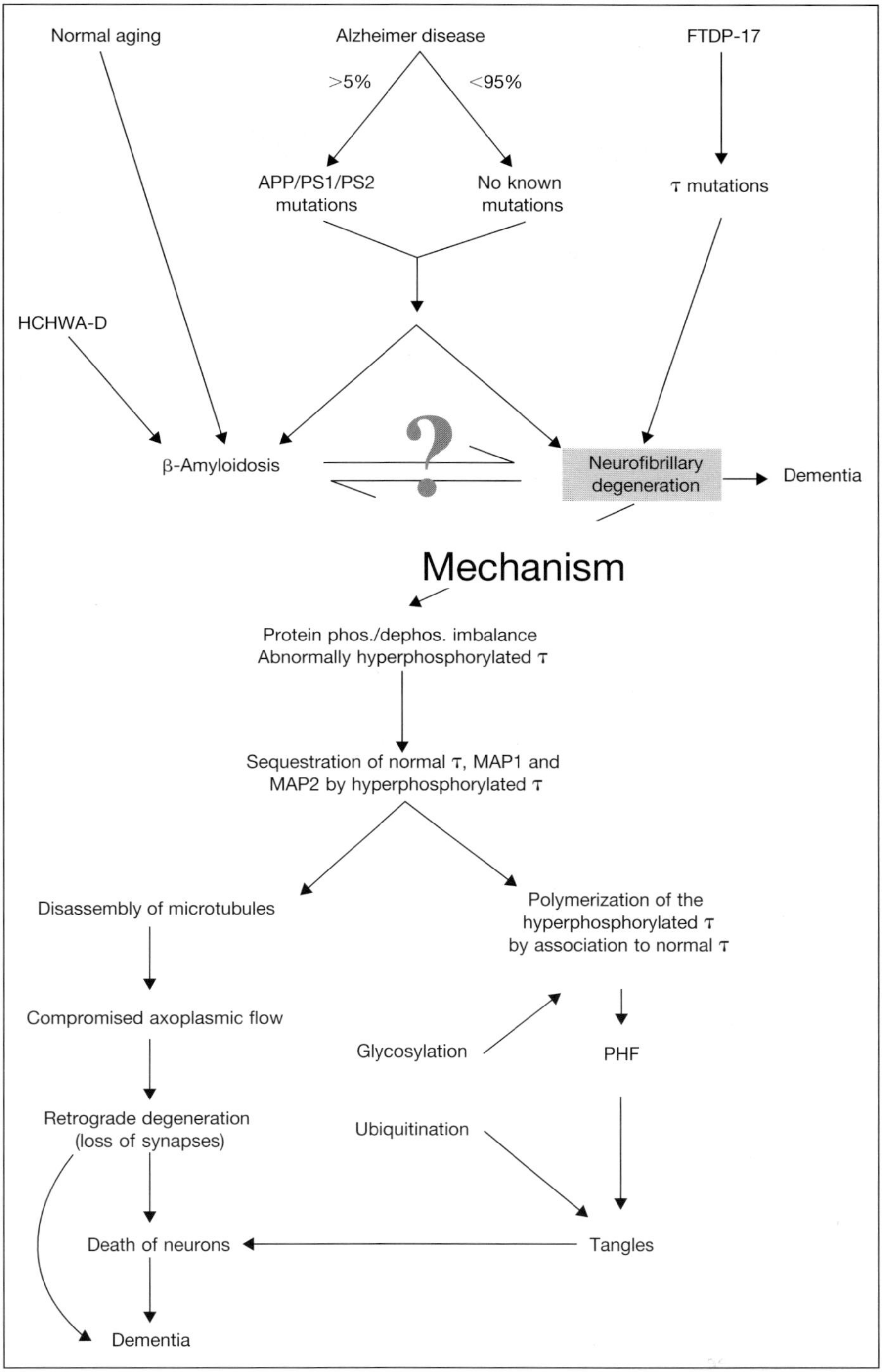
Normal aging
Alzheimer disease
FTDP-17
>5%
<95%
APP/PS1/PS2
mutations
No known
mutations
τ mutations
HCHWA-D
β-Amyloidosis
?
Neurofibrillary
degeneration
Dementia
Mechanism
Protein phos./dephos. imbalance
Abnormally hyperphosphorylated τ
Sequestration of normal τ, MAP1 and
MAP2 by hyperphosphorylated τ
Disassembly of microtubules
Polymerization of the
hyperphosphorylated τ
by association to normal τ
Compromised axoplasmic flow
Glycosylation
PHF
Retrograde degeneration
(loss of synapses)
Ubiquitination
Death of neurons
Tangles
Dementia

1

This toxic property of AD P-tau appears to be solely due to its abnormal hyperphosphorylation because dephosphorylation by alkaline phosphatase, protein phosphatase (PP)-2A, PP-2B and to a lesser degree by PP-1 converts the abnormal tau into a normal-like protein in promoting the microtubule assembly in vitro [25–30].

The six human tau isoforms, τ4RL (4R, 2N), τ4S (4R, 1N), τ4 (4R, no N), τ3RL (3R, 2N), τ3RS (3R, 1N), and τ3, also called fetal tau (3R, no N), are differentially sequestered by AD P-tau, in vitro [31]. The association of AD P-tau to normal tau, human brain recombinant taus is τ4RL > τ4RS > τ4R and τ3RL > τ3RS > τ3, and that of τ4RL > τ3RL. AD P-tau also inhibits the assembly and disrupts microtubules preassembled with each tau isoform with an efficiency which corresponds directly to the degree of interaction with these isoforms. In vitro hyperphosphorylation of recombinant tau converts it into an AD P-tau-like state in sequestering normal tau and inhibiting microtubule assembly. The preferential sequestration of 4R taus and taus with amino terminal inserts explains both (1) why fetal brain (fetal tau is with 3R and no N) is protected from Alzheimer neurofibrillary pathology and (2) why intronic mutations seen in certain inherited cases of FTDP-17, which results in alternate splicing of tau mRNA, and consequently an increase in 4R:3R ratio, lead to neurofibrillary degeneration and the disease.

The abnormal hyperphosphorylation of tau makes it resistant to proteolysis by the calcium-activated neutral protease [26, 30] and most likely it is because of this reason the levels of tau are several-fold increased in AD [21, 22]. It is likely that to neutralize the ability of AD P-tau to sequester normal MAPs and cause disassembly of microtubules the affected neurons promote the self-assembly of the abnormal tau into tangles of PHF. The fact that the tangle-bearing neurons seem to survive for many years [32] is consistent with such a self-defense role of the formation of tangles. The AD P-tau readily self-assembles into tangles of PHF/SF in vitro under physiological conditions of protein concentration, pH, ionic strength and reducing conditions [33]. Furthermore, dephosphorylation inhibits the self-assembly of AD P-tau into PHF/SF, and the in vitro abnormal hyperphosphorylation of each of the six recombinant human brain tau isoforms promotes their assembly into tangles of PHF/SF. Thus, all these studies taken together demonstrate the pivotal involvement of abnormal hyperphosphorylation in neurofibrillary degeneration.

Tau Kinases and Phosphatases

The state of phosphorylation of a phosphoprotein is a function of the balance between the activities of the protein kinases and the protein phosphatases that regulate its phosphorylation. Tau, which is phosphorylated only at serine/threonine residues, is a substrate for several protein kinases such as glycogen synthase kinase-3, cyclin dependent protein kinase-5, protein kinase A, calcium and calmodulin-dependent protein kinase-II, and stress-activated protein kinases (for review, see ref. [2, 34]). However, to date, the activities of none of these protein kinases have been reproducibly shown to be upregulated in AD brains. In contrast, the activities of protein phosphatase (PP)-2A and PP-1 are compromised by ~20–30% in AD brains [35, 36], and the phosphorylation of tau that suppresses its microtubule binding and assembly activities in adult mammalian brain is regulated by PP-2A, not by PP-2B [37, 38]. PP-2A also regulates the activities of several tau kinases in brain. Inhibition of PP-2A activity by okadaic acid in metabolically active rat brain slices results in abnormal hyperphosphorylation of tau at several of the same sites as in AD, not only directly by a decrease in dephosphorylation but also indirectly by promoting the activities of CaM kinase II [37], MAP kinase kinase (MEK1/2), extracellular regulated kinase (ERK 1/2) and P70S6 kinase [Pei et al., Am J Pathol, in press].

PP-2A and PP-1 make more than 90% of the serine/threonine protein phosphatase activity in mammalian cells [39]. The intracellular activities of these enzymes are regulated by endogenous inhibitors. PP-1 activity is regulated mainly by a 18.7-kD heat-stable protein called inhibitor-1 (I-1) [40, 41]. In addition, a structurally related protein, DARPP-32 (dopamine and cAMP-regulated phosphoprotein of apparent molecular weight 32,000) is expressed predominantly in the brain [42]. I-1 and DARPP-32 are activated on phosphorylation by protein kinase A and inactivated at basal calcium level by PP-2A. Thus, inhibition of PP-2A activity would keep I-1, DARPP-32 in active form and thereby result in a decrease in PP-1 activity. In AD brain a reduction in PP-2A activity might have decreased the PP-1 activity by allowing the upregulation of I-1/DARPP-32 activity. PP-2A is inhibited in the mammalian tissue by two heat-stable proteins: (1) the I_1^{PP2A}, a 30-kD cytosolic protein that inhibits PP-2A with a ki of 30 n*M* and (2) the I_2^{PP2A}, a 39-kD nuclear protein that inhibits PP-2A at a ki of 23 n*M* [43, 44]. Both I_1^{PP2A} and I_2^{PP2A} have been cloned from human kidney [44, 45] and brain [46]. I_1^{PP2A} has been found to be the same protein as the putative histocompatibility leukocyte antigen class II-associated protein (PHAP-1). This protein, which has also been described as mapmodulin, pp32 and LANP [47], is 249 amino acids long and has an apparent molecular weight of 30 kD on SDS-PAGE. I_2^{PP2A}, which is the same as TAF-1β or PHAPII, is a nuclear protein that is a homologue of the human SETα protein [48]. In a preliminary study, we have found that the level of I_1^{PP2A} is ~20% increased in AD brains as compared with age-matched control brains.

Therapeutic Targets

The most promising therapeutic approaches to inhibit neurofibrillary degeneration and consequently AD are (1) to inhibit sequestration of normal MAPs by the AD P-tau and (2) to inhibit the abnormal hyperphosphorylation of tau. The latter can be carried out either by inhibiting the activity of a PP-2A inhibitor or restoring the PP-2A activity in the affected areas of the brain or by inhibiting the activity of one or more tau kinase activities that are critically involved in converting normal tau into an abnormal state whereby it sequesters normal MAPs. Memantine, a low to moderate affinity NMDA receptor antagonist, which improves mental function and the quality of daily living of patients with moderate to severe AD [49, 50] restores the okadaic acid-induced inhibition of PP-2A activity and the abnormal hyperphosphorylation of tau at Ser-262 in hippocampal slice cultures from adult rats [51].

Development of therapeutic targets requires (1) a compelling scientific rationale for a therapeutic target and (2) the availability of a practical outcome measure(s). Tau is primarily a neuronal protein and its level in CSF is a reliable measure of the rate of neuronal degeneration. The CSF level of this protein both as total tau and as tau abnormally hyperphosphorylated at Ser-396/404 is markedly elevated in AD [52]. The differential diagnosis between AD and vascular dementia, the two major causes of age-associated dementia can be made in the living patients by determining the ratio of the abnormally hyperphosphorylated tau to total tau in the lumbar CSF [52]. Thus, levels of CSF total tau and abnormally hyperphosphorylated tau offer excellent outcome measures to test the efficacy of therapeutic agents towards total neurodegeneration and neurofibrillary degeneration, respectively. These outcome measures can be used to test drugs that inhibit neurofibrillary degeneration either by inhibiting the sequestration of normal MAPs by the AD P-tau or by inhibiting the abnormal hyperphosphorylation of tau.

In conclusion, given the pivotal and the primary role of neurofibrillary degeneration in AD and related tauopathies, identification of the rational therapeutic targets for this lesion and the availability of the CSF total tau and abnormally hyperphosphorylated tau as outcome measures, it is now feasible to develop a new generation of drugs that can inhibit and or prevent AD and related tauopathies.

Acknowledgements

We are grateful to Janet Biegelson and Sonia Warren for secretarial assistance. Studies in our laboratories were supported in part by the New York State Office of Mental

Retardation and Developmental Disabilities and NIH grant AG19158, Alzheimer's Association (Chicago, Ill.) grant IIRG-00–2002 and a grant from the Institute for the Study of Aging (ISOA), New York, N.Y.

References

1 Finch C, Tanzi RE: Genetics of aging. Science 1997;278:407–411.
2 Iqbal K, Grundke-Iqbal I: Metabolic hypothesis, mechanism and therapeutic targets of Alzheimer neurofibrillary degeneration. Neurosci News 2000;3:14–20.
3 Coria F, Castaño B, Frangione B: Brain amyloid in normal aging and cerebral amyloid angiopathy is antigenically related to Alzheimer's disease beta-protein. Am J Pathol 1987;129:422.
4 Levy E, Carman MD, Fernandez-Madrid IJ, Power MD, Lieberburg I, van Duinen SG, Bots GT, Luyendijk W, Frangione B: Mutation of the Alzheimer's disease amyloid gene in hereditary cerebral hemorrhage, Dutch type. Science 1990;248:1124–1126.
5 van Duinen SG, Castano EM, Prelli F, Bots GT, Luyendijk W, Frangione B: Hereditary cerebral hemorrhage with amyloidosis in patients of Dutch origin is related to Alzheimer disease. Proc Natl Acad Sci USA 1987;84:5991–5994.
6 Alafuzoff I, Iqbal K, Friden H, Adolfson R, Winblad B: Histopathological criteria for progressive dementia disorders: Clinical-pathological correlation and classification by multivariate data analysis. Acta Neuropathol (Berlin) 1987;74:209–225.
7 Arrigada PA, Growdon JH, Hedley-White ET, Hyman BT: Neurofibrillary tangles but not senile plaques parallel duration and severity of Alzheimer's disease. Neurology 1992;42:631–639.
8 Dickson DW, Crystal HA, Mattiace LA, Masur DM, Blau AD, Davies P, Yen SH, Aronson M: Identification of normal and pathological aging in prospectively studied non-demented elderly humans. Neurobiol Aging 1991;13:179–189.
9 Katzman R, Terry RD, DeTeresa R, Brown R, Davies P, Fuld P, Renling X, Peck A: Clinical, pathological and neurochemical changes in dementia: A subgroup with preserved mental status and numerous neocortical plaques. Ann Neurol 1988;23:138–144.
10 Hutton M, Lendon CL, Rizzu P, Baker M, Froelich S, Houlden H, Pickering-Brown S, Chakraverty S, Isaacs A, Grover A, Hackett J, Adamson J, Lincoln S, Dickson D, Davies P, Petersen RC, Stevens M, de Graaff E, Wauters E, van Baren J, Hillebrand M, Joosse M, Kwon JM, Nowotny P, Heutink P, et al: Association of missense and 5′-splice-site mutations in tau with the inherited dementia FTDP-17. Nature 1998;393:702–705.
11 Poorkaj P, Bird TD, Wijsman E, Nemens E, Garruto RM, Anderson L, Andreadis A, Wiederholt WC, Raskind M, Schellenberg GD: Tau is a candidate gene for chromosome 17 frontotemporal dementia. Ann Neurol 1998;43:815–825.
12 Spillantini MG, Murrell JR, Goedert M, Farlow MR, Klug A, Ghetti B: Mutation in the tau gene in familial multiple system tauopathy with presenile dementia. Proc Natl Acad Sci USA 1998; 95:7737–7741.
13 Goedert M, Spillantini MG, Jakes R, Rutherford D, Crowther RA: Multiple isoforms of human microtubule-associated protein tau: Sequences and localization in neurofibrillary tangles of Alzheimer's disease. Neuron 1989;3:519–526.
14 Weingarten MD, Lockwood AH, Hwo SY, Kirschner MW: A protein factor essential for microtubule assembly. Proc Natl Acad Sci USA 1975;72:1858–1862.
15 Grundke-Iqbal I, Iqbal K, Quinlan M, Tung YC, Zaidi MS, Wisniewski HM: Microtubule-associated protein tau: A component of Alzheimer paired helical filaments. J Biol Chem 1986;261:6084–6089.
16 Grundke-Iqbal I, Iqbal K, Tung YC, Quinlan M, Wisniewski HM, Binder LI: Abnormal phosphorylation of the microtubule associated protein tau in Alzheimer cytoskeletal pathology. Proc Natl Acad Sci USA 1986;83:4913–4917.
17 Iqbal K, Grundke-Iqbal I, Zaidi T, Merz PA, Wen GY, Shaikh SS, Wisniewski HM, Alafuzoff I, Winblad B: Defective brain microtubule assembly in Alzheimer's disease. Lancet 1986; ii:421–426.

18 Iqbal K, Grundke-Iqbal I, Smith AJ, George L, Tung YC, Zaidi T: Identification and localization of a tau peptide to paired helical filaments of Alzheimer disease. Proc Natl Acad Sci USA 1989; 86:5646–5650.
19 Lee VMY, Balin BJ, Otvos L Jr, Trojanowski JQ: A68:A major subunit of paired helical filaments and derivitized forms of normal tau. Science 1991;251:675–678.
20 Mah VH, Eskin TA, Kazee AM, Lapham L, Higgins GA: In situ hybridization of calcium/calmodulin dependent protein kinase II and tau mRNAs; species differences and relative preservation in Alzheimer's disease. Brain Res Mol Brain Res 1992;12:85–94.
21 Khatoon S, Grundke-Iqbal I, Iqbal K: Brain levels of microtubule-associated protein tau are elevated in Alzheimer's disease: A radioimmuno-slot-blot assay for nanograms of the protein. J Neurochem 1992;59:750–753.
22 Khatoon S, Grundke-Iqbal I, Iqbal K: Levels of normal and abnormally phosphorylated tau in different cellular and regional compartments of Alzheimer disease and control brains. FEBS Lett 1994;351:80–84.
23 Bancher C, Brunner C, Lassmann H, Budka H, Jellinger K, Wiche G, Seitelberger F, Grundke-Iqbal I, Iqbal K, Wisniewski HM: Accumulation of abnormally phosphorylated tau precedes the formation of neurofibrillary tangles in Alzheimer's disease. Brain Res 1989;477:90–99.
24 Köpke E, Tung YC, Shaikh S, Alonso A del C, Iqbal K, Grundke-Iqbal I: Microtubule associated protein tau: Abnormal phosphorylation of non-paired helical filament pool in Alzheimer disease. J Biol Chem 1993;268:24374–24384.
25 Iqbal K, Zaidi T, Bancher C, Grundke-Iqbal I: Alzheimer paired helical filaments: Restoration of the biological activity by dephosphorylation. FEBS Lett 1994;349:104–108.
26 Wang JZ, Gong CX, Zaidi T, Grundke-Iqbal I, Iqbal K: Dephosphorylation of Alzheimer paired helical filaments by protein phosphatase-2A and -2B. J Biol Chem 1995;270:4854–4860.
27 Alonso A del C, Zaidi T, Grundke-Iqbal I, Iqbal K: Role of abnormally phosphorylated tau in the breakdown of microtubules in Alzheimer disease. Proc Natl Acad Sci USA 1994;91:5562–5566.
28 Alonso A del C, Grundke-Iqbal I, Iqbal K: Alzheimer's disease hyperphosphorylated tau sequesters normal tau into tangles of filaments and disassembles microtubules. Nat Med 1996;2:783–787.
29 Alonso A del C, Grundke-Iqbal I, Barra HS, Iqbal K: Abnormal phosphorylation of tau and the mechanism of Alzheimer neurofibrillary degeneration: Sequestration of MAP1 and MAP2 and the disassembly of microtubules by the abnormal tau. Proc Natl Acad Sci USA 1997;94:298–303.
30 Wang JZ, Grundke-Iqbal I, Iqbal K: Restoration of biological activity of Alzheimer abnormally phosphorylated by dephosphorylation with protein phosphatase-2A, -2B and -1. Mol Brain Res 1996;38:200–208.
31 Alonso A del C, Zaidi T, Novak HS, Barra HS, Grundke-Iqbal I, Iqbal K: Interaction of tau isoforms with Alzheimer's disease abnormally hyperphosphorylated tau and in vitro phosphorylation into the disease-like protein. J Biol Chem 2001;276:37967–37973.
32 Morsch R, Simon W, Coleman PD: Neurons may live for decades with neurofibrillary tangles. J Neuropathol Exp Neurol 1999;58:188–197.
33 Alonso A del C, Zaidi T, Novak M, Grundke-Iqbal I, Iqbal K: Hyperphosphorylation induces self-assembly of tau into tangles of paired helical filaments/straight filaments. Proc Natl Acad Sci USA 2001;98:6923–6928.
34 Iqbal K, Alonso A del C, Gondal JA, Gong CX, Haque N, Khatoon S, Sengupta A, Wang JZ, Grundke-Iqbal I: Mechanism of neurofibrillary degeneration and pharmacologic therapeutic approach. J Neural Transm 2000;59:213–222.
35 Gong CX, Singh TJ, Grundke-Iqbal I, Iqbal K: Phosphoprotein phosphatase activities in Alzheimer disease brain. J Neurochem 1993;61:921–927.
36 Gong CX, Shaikh S, Wang JZ, Zaidi T, Grundke-Iqbal I, Iqbal K: Phosphatase activity towards abnormally phosphorylated τ: Decrease in Alzheimer disease brain. J Neurochem 1995;65:732–738.
37 Bennecib M, Gong CX, Grundke-Iqbal I, Iqbal K: Inhibition of PP-2A upregulates CaMKII in rat forebrain and induces hyperphosphorylation of tau at Ser 262/356. FEBS Lett 2001;490: 15–22.
38 Gong CX, Lidsky T, Wegiel J, Zuck L, Grundke-Iqbal I, Iqbal K: Phosphorylation of microtubule-associated protein tau is regulated by protein phosphatase 2A in mammalian brain. J Biol Chem 2000;275:5535–5544.

39 Oliver CJ, Shenolikar S: Physiologic importance of protein phosphatase inhibitors. Frontiers Biosci 1998;3:961–972.
40 Cohen P, Alemany S, Hemmings BA, Resink TJ, Stralfors P, Lim Tung HY: Protein phosphatase-1 and protein phosphatase-2A from rabbit skeletal muscle. Methods Enzymol 1988;159:390–408.
41 Cohen P: The structure and regulation of protein phosphatases. Annu Rev Biochem 1989; 58:453–508.
42 Walaas SI, Greengard P: Protein phosphorylation and neuronal function. Pharmacol Rev 1991; 43:299–349.
43 Li M, Guo H, Damuni Z: Purification and characterization of two potent heat-stable protein inhibitors of protein phosphatase 2A from bovine kidney. Biochemistry 1995;34:1988–1996.
44 Li M, Makkinje A, Damuni Z: Molecular identification of I_1^{PP2A}, a novel potent heat-stable inhibitor protein of protein phosphatase 2A. Biochemistry 1996;35:6998–7002.
45 Li M, Makkinje A, Damuni Z: The Myeloid Leukemia-associated protein SET is a potent inhibitor of protein phosphatase 2A. J Biol Chem 1996;271:11059–11062.
46 Tsujio T, Xu J, Kotula L, Grundke-Iqbal I, Iqbal K: The structures and activities of the endogenous inhibitors of PP-2A in brain. Neurobiol Aging 2002;23:S498(1814).
47 Ulitzur N, Rancano C, Pfeffer SR: Biochemical characterization of mapmodulin, a protein that binds microtubule-associated proteins. J Biol Chem 1997;272:30577–30582.
48 von Lindern M, van Baal S, Wiegant J, Raap A, Hagemeijer A, Grosveld G: *can*, a putative oncogene associated with myeloid leukemogenesis, may be activated by fusion of its 3′ half to different genes: Characterization of the *set* gene. Mol Cell Biol 1992;12:3346–3355.
49 Reisberg B, Ferris S, Mobius HJ, Stoffler A, Schmitt F, Doody R: Long-term treatment with the NMDA antagonist memantine: Results of a 24-week, open-label extension study in Alzheimer's disease. Neurobiol Aging 2002;23:S555(2039).
50 Winblad B, Poritis N: Memantine in severe dementia: Results of the 9M-Best Study (Benefit and efficacy in severely demented patients during treatment with memantine). Int J Geriatr Psychiatry 1999;14:135–146.
51 Li L, Sengupta A, Grundke-Iqbal I, Iqbal K: Memantine restores the okadaic acid-induced changes in the activities of protein phosphatase-2A and calcium, calmodulin-protein kinase II and hyperphosphorylation of tau in rat hippocampal slices in culture. Neurobiol Aging 2002;23: S111(421).
52 Hu YY, He SS, Wang X, Duan QH, Grundke-Iqbal I, Iqbal K, Wang JZ: Levels of nonphosphorylated and phosphorylated tau in CSF of Alzheimer disease patients: An ultrasensitive bienzyme-substrate-recycle ELISA. Am J Pathol 2002;160:1269–1278.

Khalid Iqbal, PhD, Chairman
Department of Neurochemistry, Head, Chemical Neuropathology Laboratory
New York State Institute for Basic Research in Developmental Disabilities
1050 Forest Hill Road, Staten Island, New York, NY 10314–6399 (USA)
Tel. +1 718 494 5259, Fax +1 718 494 1080, E-Mail iqbalk@worldnet.att.net

Takeda M, Tanaka T, Cacabelos R (eds): Molecular Neurobiology of Alzheimer Disease and Related Disorders. Basel, Karger, 2004, pp 52–61

Tau Pathology of Sporadic Tauopathies

Tetsuaki Arai[a], *Haruhiko Akiyama*[a], *Kuniaki Tsuchiya*[b], *Shuji Iritani*[b], *Koichi Ishiguro*[c], *Saburo Yagishita*[d], *Tatsuro Oda*[e], *Toshinari Odawara*[f], *Eizo Iseki*[f], *Kenji Ikeda*[a]

[a] Department of Psychogeriatrics, Tokyo Institute of Psychiatry, and
[b] Tokyo Metropolitan Matsuzawa Hospital, Tokyo,
[c] Mitsubishi Kasei Institute of Life Sciences, Tokyo,
[d] Division of Pathology, Kanagawa Rehabilitation Center, Kanagawa,
[e] Department of Neuropsychiatry, National Shimofusa Mental Hospital, Chiba,
[f] Department of Psychiatry, Yokohama City University School of Medicine, Yokohama, Japan

Tau belongs to a family of microtubule-associated proteins that are involved in microtubule assembly and stabilization. In the human brain, alternative splicing causes production of six tau isoforms, three with three microtubule-binding repeats in the carboxyl-terminal region (3Rtau) and the other three with four microtubule-binding repeats (4Rtau), containing the insertion of an exon 10-derived 31-amino acid fragment. Tau aggregates as cytoplasmic inclusions in neurons and glial cells in a variety of neurodegenerative diseases, which are now collectively referred to as tauopathies.

Progressive supranuclear palsy (PSP), corticobasal degeneration (CBD), argyrophilic grain disease (AGD) and Pick's disease (PiD) belong to sporadic tauopathies. At present, biochemical features of tau that characterize each disease are isoform composition and phosphorylation states of aggregated tau. It has been reported that aggregated tau in PSP, CBD and AGD consists of 4Rtau [1–5], while that in PiD is composed of 3Rtau [2, 6]. However, Zhukareva et al. [7] have recently reported that in some PiD, aggregated tau in Pick bodies (PiB) consists of 4Rtau. Although phosphorylation at serine (Ser) 262 of tau has been consistently found in aggregated tau in PSP and CBD [1], that in PiB and argyrophilic grains remains controversial [4, 6–11]. Moreover, it is unclear whether biochemical differences of aggregated tau exist between PSP and CBD.

Table 1. Patients examined in this study

Case no.	Age at death (years)	Gender	Duration (years)	BW (g)	Region
PSP					
1	84	M	3	1,030	Prec/Poc
2	70	M	3	–	Striatum
3	72	M	7	1,290	Prec/Poc
CBD					
4	67	M	2	1,160	Prec/Poc
5	–	–	–	–	Prec/Poc
6	78	F	5	1,055	Fr
AGD					
7	80	F	10	1,220	T
8	81	M	–	–	Ent
9	84	M	7	1,300	Ent
PiD					
10	–	–	–	1,000	T
11	71	M	15	680	Fr
12	72	M	9	920	Fr
13	75	M	10	1,080	Fr
14	67	M	14	–	T

PSP, progressive supranuclear palsy; CBD, corticobasal degeneration; AGD, argyrophilic grain disease; PiD, Pick's disease; –, not available; M, male; F, female; BW, brain weight; Prec, precentral; Poc, postcentral; Fr, frontal; T, temporal; Ent, entorhinal.

To elucidate these issues, we examined the brains from three cases of PSP, CBD and AGD and five cases of PiD. The age, gender, disease duration, brain weight and brain regions examined are given in table 1. A fresh-frozen brain block was taken from each case and cut into two pieces. One piece was homogenized and used for immunoblot analysis. The other piece was fixed in 70% ethanol containing 150 m*M* NaCl or 4% paraformaldehyde in 0.1 *M* phosphate buffer (pH 7.4) for 2 days and used for immunohistochemistry. Isoform composition was investigated by dephosphorylation of sarkosyl-insoluble tau [12] and immunohistochemistry with rabbit anti-tau antibody specific for 4Rtau. The phosphorylation state of aggregated tau was examined using immunohistochemistry and immunoblotting with anti-tau antibodies specific to phosphorylated Ser262. Moreover, we investigated biochemical differences of aggregated tau between PSP and CBD using immunoblotting of sarkosyl-insoluble tau.

Differences of Isoform Composition of Aggregated Tau

Immunoblot analysis with HT7 (Innogenetics, Belgium), a phosphorylation-independent monoclonal antibody to tau, showed that sarkosyl-insoluble tau extracted from all cases of PSP, CBD and AGD appeared as a triplet of 64, 68 and 72 kD. In case 9 of AGD, in which numerous neurofibrillary tangles (NFT) occurred in the entorhinal region, an additional band of 60 kD appeared (fig. 1a). On dephosphorylation, insoluble tau consisted predominantly of 4Rtau in all cases of PSP and CBD and in two cases of AGD (cases 7 and 8). Dephosphorylated tau comigrated with both 4Rtau and 3Rtau isoforms showing an equal ratio in AGD case 9 (fig. 1b). In PiD, nondephosphorylated tau from all cases appeared as a triplet of 60, 64 and 68 kD (fig. 1c). Dephosphorylated tau contained both 3Rtau and 4Rtau, with 3Rtau predominating over 4Rtau in all cases (fig. 1d).

For immunohistochemical examinations, we developed a polyclonal anti-tau antibody, Ex10, by immunizing rabbits with synthetic peptides conjugated to keyhole limpet hemocyanin [2]. Ex10 was raised against the first 16 amino acids of exon 10 (KVQIINKKLDLSNVQS) so that it recognizes 4Rtau but not 3Rtau. AT8 (Innogenetics), a monoclonal antibody specific to phosphorylated Ser199/202, was used to identify tau pathology in tissue sections.

Ex10 labeled neuronal and glial tau-positive structures in PSP, CBD and AGD. They include NFT, pretangles, tuft-shaped astrocytes (fig. 2a), oligodendroglial coiled bodies (fig. 2b), astrocytic plaques (fig. 2c) and argyrophilic grains (fig. 2d). Ex10-positive structures in these diseases were comparable in both number and morphology with those positive for AT8 in all cases examined. In PiD, Ex10 demonstrated fewer positive structures than those seen in adjacent AT8-stained sections. Therefore, further analysis was performed with double immunostaining for Ex10 and AT8, where Ex10 was stained purple and AT8 brown. In 4 of 5 cases, the vast majority of PiB was negative for Ex10 and, therefore, stained brown for AT8 (fig. 2e). Only case 13 showed a considerable number of Ex10-positive PiB, although there were fewer Ex10-positive PiB than AT8-positive PiB (data not shown). In the white matter, a large number of oligodendroglial coiled bodies were seen in all PiD cases. They were positive for AT8 but negative for Ex10, indicating that they lack 4Rtau (fig. 2f). In the cortex, Ex10 stained many AT8-positive astrocytes (fig. 2g). Such astrocytic inclusions were found in all PiD cases examined but the frequency varied from case to case.

In summary, in PSP, CBD and AGD, the isoform composition of insoluble tau was 4Rtau dominant, and all neuronal and glial inclusions consisted of 4Rtau. Insoluble tau from AGD case 9 contained both 3Rtau and 4Rtau with an equal ratio. This is not surprising since a number of NFT in the hippocampal region was observed in this case and tau in NFT consists of all six tau isoforms [12]. We have previously reported a case of AGD concomitant with numerous NFT in the

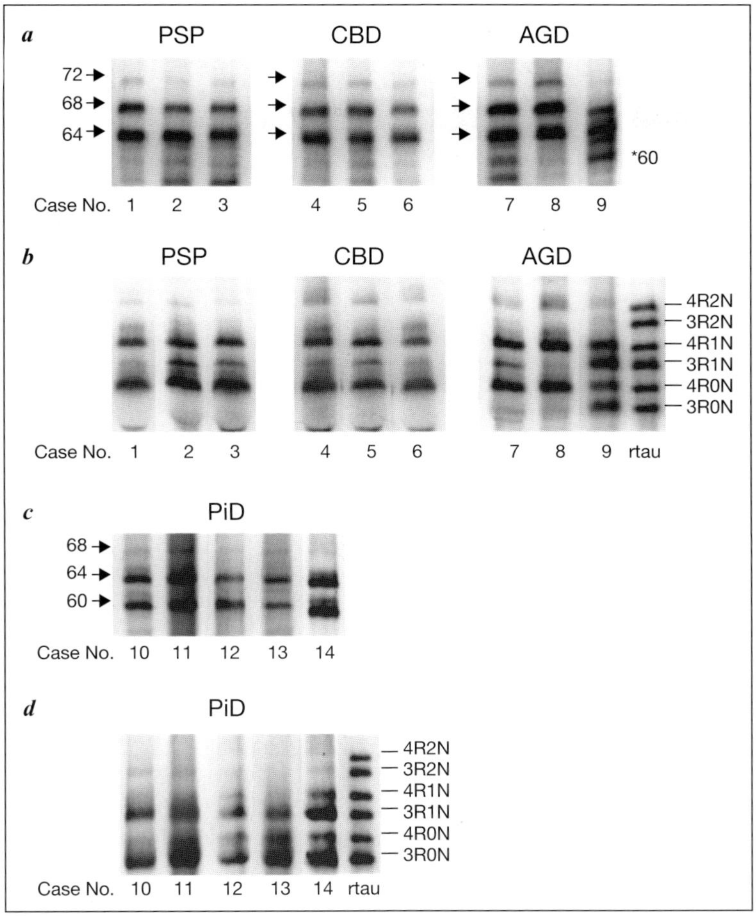

Fig. 1. Immunoblot analysis of sarkosyl-insoluble tau from the brains of PSP, CBD AGD (***a***, ***b***) and PiD (***c***, ***d***) stained with HT7, a phosphorylation-independent monoclonal antibody to τ. ***a*** Nondephosphorylated tau from all cases of PSP and CBD and two cases of AGD (cases 7 and 8) migrates as two major bands of 64 and 68 kD and a minor band of 72 kD (arrows). In AGD case 9, which showed numerous NFT in the entorhinal region, an additional 60-kD band appears. ***b*** Dephosphorylated τ is stained as two major and one minor bands that align with all recombinant 4Rtau and one minor band corresponding to 3Rτ with one amino terminal insert in all cases of PSP and CBD and two cases of AGD (cases 7 and 8). In AGD case 9, dephosphorylated tau comigrates with both 4Rtau and 3Rtau isoforms. ***c*** Nondephosphorylated tau from all cases of PiD migrates as two major bands of 60 and 64 kD and a minor band of 68 kD (arrows). ***d*** Dephosphorylated tau from all cases of PiD appears as two major and one minor bands that align with all recombinant 3Rtau isoforms and one or two minor bands corresponding to 4Rtau with zero or one amino terminal insert. Recombinant tau isoforms (rtau) are indicated to the right of ***b*** and ***d***.

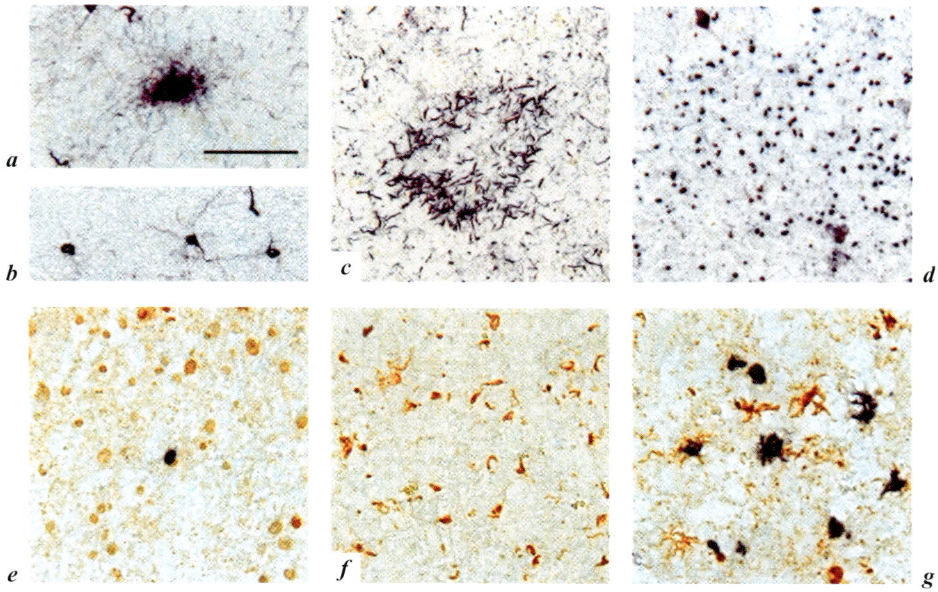

Fig. 2. ***a–d*** Single immunostaining for Ex10. ***a*** Tuft-shaped astrocyte in the frontal cortex of PSP. ***b*** Oligodendroglial coiled bodies in the cerebral white matter of PSP. ***c*** An astrocytic plaque in the frontal cortex of CBD. ***d*** Argyrophilic grains in the entorhinal cortex of AGD. ***e–g*** Double immunostaining for Ex10 (purple) and AT8 (brown) in the temporal lobe of PiD. ***e*** The majority of PiB are stained brown, indicating the absence of 4Rtau. Occasional Ex10-positive round inclusions are seen among Ex10-negative PiB (case 10). ***f*** Most oligodendroglial coiled bodies in the white matter are stained brown, indicating the absence of 4Rtau (case 10). ***g*** AT8-positive astrocytes are labeled with Ex10 (case 12). Bar in ***a*** = 60 μm. ***a–g*** are at the same magnification.

hippocampal region [13]. In PiD, the isoform composition of insoluble tau was 3Rtau dominant. Many PiB and oligodendroglial coiled bodies consist of 3Rtau. Astrocytic tangles, part of PiB, occasional NFT and neuropil threads contain 4Rtau. Such results suggest that, in neurons and oligodendroglia, tau isoforms involved in the pathological processes differ between PSP/CBD/AGD and PiD, and are thus disease specific. This contrasts with astrocytic tau isoforms that accumulate similarly in PSP, CBD and PiD.

Differences of the Phosphorylation State of Aggregated Tau

We examined tau-positive structures of each disease using three anti-tau antibodies specific to phosphorylated Ser262, consisting of 12E8 [14], anti-PS262 [15] and pS^{262} (Biosource, USA).

Table 2. Immunoreactivities of tau-positive structures of each tauopathy with or without formic acid treatment

	12E8		anti-PS262, pS^{262}	
	FA(−)	FA(+)	FA(−)	FA(+)
PSP	+++	+++	+	+++
CBD	+++	+++	+	+++
AGD	+++	+++	+	+++
PiD	+~++	+++	+	+++

FA(−), without formic acid treatment; FA(+), with formic acid treatment; PSP, progressive supranuclear palsy; CBD, corticobasal degeneration; AGD, Argyrophilic grain disease; PiD, Pick's disease; −, negative; +, slight; ++, moderate; +++, abundant.

Immunohistochemical examinations in PiD showed relatively weak staining of PiB for 12E8 without pretreatment. Pretreatment of tissue sections with formic acid markedly enhanced 12E8 staining of PiB. In PSP, CBD and AGD; however, 12E8 intensely stained all tau-positive structures even without formic acid pretreatment. Anti-PS262 and pS^{262} showed similar reactivity in all diseases. Immunostaining of tau-positive structures of all diseases was weak without pretreatment and markedly enhanced after formic acid pretreatment (table 2).

Figure 3 shows immunoblots of sarkosyl-insoluble tau from brains of each disease stained with HT7 (fig. 3a), 12E8 (fig. 3b) and anti-PS262 (fig. 3c). In PSP, CBD and AGD, the reactivity of insoluble tau to three antibodies was comparable. In PiD, however, reactivity to 12E8 and anti-PS262 was weaker than reactivity to HT7 except for case 12, in which insoluble tau was similarly stained with 12E8 and HT7.

In summary, in both immunohistochemistry and immunoblot, immunoreactivity of aggregated tau to antibodies against tau phosphorylated at Ser262 tended to be weaker in PiD than in any other diseases. These results suggest that tau in PiB is partly phosphorylated at Ser262, but the proportion of tau phosphorylated at Ser262 is smaller than other abnormal tau structures. Argyrophilic grains of AGD, however, were phosphorylated at Ser262 comparable to that in PSP and CBD.

Differences in Proteolytic Processing of Aggregated Tau

To investigate biochemical differences of aggregated tau between PSP and CBD, we employed immunoblot analysis of sarkosyl-insoluble tau and the results

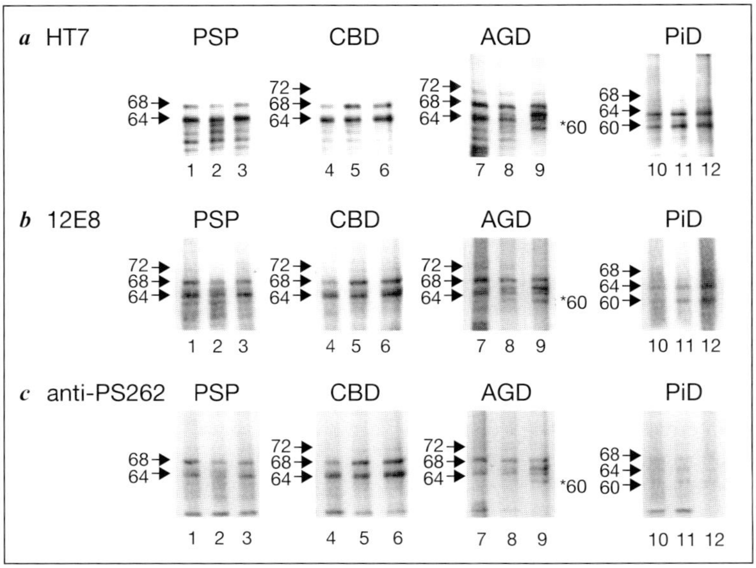

Fig. 3. Immunoblotting of the sarkosyl-insoluble tau from brains of PSP, CBD, AGD and PiD stained for HT7 (***a***), 12E8 (***b***), and anti-PS262 (***c***). HT7 stains a major doublet of 64 and 68 kD in PSP, CBD and AGD, and a major doublet of 60 and 64 kD in PiD. In AGD case 9, an additional 60-kD band appears. The immunoreactivity of these insoluble tau bands with HT7 is similar among diseases (***a***). In PiD, comparing immunoreactivities of insoluble τ to HT7, those to 12E8 (***b***) and anti-PS262 (***c***) are weaker than in any other diseases except for case 12, in which 12E8 similarly stained insoluble tau to HT7.

were compared between diseases [16]. Antibodies used in this study and the tau epitopes recognized by these antibodies were Alz-50 (2–10) [17], HT7 (159–163; Innogenetics), AT8 (189–207; Innogenetics), anti-PS262 [14], PHF-1 (396–404) [18], and T46 (404–441; Zymed, USA). The sequence number corresponds to that of the longest human isoform. Immunoreactivity of Alz-50, HT7 and T46 is independent of tau phosphorylation. AT8, anti-PS262 and PHF-1 are specific to tau phosphorylated at Ser199/202, Ser262 and Ser396/404, respectively.

Figure 4 shows immunoblot of sarkosyl-insoluble tau of PSP and CBD stained with PHF-1 (fig. 4a) and T46 (fig. 4b). Typical triplet of 64, 68 and 72 kD was commonly observed in both diseases. The pattern of low-molecular tau fragments, however, differed between the two diseases. The low-molecular fragments in PSP demonstrated a prominent 33-kD fragment, while those in CBD predominantly showed two closely related bands of approximately 37 kD. In addition to PHF-1 and T46, these low-molecular bands were positive for anti-PS262 and AT8, but were negative for HT7 and Alz-50 (data not shown). Such a profile suggests that these low-molecular bands are composed of amino-terminally

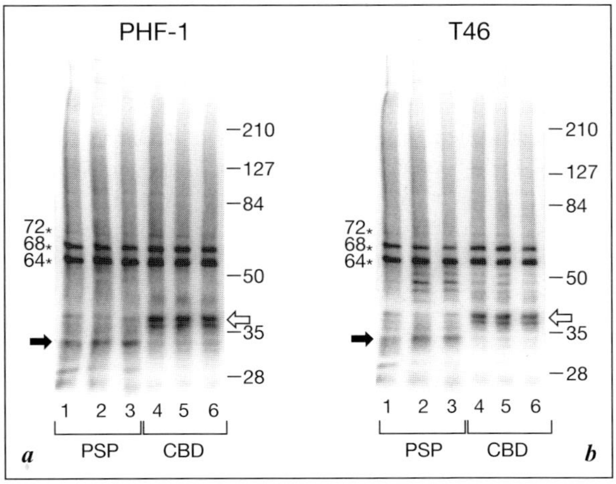

Fig. 4. Immunoblot analysis of sarkosyl-insoluble tau from the brains of PSP and CBD stained with PHF-1 (***a***) and T46 (***b***). A typical triplet of 64, 68 and 72 kD is seen in both PSP and CBD (asterisks). Low-molecular fragments in PSP demonstrate a prominent 33-kD fragment (closed arrow), while those in CBD demonstrate two closely associated bands of approximately 37 kD (open arrow). Molecular weight markers are shown at the right.

truncated tau fragments. The difference in the molecular sizes of these fragments indicates that the cleavage site varies between the HT7 epitope and the AT8 epitope. While the significance of such a difference in tau fragments is not clear, there is a possibility that posttranslational modifications of tau such as phosphorylation, glycosylation and ubiquitination differ between PSP and CBD, resulting in different amino-terminal processing of aggregated tau between these diseases. For instance, it has been reported that pathological tau in CBD is more acidic than in PSP [1]. Such biochemical differences may be related to the neuropathological features of these diseases.

Conclusion

The pathology of PiD is characterized by aggregation of tau in neurons and oligodendroglia, and the aggregated tau in these cells is 3Rtau dominant and less phosphorylated at Ser262. Tau in argyrophilic grains of AGD was phosphorylated at Ser262 similarly to PSP and CBD. Amino-terminal processing of aggregated tau differs between PSP and CBD although the isoform pattern is identical in these diseases. It is suggested that the differences of isoform composition, phosphorylation state, and proteolytic processing characterize the abnormal tau of each sporadic tauopathy.

Acknowledgement

We are grateful to Dr. M. Goedert (MRC Laboratory of Molecular Biology, Cambridge, UK) and Dr. M. Hasegawa (Department of Molecular Neurobiology, Tokyo Institute of Psychiatry, Japan) for providing tau cDNA and advice. We thank Dr. P. Davies (Albert Einstein College of Medicine, USA) for Alz-50 and PHF-1, and Elan Pharmaceuticals for 12E8. We also thank Drs. T. Beach and J. Rogers (Sun Health Research Institute, USA) for providing brain tissues. This study was supported by a grant-in-aid for encouragement of young scientists from the Ministry of Education, Science, Sports and Culture of Japan and the Welfide Medicinal Research Foundation.

References

1 Sergeant N, Wattez A, Delacourte A: Neurofibrillary degeneration in progressive supranuclear palsy and corticobasal degeneration: Tau pathologies with exclusively "exon 10" isoforms. J Neurochem 1999;72:1243–1249.
2 Arai T, Ikeda K, Akiyama H, Shikamoto Y, Tsuchiya K, Yagishita S, Beach T, Rogers J, Schwab C, McGeer PL: Distinct isoforms of tau aggregated in neurons and glial cells in brains of patients with Pick's disease, corticobasal degeneration and progressive supranuclear palsy. Acta Neuropathol (Berl) 2001;101:167–173.
3 Togo T, Sahara N, Yen SH, Cookson N, Ishizawa T, Hutton M, de Silva R, Lees A, Dickson DW: Argyrophilic grain disease is a sporadic 4-repeat tauopathy. J Neuropathol Exp Neurol 2002;61: 547–556.
4 Tolnay M, Sergeant N, Ghestem A, Chalbot S, De Vos RA, Jansen Steur EN, Probst A, Delacourte A: Argyrophilic grain disease and Alzheimer's disease are distinguished by their different distribution of tau protein isoforms. Acta Neuropathol (Berl) 2002;104:425–434.
5 Zhukareva V, Shah K, Uryu K, Braak H, Del Tredici K, Sundarraj S, Clark C, Trojanowski JQ, Lee VM: Biochemical analysis of tau proteins in argyrophilic grain disease, Alzheimer's disease, and Pick's disease: A comparative study. Am J Pathol 2002;161:1135–1141.
6 Delacourte A, Sergeant N, Wattez A, Gauvreau D, Robitaille Y: Vulnerable neuronal subsets in Alzheimer's and Pick's disease are distinguished by their τ isoform distribution and phosphorylation. Ann Neurol 1998;43:193–204.
7 Zhukareva V, Mann D, Pikering-Brown S, Uryu K, Shunk T, Shah K, Grossman M, Miller BL, Hulette CM, Feinstein SC, Trojanowski JQ, Lee VMY: Sporadic Pick's disease: A tauopathy characterized by a spectrum of pathological τ isoforms in gray and white matter. Ann Neurol 2002; 51:730–739.
8 Probst A, Tolnay M, Langui D, Goedert M, Spillantini MG: Pick's disease: Hyperphosphorylated tau protein segregates to the somatoaxonal compartment. Acta Neuropathol (Berl) 1996;92: 588–596.
9 Tolnay M, Spillantini MG, Goedert M, Ulrich J, Langui D, Probst A: Argyrophilic grain disease: Widespread hyperphosphorylation of tau protein in limbic neurons. Acta Neuropathol (Berl) 1997;93:477–484.
10 Ferrer I, Barrachina M, Puig B: Anti-tau phospho-specific Ser(262) antibody recognizes a variety of abnormal hyper-phosphorylated tau deposits in tauopathies including Pick bodies and argyrophilic grains. Acta Neuropathol (Berl) 2002;104:658–664.
11 Arai T, Ikeda K, Akiyama H, Tsuchiya K, Iritani S, Ishiguro K, Yagishita S, Oda T, Odawara T, Iseki E: Different immunoreactivities of the microtubule-binding region of tau and its molecular basis in brains from Alzheimer's disease, Pick's disease, progressive supranuclear palsy and corticobasal degeneration. Acta Neuropathol (Berl) 2003;105:489–498.
12 Goedert M, Spillantini MG, Cairns NJ, Crowther RA: Tau proteins of Alzheimer paired helical filaments: abnormal phosphorylation of all six brain isoforms. Neuron 1992;8:159–168.

13 Oshima K, Tsuchiya K, Iritani S, Ueno H, Niizato K, Nakamura R, Arai T, Akiyama H, Ikeda K: An autopsy case of argyrophilic grain dementia with abundant neurofibrillary tangles. Brain Nerve (Tokyo) 2003;55:133–138.
14 Seubert P, Mawal-Dewan M, Barbour R, Jakes R, Goedert M, Johnson GVW, Literski JM, Schenk D, Lieberburg I, Trojanowski JQ, Lee VM-Y: Detection of phosphorylated Ser262 in fetal tau, adult tau and paired helical filament-tau. J Biol Chem 1995;270:18917–18922.
15 Ishiguro K, Sato K, Takamatsu M, Park J, Uchida T, Imahori K: Analysis of phosphorylation of tau with antibodies specific for phosphorylation site. Neurosci Lett 1995;202:81–84.
16 Arai T, Ikeda K, Akiyama H, Tsuchiya K, Yagishita S, Takamatsu J: Intracellular processing of aggregated tau differs between corticobasal degeneration and progressive supranuclear palsy. Neuroreport 2001;12:935–938.
17 Wolozin GL, Pruchnicki A, Dickson DW, Davies P: A neuronal antigen in the brain of Alzheimer's patients. Science 1986;232:648–650.
18 Greenberg SG, Davies P, Schein JD, Binder LI : Hydrofluoric acid-treated tau PHF proteins display the same biochemical properties as normal tau. J Biol Chem 1992;267:564–569.

Dr. Tetsuaki Arai
Department of Psychogeriatrics
Tokyo Institute of Psychiatry
2–1–8, Kamikitazawa, Setagaya-ku, Tokyo 156 (Japan)
Tel. +81 3 3304 5701, Fax +81 3 3329 8035, E-Mail arai@prit.go.jp

Takeda M, Tanaka T, Cacabelos R (eds): Molecular Neurobiology of Alzheimer Disease and Related Disorders. Basel, Karger, 2004, pp 62–70

Deregulation of GSK-3β and JNK in a Mouse Model of Tauopathy: A Kinase Combination That Induces Alzheimer-Type Tau Hyperphosphorylation

Yoshitaka Tatebayashi[a], *Shinji Sato*[a], *Takumi Akagi*[b], *De-Hua Chui*[a], *Tomohiro Miyasaka*[a], *Emmanuel Planel*[a], *Miyuki Murayama*[a], *Akihiko Takashima*[a]

Laboratories for [a]Alzheimer's Disease and [b]Neural Architecture, Brain Science Institute, RIKEN, Saitama, Japan

Alzheimer disease (AD) is a neurodegenerative disorder clinically characterized by dementia. The main pathological features of AD include progressive loss of neurons, amyloid plaques and cytoplasmic filamentous materials known as neurofibrillary tangles (NFTs). Many studies have led to the conclusion that aberrantly hyperphosphorylated forms of the microtubule-associated protein tau are the main constituents of NFTs [1, 2]. Although the significance of the NFT formation in AD has remained elusive, findings of tau gene mutations in patients with familial frontotemporal dementia and parkinsonism linked to chromosome-17 (FTDP-17), another type of neurodegenerative disorder with NFT formation, have led to the idea that tau abnormality alone can cause neurodegeneration [2–10]. Thus, study of the molecular mechanisms of NFT formation may provide significant insight into the process of neurodegeneration in AD.

To date, more than 26 intronic and exonic tau gene mutations have been identified in FTDP-17 [3]. Among them, R406W is unique since this mutation is only associated with AD-like clinical symptoms (with the exception of a 47-year-old man in one US family, who first presented with psychiatric symptoms followed by overt dementia) [6, 11, 12]. Several in vitro studies have revealed that the R406W mutation exerts detrimental effects on the known biological features of tau, including those of assembling and binding to microtubules, being degraded, and self-assembling into taufilaments; however, it appears to be relatively minor when compared to other exonic FTDP-17 mutations [6, 13–17]. Furthermore, detergent-insoluble filaments isolated from the brains of patients with this mutation are indistinguishable from those from AD brains – paired helical and straight filaments composed of both mutant and wild-type hyperphosphorylated tau [6, 18, 19]. Therefore, it is likely that, in addition to the direct mutant effects, the environmental alterations associated with the mutation may also be responsible for NFT formation in R406W FTDP-17.

Imbalance between protein kinases and phosphatases has long been suspected as one of such environmental alterations, since tau is always aberrantly hyperphosphorylated in AD [20]. It is possible that hyperphosphorylation disengages tau from microtubules, thereby increasing the pool of unbound hyperphosphorylated tau which may be more prone to self-aggregate than microtubule-bound tau. Therefore, a large number of serine/threonine protein kinases have been suggested to play a role in NFT formation in vivo [21]. These kinases include mitogen-activated protein kinase [22, 23], glycogen synthase kinase 3β (GSK-3β) [24, 25], cyclin-dependent kinase 2 (cdk2) [26], cdk5 [26], cAMP-dependent protein kinase [27], calcium/calmodulin-dependent protein kinase II [28], and MT-affinity-regulating kinase [29]. In addition, several members of the family of stress-activated protein kinases also phosphorylate tau at multiple sites [30–32]. Nonetheless, many of these studies provide only in vitro evidence to implicate specific kinases. Therefore, it remains unclear which kinases are involved and what role they play in in vivo tau hyperphosphorylation and aggregation.

In the present study, we first focus on the expression of kinases during neurodegeneration in Tg mice expressing the longest human tau isoform with the R406W mutation [33]. As presented by Miyasaka et al. [34] in this issue, these mice develop NFTs-like tau inclusions when they are aged. We found deregulation of GSK-3β and JNK but not apparently other kinases associated with NFT-like tau inclusions in the aged Tg hippocampus. Coexpression of tau, GSK-3β, JNK3, and its activator, MEKK in COS-7 produces most of the pathological phosphorylation-epitopes of tau including AT100 and phospho-Ser422 [35, 36]. Furthermore, this coexpression resulted in the formation of tau aggregates having short fibrils that were detergent insoluble. Taken together, these data suggest a possible involvement of these kinases in the formation of AD-type tau hyperphosphorylation and probably NFTs in vivo.

Experimental Procedures

Antibodies

The following antibodies were used: mouse monoclonal anti-myc (clone 9E10; BAbCO), rabbit polyclonal anti-tau JM [37], anti-tauC (specific to tau at residues, 422–438) [38], phosphorylation-dependent mouse monoclonal anti-tau AT8 (specific to tau phosphorylated at Ser202 and Thr205), AT100 (Thr212 and Ser214), AT180 (Thr231 and Ser235), AT270 (Thr181) (AT series were all purchased from Innogenetics), tau-1 (specific to tau dephosphorylated tau at Ser195, 198, 199 and 202, Boehringer), PHF-1 (specific to tau phosphorylated at Ser396 and 404, generously donated by Dr. P. Davies, Albert Einstein College of Medicine), phosphorylation-dependent rabbit polyclonal anti-tau antibodies PS199, PT205, PS396, PS404, PS413, PS422, which recognize tau phosphorylated at each corresponding site [38], rabbit polyclonal antibodies generated against mutant R406W (AW406) and corresponding wild type tau (AR406) (generously donated by Dr. Y. Ihara, University of Tokyo) [19], rabbit polyclonal anti-total JNK (Cell signaling), anti-JNK3/SAPKββ (Upstate Biotechnology UK, LTD), anti-phospho-JNK (Thr183/Tyr185; Promega), anti-Erk1/2 (cell signaling), anti-p38 (cell signaling), anti-cdk5 (Santa Cruz Biotechnology), anti-phospho-CaMKII (Promega), anti-phospho-GSK-3β (Tyr216; Bio Source), mouse monoclonal anti-phospho-JNK (Thr183/Tyr185; cell signaling), anti-GSK-3β (Transduction Laboratories), anti-phospho-GSK-3β (Tyr216; Transduction Laboratories).

Immunoblot and Immunohistochemical Procedures

Immunoblot analysis was performed as described previously [35]. The enhanced chemiluminescence (ECL) system (Amersham Biosciences) was used for the visualization. Images were documented with a LAS-1000 luminescent image analyzer (Fuji Films). For immunohistochemistry, paraffin embedded brain sections were used [33]. Deparaffinized sections (4–6 μm) were treated in Target Retrieval Solution (DAKO) for 20 min at 80°C, blocked in 0.1% bovine serum albumin/TBS (BSA/TBS) and incubated with primary antibodies in 0.1% BSA/TBS overnight at 4°C. Radiance 2000 KR3 confocal microscope (BIO-RAD, UK) was used for the observations.

Generation and Infection of Recombinant Adenoviruses

All cosmid clones used were generated as described previously [35]. Recombinant adenoviruses were generated from the corresponding cosmid clone using the COS-TPC method. COS-7 cells were exposed to recombinant adenovirus at various multiplicities of infection in Dulbecco's modified Eagle's medium containing 5% fetal bovine serum. Forty-eight hours after infection, cells were lysed and the resulting cell lysates were analyzed as described previously [35].

Immunoelectron Microscopy of RIPA-Insoluble Materials and Tau-Expressing Cells

COS-7 cells infected with recombinant adenoviruses were lysed with RIPA buffer containing 1% SDS. The cell lysates were centrifuged (20 min at 100,000 *g* at 4°C) and the resulting pellets were rehomogenized in the RIPA buffer using a sonic homogenizer. The resulting cell lysates were treated four times as above. The final insoluble pellets were solubilized in 70% formic acid for the immunoblot analysis, or resuspended in 100 m*M* Tris-HCl (pH 8.3) for electron microscopy. For immunoelectron microscopic observation, the RIPA-insoluble materials resuspended in Tris-HCl as well as the infected cells were treated as described previously [35]. The final samples were examined with an electron microscope (LEO 912AB, LEO Electron Microscopy, Ltd.).

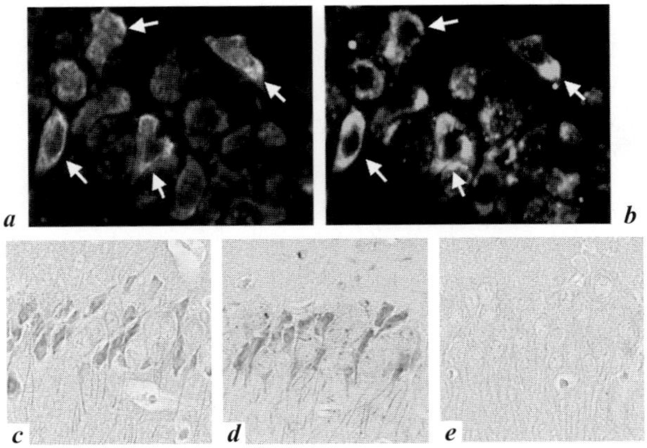

Fig. 1. Accumulation of activated JNK and GSK-3β in the mutant tau-positive degenerating neurons in aged Tg mice. Immunostaining of the hippocampal region of a 19-month-old Tg mouse with anti-myc antibody (***a***), anti-phospho-JNK (***b***), anti-PY216 (active form of GSK-3β) (***c***), and AW406 (specific to mutant tau) (***d***), and of a non-Tg littermate with anti-PY216 (***e***). Activated JNK and GSK-3β were accumulated mainly in the cytoplasm of the mutant tau positive neurons. Arrows in ***a*** and ***b*** indicate neurons with tau inclusions.

Results and Discussion

We produced Tg mice expressing R406W mutant human tau that exhibit a disease phenotype mimicking human R406W FTDP-17 when they are aged. As with human cases, tau inclusions in aged Tg mice were composed of both mutant and endogenous wild-type hyperphosphorylated tau. To investigate the potential involvement of kinases in R406W-associated neurodegeneration, we have compared the distribution of various kinases in the hippocampus of the aged R406W Tg mice and age-matched non-Tg littermate controls by immunohistochemistry. Phosphodependent antibodies recognizing activated forms of p38 and CaMK-II and phospho-independent antibodies recognizing Erk1/2 and cdk5 did not reveal any major immunostaining differences in hippocampal neurons between R406W Tg mice and controls (data not shown). In contrast, a striking difference was observed in the immunostaining with antibodies against active GSK-3β and JNK (fig. 1). In normal mice, anti-active GSK-3β and JNK antibodies yielded very poor immunoreactivity in hippocampal neurons, whereas intense immunostaining with these antibodies was observed in affected neurons of R406W Tg mice. Moreover, whereas active JNK is normally targeted to the nucleus although only very weakly, significantly increased levels of active JNK were detected mainly in the cytoplasm of affected neurons in association with tau accumulation in R406W Tg mice. These findings suggest

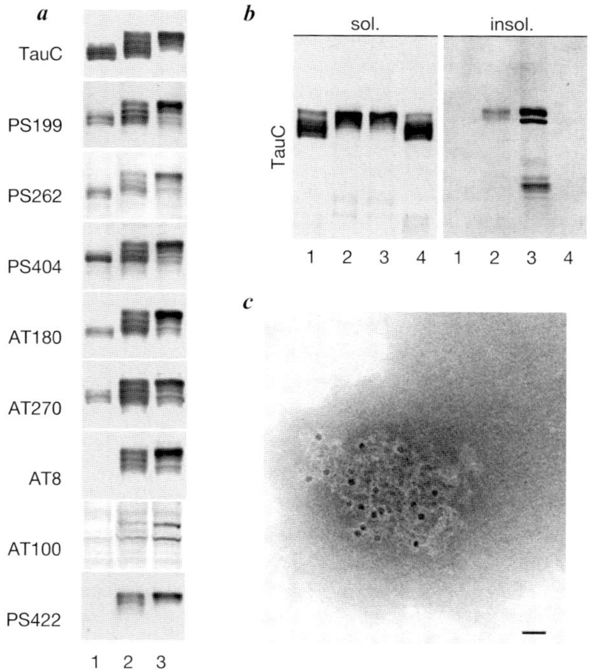

Fig. 2. Tau hyperphosphorylation and detergent-insoluble tau aggregate formation by JNK3 and GSK-3β in COS-7 cells. ***a*** COS-7 cells co-infected with tau (10 MOI) with LacZ (110 MOI, in lane 1), with JNK3 (10 MOI), ΔMEKK (50 MOI) and LacZ (50 MOI, in lane 2), and with JNK3 (10 MOI), ΔMEKK (50 MOI) and GSK3β (50 MOI in lane 3) were lysed and the lysates were subjected to immunoblot analysis. Blots were immunostained with indicated antibodies. ***b*** Immunoblots showing RIPA-soluble (sol.) and -insoluble tau (insol.) from COS-7 cells infected with tau with JNK3 + ΔMEKK (lane 1), with GSK-3β (lane 2), with JNK3 + ΔMEKK + GSK-3β (lane 3), or with LacZ (lane 4). The blots were stained with the phosphorylation-independent antibody TauC. ***c*** Fibrillar tau aggregates in the RIPA-insoluble pellets isolated from quadruple-expressing COS-7 cells. Aggregates were immunolabeled with AT100 antibody followed by 5-nm gold particles conjugated secondary antibody. Similar immunogold-positive tau aggregates were observed in quadruple-expressing COS-7 cells but not in control cells expressing tau and LacZ. Scale bar: 20 nm.

that deregulation of these two kinases might be involved in the neurodegeneration in R406W Tg mice.

We then studied the effect of overexpression of these kinases on the nature of tau in living cells. We expressed tau, ΔMEKK, JNK3, and GSK-3β in COS-7 cells, and examined tau phosphorylation using various phosphodependent anti-tau antibodies. We found that adenovirus-mediated gene expression of ΔMEKK, JNK3, and GSK-3β induced AD-type tau hyperphosphorylation in COS-7 cells (fig. 2a). Such hyperphosphorylation includes AT100 epitopes

(Ser214 and Thr217) and phosphorylation at Ser422. Furthermore, significantly higher levels of tau were recovered in 1% SDS-insoluble fractions from quadruple-expressing cells (fig. 2b). Immunoelectron microscopic observation further revealed that these insoluble materials were composed of amorphous tau aggregates having short fibrils (fig. 2c). Thus, these results suggested that tau hyperphosphorylation by GSK-3β and active JNK could form tau aggregates in the cytoplasm of living cells.

In this study, we provide in vivo evidence that activated JNK and GSK-3β are altered immunohistochemically in degenerating neurons of aged R406W Tg mice. Given that R406W tau mutation is associated with AD-like clinical symptoms [11] and AD-type tau filament formation in humans [6], this type of environmental alteration observed in our Tg mice might also occur in AD neurons. In fact, these kinases have been reported to colocalize with NFTs in AD brains [39–43]. Furthermore, we demonstrate that overexpression of these kinases with wild-type tau in COS-7 cells induces AD-like tau hyperphosphorylation and the formation of amorphous tau aggregates. Overexpression of single kinase with tau, however, did not produce such changes but only lesser levels of tau phosphorylation, suggesting that this combination may be important for the in vivo hyperphosphorylation of tau and that tau hyperphosphorylation may have a role in the NFTs formation. However, only short fibrils but not PHFs or straight filaments were recovered as the detergent-insoluble materials from quadruple-expressing COS-7 cells, suggesting that other factors are also required for NFT formation.

Although our data failed to uncover how R406W tau expression deregulates these two kinases, one possible explanation might be that the particular cellular stresses associated with this mutation, such as oxidative stress and endoplasmic reticulum(ER) stress induce such activation. In fact, JNK has been shown to become activated by these stresses [44–48]. More recently, ER stress also has been shown to activate GSK-3β [49]. It is possible that expression of mutant tau impairs some neuronal functions such as axonal transport, resulting in the chronic accumulation of these stresses in neurons. Further study is necessary to elucidate this possibility.

Acknowledgement

This work is partly supported by CREST (Japan Science and Technology, JST), and a grant-in-aid for Scientific Research (11680746, the Japanese Ministry of Education, Science and Culture).

References

1 Mandelkow EM, Mandelkow E: Tau in Alzheimer's disease. Trends Cell Biol 1998;8:425–427.
2 Spillantini MG, Goedert M: Tau protein pathology in neurodegenerative diseases. Trends Neurosci 1998;21:428–433.
3 Reed LA, Wszolek ZK, Hutton M: Phenotypic correlations in FTDP-17. Neurobiol Aging 2001;22:89–107.
4 Heutink P: Untangling tau-related dementia. Hum Mol Genet 2000;9:979–986.
5 Lee VM, Trojanowski JQ: Neurodegenerative tauopathies: Human disease and transgenic mouse models. Neuron 1999;24:507–510.
6 van Swieten JC, Stevens M, Rosso SM, Rizzu P, Joosse M, de Koning I, Kamphorst W, Ravid R, Spillantini MG, Niermeijer, Heutink P: Phenotypic variation in hereditary frontotemporal dementia with tau mutations. Ann Neurol 1999;46:617–626.
7 Hutton M, Lendon CL, Rizzu P, Baker M, Froelich S, Houlden H, Pickering-Brown S, Chakraverty S, Isaacs A, Grover A, Hackett J, Adamson J, Lincoln S, Dickson D, Davies P, Petersen RC, Stevens M, de Graaff E, Wauters E, van Baren J, Hillebrand M, Joosse M, Kwon JM, Nowotny P, Heutink P, et al: Association of missense and 5′-splice-site mutations in tau with the inherited dementia FTDP-17. Nature 1998;393:702–705.
8 Poorkaj P, Bird TD, Wijsman E, Nemens E, Garruto RM, Anderson L, Andreadis A, Wiederholt WC, Raskind M, Schellenberg GD: Tau is a candidate gene for chromosome 17 frontotemporal dementia. Ann Neurol 1998;43:815–825.
9 Spillantini MG, Murrell JR, Goedert M, Farlow MR, Klug A, Ghetti B: Mutation in the tau gene in familial multiple system tauopathy with presenile dementia. Proc Natl Acad Sci USA 1998; 95:7737–7741.
10 Goedert M, Spillantini MG, Crowther RA, Chen SG, Parchi P, Tabaton M, Lanska DJ, Markesbery WR, Wilhelmsen KC, Dickson DW, Petersen RB, Gambetti P: Tau gene mutation in familial progressive subcortical gliosis. Nat Med 1999;5:454–457.
11 Reed LA, Grabowski TJ, Schmidt ML, Morris JC, Goate A, Solodkin A, Van Hoesen GW, Schelper RL, Talbot CJ, Wragg MA, Trojanowski JQ: Autosomal dominant dementia with widespread neurofibrillary tangles. Ann Neurol 1997;42:564–572.
12 Saito Y, Geyer A, Sasaki R, Kuzuhara S, Nanba E, Miyasaka T, Suzuki K, Murayama S: Early-onset, rapidly progressive familial tauopathy with R406W mutation. Neurology 2002;58:811–813.
13 Hasegawa M, Smith MJ, Goedert M: Tau proteins with FTDP-17 mutations have a reduced ability to promote microtubule assembly. FEBS Lett 1998;437:207–210.
14 Dayanandan R, Van Slegtenhorst M, Mack TG, Ko L, Yen SH, Leroy K, Brion JP, Anderton BH, Hutton M, Lovestone S: Mutations in tau reduce its microtubule binding properties in intact cells and affect its phosphorylation. FEBS Lett 1999;446:228–232.
15 Nacharaju P, Lewis J, Easson C, Yen S, Hackett J, Hutton M, Yen SH: Accelerated filament formation from tau protein with specific FTDP-17 missense mutations. FEBS Lett 1999;447:195–199.
16 DeTure M, Ko LW, Yen S, Nacharaju P, Easson C, Lewis J, van Slegtenhorst M, Hutton M, Yen SH: Missense tau mutations identified in FTDP-17 have a small effect on tau-microtubule interactions. Brain Res 2000;853:5–14.
17 Barghorn S, Zheng-Fischhofer Q, Ackmann M, Biernat J, von Bergen M, Mandelkow EM, Mandelkow E: Structure, microtubule interactions, and paired helical filament aggregation by tau mutants of frontotemporal dementias. Biochemistry 2000;39:11714–11721.
18 Hong M, Zhukareva V, Vogelsberg-Ragaglia V, Wszolek Z, Reed L, Miller BI, Geschwind DH, Bird TD, McKeel D, Goate A, Morris JC, Wilhelmsen KC, Schellenberg GD, Trojanowski JQ, Lee VM: Mutation-specific functional impairments in distinct tau isoforms of hereditary FTDP-17. Science 1998;282:1914–1917.
19 Miyasaka T, Morishima-Kawashima M, Ravid R, Heutink P, van Swieten JC, Nagashima K, Ihara Y: Molecular analysis of mutant and wild-type tau deposited in the brain affected by the FTDP-17 R406W mutation. Am J Pathol 2001;158:373–379.
20 Iqbal K, Grundke-Iqbal I: Molecular mechanism of Alzheimer's neurofibrillary degeneration and therapeutic intervention. Ann NY Acad Sci 1996;777:132–138.

21 Billingsley ML, Kincaid RL: Regulated phosphorylation and dephosphorylation of tau protein: Effects on microtubule interaction, intracellular trafficking and neurodegeneration. Biochem J 1997;323(pt 3):577–591.
22 Drewes G, Lichtenberg-Kraag B, Doring F, Mandelkow EM, Biernat J, Goris J, Doree M, Mandelkow E: Mitogen activated protein (MAP) kinase transforms tau protein into an Alzheimer-like state. Embo J 1992;11:2131–2138.
23 Drechsel DN, Hyman AA, Cobb MH, Kirschner MW: Modulation of the dynamic instability of tubulin assembly by the microtubule-associated protein tau. Mol Biol Cell 1992;3:1141–1154.
24 Hanger DP, Hughes K, Woodgett JR, Brion JP, Anderton BH: Glycogen synthase kinase-3 induces Alzheimer's disease-like phosphorylation of tau: Generation of paired helical filament epitopes and neuronal localisation of the kinase. Neurosci Lett 1992;147:58–62.
25 Mandelkow EM, Drewes G, Biernat J, Gustke N, Van Lint J, Vandenheede JR, Mandelkow E: Glycogen synthase kinase-3 and the Alzheimer-like state of microtubule-associated protein tau. FEBS Lett 1992;314:315–321.
26 Baumann K, Mandelkow EM, Biernat J, Piwnica-Worms H, Mandelkow E: Abnormal Alzheimer-like phosphorylation of tau-protein by cyclin-dependent kinases cdk2 and cdk5. FEBS Lett 1993; 336:417–424.
27 Litersky JM, Johnson GV: Phosphorylation by cAMP-dependent protein kinase inhibits the degradation of tau by calpain. J Biol Chem 1992;267:1563–1568.
28 Baudier J, Cole RD: Phosphorylation of tau proteins to a state like that in Alzheimer's brain is catalyzed by a calcium/calmodulin-dependent kinase and modulated by phospholipids. J Biol Chem 1987;262:17577–17583.
29 Drewes G, Ebneth A, Preuss U, Mandelkow EM, Mandelkow E: MARK, a novel family of protein kinases that phosphorylate microtubule-associated proteins and trigger microtubule disruption. Cell 1997;89:297–308.
30 Goedert M, Hasegawa M, Jakes R, Lawler S, Cuenda A, Cohen P: Phosphorylation of microtubule-associated protein tau by stress-activated protein kinases. FEBS Lett 1997;409:57–62.
31 Reynolds CH, Utton MA, Gibb GM, Yates A, Anderton BH: Stress-activated protein kinase/c-jun N-terminal kinase phosphorylates tau protein. J Neurochem 1997;68:1736–1744.
32 Reynolds CH, Nebreda AR, Gibb GM, Utton MA, Anderton BH: Reactivating kinase/p38 phosphorylates tau protein in vitro. J Neurochem 1997;69:191–198.
33 Tatebayashi Y, Miyasaka T, Chui DH, Akagi T, Mishima K, Iwasaki K, Fujiwara M, Tanemura K, Murayama M, Ishiguro K, Planel E, Sato S, Hashikawa T, Takashima A: Tau filament formation and associative memory deficit in aged mice expressing mutant (R406W) human tau. Proc Natl Acad Sci USA 2002;99:13896–13901.
34 Myasaka T, Tatebayashi Y, Chui DH, Akagi T, Mishima KI, Iwasaki K, Fujiwara M, Tanemura K, Murayama M, Ishiguro K, Planel E, Sato S, Hashikawa T, Takashima A: Tau filament formation and associative memory deficit in aged mice expressing mutant (R406W) human tau; in Takeda M (ed): Molecular Neurobiology of Alzheimer Disease and Related Disorders. Basel, Karger, 2003.
35 Sato S, Tatebayashi Y, Akagi T, Chui DH, Murayama M, Miyasaka T, Planel E, Tanemura K, Sun X, Hashikawa T, Yoshioka K, Ishiguro K, Takashima A: Aberrant tau phosphorylation by glycogen synthase kinase-3beta and JNK3 induces oligomeric tau fibrils in COS-7 cells. J Biol Chem 2002; 277:42060–42065.
36 Matsuo ES, Shin RW, Billingsley ML, Van deVoorde A, O'Connor M, Trojanowski JQ, Lee VM: Biopsy-derived adult human brain tau is phosphorylated at many of the same sites as Alzheimer's disease paired helical filament tau. Neuron 1994;13:989–1002.
37 Takashima A, Murayama M, Murayama O, Kohno T, Honda T, Yasutake K, Nihonmatsu N, Mercken M, Yamaguchi H, Sugihara S, Wolozin B: Presenilin 1 associates with glycogen synthase kinase-3beta and its substrate tau. Proc Natl Acad Sci USA 1998;95:9637–9641.
38 Ishiguro K, Sato K, Takamatsu M, Park J, Uchida T, Imahori K: Analysis of phosphorylation of tau with antibodies specific for phosphorylation sites. Neurosci Lett 1995;202:81–84.
39 Mohit AA, Martin JH, Miller CA: p493F12 kinase: a novel MAP kinase expressed in a subset of neurons in the human nervous system. Neuron 1995;14:67–78.

40 Zhu X, Castellani RJ, Takeda A, Nunomura A, Atwood CS, Perry G, Smith MA: Differential activation of neuronal ERK, JNK/SAPK and p38 in Alzheimer disease: the 'two hit' hypothesis. Mech Ageing Dev 2001;123:39–46.
41 Yamaguchi H, Ishiguro K, Uchida T, Takashima A, Lemere CA, Imahori K: Preferential labeling of Alzheimer neurofibrillary tangles with antisera for tau protein kinase (TPK) I/glycogen synthase kinase-3 beta and cyclin-dependent kinase 5, a component of TPK II. Acta Neuropathol (Berl) 1996;92:232–241.
42 Pei JJ, Braak E, Braak H, Grundke-Iqbal I, Iqbal K, Winblad B, Cowburn RF: Localization of active forms of C-jun kinase (JNK) and p38 kinase in Alzheimer's disease brains at different stages of neurofibrillary degeneration. J Alzheimers Dis 2001;3:41–48.
43 Pei JJ, Tanaka T, Tung YC, Braak E, Iqbal K, Grundke-Iqbal I: Distribution, levels, and activity of glycogen synthase kinase-3 in the Alzheimer disease brain. J Neuropathol Exp Neurol 1997; 56:70–78.
44 Clerk A, Michael A, Sugden PH: Stimulation of multiple mitogen-activated protein kinase subfamilies by oxidative stress and phosphorylation of the small heat shock protein, HSP25/27, in neonatal ventricular myocytes. Biochem J 1998;333(pt 3):581–589.
45 Wang X, Martindale JL, Liu Y, Holbrook NJ: The cellular response to oxidative stress: Influences of mitogen-activated protein kinase signalling pathways on cell survival. Biochem J 1998;333(pt 2): 291–300.
46 Srivastava RK, Sollott SJ, Khan L, Hansford R, Lakatta EG, Longo DL: Bcl-2 and Bcl-X(L) block thapsigargin-induced nitric oxide generation, c-Jun NH(2)-terminal kinase activity, and apoptosis. Mol Cell Biol 1999;19:5659–5674.
47 Kyriakis JM, Banerjee P, Nikolakaki E, Dai T, Rubie EA, Ahmad MF, Avruch J, Woodgett JR: The stress-activated protein kinase subfamily of c-Jun kinases. Nature 1994;369:156–160.
48 Urano F, Wang X, Bertolotti A, Zhang Y, Chung P, Harding HP, Ron D: Coupling of stress in the ER to activation of JNK protein kinases by transmembrane protein kinase IRE1. Science 2000; 287:664–666.
49 Song L, De Sarno P, Jope RS: Central role of glycogen synthase kinase-3beta in endoplasmic reticulum stress-induced caspase-3 activation. J Biol Chem 2002;277:44701–44708.

Dr. Akihiko Takashima
Laboratory for Alzheimer's Disease, Brain Science Institute
RIKEN, 2–1 Hirosawa, Wako-shi, Saitama 351–0198 (Japan)
Tel. +81 48 467 9632, Fax +81 48 467 5916, E-Mail kenneth@brain.riken.go.jp

Takeda M, Tanaka T, Cacabelos R (eds): Molecular Neurobiology of Alzheimer Disease and Related Disorders. Basel, Karger, 2004, pp 71–78

Clinical Assessment of the Genetic Risk Functions in Alzheimer's Disease

Kouzin Kamino, Tomohiro Kida, Masatoshi Takeda

Division of Psychiatry and Behavioral Proteomics, Department of Post-Genomics and Diseases, Course of Advanced Medicine, Osaka University Graduate School of Medicine, Yamadaoka, Osaka, Japan

From the end of the 20th century, late-onset Alzheimer's disease (LOAD) has been the most prevalent cause of dementia in the elderly in Japan [1]. The prominent features of LOAD are microscopically characterized by neuritic plaques in the mesenchyme, and neurofibrillary tangles in the neuronal cell bodies, accompanied with neuronal cell loss in the cerebral cortex. Neuritic plaques contain β-amyloid, α_1-antichymotrypsin and other substances (table 1). Lipid-related proteins, such as apolipoprotein E (ApoE) and low-density lipoprotein (LDL)-receptor-related protein (LRP), are also found in neuritic plaques [2, 3]. Several studies pointed out the alterations of plasma cholesterol levels in patients with LOAD, and among those results, decreased plasma levels of high-density lipoprotein cholesterol (HDL-C) were notable [4, 5]. In an epidemiological study, elevated serum HDL-C levels were associated with a significantly decreased risk of dementia [6]. On the other hand, it was shown that hydroxymethylglutaryl coenzyme A (HMG-CoA) reductase inhibitor was effective in preventing dementia, especially Alzheimer's disease (AD), in the elderly [7, 8]. The HMG-CoA reductase inhibitor is known to lower the level of plasma LDL cholesterol (LDL-C) and increase the levels of plasma HDL-C. The well-known genetic risk for LOAD is the ε4 allele of the apolipoprotein E gene (APOE-ε4) [9, 10]. It was noted that the APOE-ε4 allele increases plasma LDL-C levels in normolipidemic subjects, and this effect was also shown in patients with dementia [11, 12]. Thus, the alterations of plasma cholesterol levels in patients with LOAD are similar to those in atherosclerosis, for which both high plasma LDL-C and low plasma HDL-C level increase the risk. It remains undetermined whether these lipid alterations are vascular risk factors leading to decreased brain circulation, or are neuronal factors

Table 1. The major components of pathological hallmarks of LOAD

Neuritic plaque
β–Amyloid (APP)
α_1-Antichymotrypsin
ApoE
α_2-Macroglobulin
Complement C1q, etc.
Ferritin
Serum amyloid P component
Aluminum
LRP
Acetylcholinesterase
CLAC-P/collagen type XXV
Neurofibrillary tangle
Microtubule associated protein τ
Ubiquitin
Ubiquitin-binding protein p62
Collapsin response mediator protein-2
Lysosome-associated antigen CD68

affecting synaptic plasticity in the brain. Nonetheless, when protein exists in plasma as well as in the brain, the polymorphism could be functionally assessed by measuring the target product in relation to its allele dose. Based on this assumption, we examined the effect of the known risk alleles on plasma lipid levels.

Plasma Cholesterol Levels in Late-Onset Alzheimer's Disease

We compared plasma cholesterol levels between patients with LOAD and control subjects over 65 years old (table 2), since, as evidenced in epidemiological studies, the genetic effect on LOAD is prominent in this age group [13]. We assessed 26 patients with LOAD and 26 sex- and age-matched nondemented controls and found that plasma total cholesterol and LDL-C levels were significantly increased in patients with LOAD, whereas plasma HDL-C levels were not significantly different though plasma HDL-C levels tended to be lower in patients with LOAD compared to nondemented controls (table 2). These results support that increased LDL-C and decreased HDL-C levels are the major trend in patients with LOAD.

Genetic Factors Related to Lipid Metabolism

APOE-ε4 is the major genetic risk factor for LOAD [9] and it was also evidenced that the LRP-C allele of the C766T silent polymorphism at exon 3

Table 2. Plasma lipid alterations in AD [unpubl. data]

	Plasma levels (mean ± SD), mg/dl	
	LOAD (n = 26)	Control (n = 26)
Total cholesterol	215.8 ± 36.3*	190.2 ± 35.9
HDL-C	55.8 ± 14.7	57.8 ± 15.1
LDL-C	138.3 ± 33.9**	110.4 ± 28.3
Triglycerides	108.7 ± 41.7	109.9 ± 41.2
ApoA-I	111.4 ± 24.7*	130.7 ± 28.0
ApoA-II	18.2 ± 9.4***	26.5 ± 7.1
ApoB	103.7 ± 22.5*	90.4 ± 18.4
ApoC-II	3.0 ± 1.8	3.1 ± 1.3
ApoC-III	6.6 ± 3.4	8.4 ± 4.0
ApoE	5.1 ± 1.4	5.3 ± 1.4

*$p < 0.05$, **$p < 0.01$, ***$p < 0.001$.

of the LRP gene is a genetic risk factor for AD [14, 15]. The APOE gene is located at 19q13.2, and this locus forms a gene cluster of apolipoproteins [16]. Pericak-Vance et al. [17] indicated the existence of the risk locus at chromosome 12; the LRP gene is located at 12q13.1–13.3 [18]. We examined the risk effect of the LRP-C allele in LOAD (table 3). Among APOE-ε4 carriers, the CRP-C allele was a significant risk factor for LOAD (OR = 3.4, p = 0.002), but no effect was supported for non-APOE-ε4 carriers. When all subjects were combined into one group, we did not find any significant risk effect on LOAD.

In polygenetic diseases, combination of the risk allele results in the development of the disease. We can examine this combinative risk effect by logistic regression statistics (table 4). Among APOE-ε4 carriers, we detected a significant risk effect of the dose of the LRP-C allele (OR = 3.1, 95% confidence interval: 1.42–6.96), but not of the dose of the APOE-ε4 allele. As a reference, we examined the risk effect of the APOE-ε4, LRP-C, and the defective allele of the mitochondrial aldehyde dehydrogenase (ALDH2) gene (ALDH2-K) by multiple logistic regression [19]. While the APOE-ε4 and ALDH2-K allele showed significant risk effects, the LRP-C allele did not. Thus, the LRP-C allele is a risk factor for LOAD in APOE-ε4 carriers.

Plasma Cholesterol Level and the Genetic Risks

It was demonstrated previously that the APOE-ε4 allele inversely correlated with plasma HDL-C level [20]. This study examined the effect of the dose of

Table 3. CC genotype of the C766T polymorphism of the LRP gene associates with sporadic LOAD in APOE-ε4 carriers

	LRP genotype		OR
	CC	CT and TT	
APOE-ε4 carriers			
LOAD	93 (0.869)	14 (0.131)	3.4 (p = 0.002)
Control	37 (0.661)	19 (0.339)	
Non-APOE-ε4 carriers			
LOAD	70 (0.737)	25 (0.263)	0.6 (n.s.)
Control	210 (0.827)	44 (0.173)	
All subjects			
LOAD	163 (0.807)	39 (0.193)	
Control	247 (0.797)	63 (0.203)	0.8 (n.s.)

n.s. = Not significant.

Table 4. Logistic regressions of the APOE-ε4 and the LRP-C dose on the risk for LOAD

Allele dose	Coefficient (p value)	OR (95% CI)
APOE-ε4 carriers		
APOE-ε4	0.953 (n.s.)	2.6 (0.55–12.2)
LRP-C	1.144 (0.0048)	3.1 (1.42–6.96)
Non-APOE-ε4 carriers		
LRP-C	−0.550 (0.0387)	0.6 (0.34–0.97)
All subjects		
APOE-ε4	1.541 (<0.0001)	4.7 (3.18–6.85)
LRP-C	0.021 (n.s.)	1.0 (0.65–1.61)
ALDH2-K	0.517 (0.0007)	1.7 (1.24–2.26)

the APOE-ε4 and LRP-C alleles in patients with LOAD and nondemented elderly subjects (table 5, unpubl. data). Albumin reflects the nutritional effect, and the dose of X chromosome is likely to modify plasma lipid levels by altering the level of sex hormones. To incorporate the effect of age, multiple regression statistics of plasma HDL-C and LDL-C levels was done against age, dose of X chromosome, plasma albumin level, dose of the APOE-ε4 allele, and dose of the LRP-C allele. Age was not a factor that significantly modified either plasma HDL-C or LDL-C levels. Plasma albumin level correlated with both plasma HDL-C and LDL-C levels. The dose of X chromosome increased

Table 5. The effect of the APOE and LRP gene on plasma HDL and LDL cholesterol levels [unpubl. data]

	Coefficient of parameters				
	age	sex	albumin	APOE-ε4	LRP-C
Plasma LDL	n.s.	0.205	0.388	n.s.	0.246
Plasma HDL	n.s.	n.s.	0.283	−0.219	n.s.
HDL/LDL	n.s.	−0.188	n.s.	−0.212	−0.274

All values shown are significant at $p < 0.05$; n.s. = not significant.

plasma LDL-C level. The dose of APOE-ε4 allele was found to inversely correlate with plasma HDL-C level, but not with plasma LDL-C level. The dose of the LRP-C allele was found to correlate with plasma LDL-C level, but not with plasma HDL-C level. The ratio of plasma HDL-C to LDL-C level inversely correlated with the dose of X chromosome, and the APOE-ε4 and LRP-C alleles. Thus, the effect of genetic risk was summarized as follows: the factors that decrease the ratio of plasma HDL-C to LDL-C increase the risk for LOAD. These results suggest that the genetic predisposition to LOAD is the same as that to atherosclerosis, and it seems plausible that life-style measures for the prevention of atherosclerosis are also effective in preventing LOAD.

Functions of the LRP Gene in Late-Onset Alzheimer's Disease

Ligand-binding sites of LRP are composed of 4 regions, and exon 3 of the LRP gene encodes cluster I [21]. The LRP-C allele was shown to relate to lower brain LRP expression and to a higher amount of brain β-amyloid deposition [22, 23]. Clusters I and II of LRP are required to bind α_2-macroglobulin, which is also deposited in neuritic plaques [24]. In an exon-trapping assay, the LRP-C allele showed a significantly decreased splicing efficiency compared to the LRP-T allele, leading to the production of LRP lacking the exon 3 sequence-encoding cluster I [unpubl. data]. α_2-Macroglobulin complexes with and mediates β-amyloid endocytosis via LRP [25], and was shown to prevent the aggregation of β-amyloid. Based on this evidence, LRP could mediate the protective effect of α_2-macroglobulin in β-amyloid deposition. Moreover, it was also shown that presenilin-1 downregulates the expression of LRP [26]. Therefore, LRP, in itself, is a preventive factor for LOAD through the β-amyloid clearance pathway. On the other hand, it was shown in a cell culture study that

LRP mediates ApoE3-dependent neurite outgrowth [27], and that APOE3-enriched HDL enhanced neurite outgrowth more than ApoE4-enriched HDL [28]. Thus, LRP and APOE could participate in synaptic plasticity. We found the risk effect of the LRP-C allele only in APOE-ε4 carriers, suggesting that the risk effect of the LRP-C allele in the population differs between APOE-ε4 and non-APOE-ε4 carriers. In fact, the LRP-T allele increases the risk for coronary artery disease in subjects harboring the 5G/5G genotype of the 4G/5G promoter polymorphism of plasminogen activator inhibitor-1 gene [29]. LRP is expressed in the macrophage lineage, and also functions as the second major receptor for APOE [30, 31]. In summary, LRP participates in cholesterol clearance as well as lipid homeostasis, and the risk effect of the LRP-C allele for LOAD could be complicated in the elderly population.

Conclusions

The LRP-C allele is a risk factor for LOAD in APOE-ε4 carriers, and increases the plasma LDL-C level. LRP expression could be protective against the development of LOAD. Since LRP is also related to systemic lipid metabolism and to atherosclerosis, induction of LRP expression restricted to the brain could be therapeutic and protective against LOAD. However, the application of LRP-C genotyping in diagnosing LOAD is not warranted.

Acknowledgement

The genome ethical committee of Osaka Univerisity approved this study, which was supported by Research for the Future Program, Japan Society for the Promotion of Science (JSPS).

References

1 Shibayama H, Kobayashi H, Nakagawa M, Marui Y, Miyachi T, Kayukawa Y, Yamada K, Iwata H, Takeuchi T, Iwai K, Kogawa S, Kagawa N, Imai M, Mizuno Y, Hashimoto N, Ibuki Y, Ogasawara S, Tadano F, Nakagawa T, Ohta T: Prevalence of dementia among the elderly in a Japanese community population – comparative study on the 1983 and 1996 survey: The Aichi Study. Psychogeriatrics 2001;1:317–325.

2 Namba Y, Tomonaga M, Kawasaki H, Otomo E, Ikeda K: Apolipoprotein E immunoreactivity in cerebral amyloid deposits and neurofibrillary tangles in Alzheimer's disease and kuru plaque amyloid in Creutzfeldt-Jakob disease. Brain Res 1991;541:163–166.

3 Rebeck GW, Reiter JS, Strickland DK, Hyman BT: Apolipoprotein E in sporadic Alzheimer's disease: Allelic variation and receptor interactions. Neuron 1993;11:575–580.

4 Kuriyama M, Takahashi K, Yamano T, Hokezu Y, Togo S, Osame M, Igakura T: Low levels of serum apolipoprotein AI and AII in senile dementia. Jpn J Psychiatry Neurol 1994;48:589–593.

5 Kawano M, Kawakami M, Otsuka M, Yashima H, Yaginuma T, Ueki A: Marked decrease of plasma apolipoprotein AI and AII in Japanese patients with late-onset non-familial Alzheimer's disease. Clin Chim Acta 1995;239:209–211.
6 Merched A, Xia Y, Visvikis S, Serot JM, Siest G: Decreased high-density lipoprotein cholesterol and serum apolipoprotein AI concentrations are highly correlated with the severity of Alzheimer's disease. Neurobiol Aging 2000;21:27–30.
7 Jick H, Zornberg GL, Jick SS, Seshadri S, Drachman DA: Statins and the risk of dementia. Lancet 2000;356:1627–1631.
8 Rockwood K, Kirkland S, Hogan DB, MacKnight C, Merry H, Verreault R, Wolfson C, McDowell I: Use of lipid-lowering agents, indication bias, and the risk of dementia in community-dwelling elderly people. Arch Neurol 2002;59:223–227.
9 Farrer LA, Cupples LA, Haines JL, Hyman B, Kukull WA, Mayeux R, Myers RH, Pericak-Vance MA, Risch N, van Duijn CM, for the APOE and Alzheimer Disease Meta Analysis Consortium: Effects of age, sex, and ethnicity on the association between apolipoprotein E genotype and Alzheimer disease. A meta-analysis. JAMA 1997;278:1349–1356.
10 Saunders AM, Strittmatter WJ, Schmechel D, St George-Hyslop PH, Pericak-Vance MA, Joo SH, Rosi BL, Gusella JF, Crapper-MacLachlan DR, Alberts MJ, Hulette C, Crain B, Goldgaber D, Roses AD: Association of apolipoprotein E allele ε4 with late-onset familial and sporadic Alzheimer's disease. Neurology 1993;43:1467–1472.
11 Sing CF, Davignon J: Role of the apolipoprotein E polymorphism in determining normal plasma lipid and lipoprotein variation. Am J Hum Genet 1985;37:268–285.
12 Wehr H, Parnowski T, Puzynski S, Bednarska-Makaruk M, Bisko M, Kotapka-Minc S, Rodo M, Wolkowska M: Apolipoprotein E genotype and lipid and lipoprotein levels in dementia. Dement Geriatr Cogn Disord 2000;11:70–73.
13 Rao VS, van Duijn CM, Connor-Lacke L, Cupples LA, Growdon JH, Farrer LA: Multiple etiologies for Alzheimer disease are revealed by segregation analysis. Am J Hum Genet 1994;55:991–1000.
14 Kang DE, Saitoh T, Chen X, Xia Y, Masliah E, Hansen LA, Thomas RG, Thal LJ, Katzman R: Genetic association of the low-density lipoprotein receptor-related protein gene (LRP), an apolipoprotein E receptor, with late-onset Alzheimer's disease. Neurology 1997;49:56–61.
15 Sanchez-Guerra M, Combarros O, Infante J, Llorca J, Berciano J, Fontalba A, Fernandez-Luna JL, Pena N, Fernandez-Viadero C: Case-control study and meta-analysis of low density lipoprotein receptor-related protein gene exon 3 polymorphism in Alzheimer's disease. Neurosci Lett 2001; 316:17–20.
16 Smit M, van der Kooij-Meijs E, Frants RR, Havekes L, Klasen EC: Apolipoprotein gene cluster on chromosome 19. Definite localization of the APOC2 gene and the polymorphic Hpa I site associated with type III hyperlipoproteinemia. Hum Genet 1988;78:90–93.
17 Rogaeva E, Premkumar S, Song Y, Sorbi S, Brindle N, Paterson A, Duara R, Levesque G, Yu G, Nishimura M, Ikeda M, O'Toole C, Kawarai T, Jorge R, Vilarino D, Bruni AC, Farrer LA, St George-Hyslop PH: Evidence for an Alzheimer disease susceptibility locus on chromosome 12 and for further locus heterogeneity. JAMA 1998;280:614–618.
18 Hilliker C, Van Leuven F, Van Den Berghe H: Assignment of the gene coding for the alpha(2)-macroglobulin receptor to mouse chromosome 15 and to human chromosome 12q13–q14 by isotopic and nonisotopic in situ hybridization. Genomics 1992;13:472–474.
19 Kamino K, Nagasaka K, Imagawa M, Yamamoto H, Yoneda H, Ueki A, Kitamura S, Namekata K, Miki T, Ohta S: Deficiency in mitochondrial aldehyde dehydrogenase increases the risk for late-onset Alzheimer's disease in the Japanese population. Biochem Biophys Res Commun. 2000;273: 192–196.
20 Hoshino T, Kamino K, Matsumoto M: Gene dose effect of the APOE-epsilon4 allele on plasma HDL cholesterol level in patients with Alzheimer's disease. Neurobiol Aging 2002;23:41–45.
21 Krieger M, Herz J: Structures and functions of multiligand lipoprotein receptors: Macrophage scavenger receptors and LDL receptor-related protein (LRP). Annu Rev Biochem 1994;63: 601–637.
22 Kang DE, Pietrzik CU, Baum L, Chevallier N, Merriam DE, Kounnas MZ, Wagner SL, Troncoso JC, Kawas CH, Katzman R, Koo EH: Modulation of amyloid beta-protein clearance and Alzheimer's disease susceptibility by the LDL receptor-related protein pathway. J Clin Invest 2000;106:1159–1166.

23 Rebeck GW, Harr SD, Strickland DK, Hyman BT: Multiple, diverse senile plaque-associated proteins are ligands of an apolipoprotein E receptor, the alpha-2-macroglobulin receptor/low-density-lipoprotein receptor-related protein. Ann Neurol 1995;37:211–217.
24 Mikhailenko I, Battey FD, Migliorini M, Ruiz JF, Argraves K, Moayeri M, Strickland DK: Recognition of alpha-2-macroglobulin by the low density lipoprotein receptor-related protein requires the cooperation of two ligand binding cluster regions. J Biol Chem 2001;276:39484–39491.
25 Narita M, Holtzman DM, Schwartz AL, Bu G: Alpha-2-macroglobulin complexes with and mediates the endocytosis of beta-amyloid peptide via cell surface low-density lipoprotein receptor-related protein. J Neurochem 1997;69:1904–1911.
26 Van Uden E, Carlson G, St George-Hyslop P, Westaway D, Orlando R, Mallory M, Rockenstein E, Masliah E: Aberrant presenilin-1 expression downregulates LDL receptor-related protein (LRP): Is LRP central to Alzheimer's disease pathogenesis? Mol Cell Neurosci 1999;14:129–140.
27 Holtzman DM, Pitas RE, Kilbridge J, Nathan B, Mahley RW, Bu G, Schwartz AL: Low density lipoprotein receptor-related protein mediates apolipoprotein E-dependent neurite outgrowth in a central nervous system-derived neuronal cell line. Proc Natl Acad Sci USA 1995;92:9480–9484.
28 Bellosta S, Nathan BP, Orth M, Dong LM, Mahley RW, Pitas RE: Stable expression and secretion of apolipoproteins E3 and E4 in mouse neuroblastoma cells produces differential effects on neurite outgrowth. J Biol Chem 1995;270:27063–27071.
29 Benes P, Muzik J, Benedik J, Elbl L, Vasku A, Siskova L, Znojil V, Vacha J: The C766T low-density lipoprotein receptor related protein polymorphism and coronary artery disease, plasma lipoproteins, and longevity in the Czech population. J Mol Med 2001;79:116–120.
30 Herz J, Hamann U, Rogne S, Myklebost O, Gausepohl H, Stanley KK: Surface location and high affinity for calcium of a 500 kd liver membrane protein closely related to the LDL-receptor suggest a physiological role as lipoprotein receptor. EMBO J 1988;7:4119–4127.
31 Kowal RC, Herz J, Goldstein JL, Esser V, Brown MS: Low density lipoprotein receptor-related protein mediates uptake of cholesteryl esters derived from apoprotein E-enriched lipoproteins. Proc Natl Acad Sci USA 1989;86:5810–5814.

Kouzin Kamino, MD, PhD
Division of Psychiatry and Behavioral Proteomics
Department of Post-Genomics and Diseases, Course of Advanced Medicine
Osaka University Graduate School of Medicine
2–2, D3, Yamadaoka, Suita, Osaka 565–0871 (Japan)
Tel. +81 6 6879 3051, Fax +81 6 6879 3059, E-Mail kkamino@psy.med.osaka-u.ac.jp

Takeda M, Tanaka T, Cacabelos R (eds): Molecular Neurobiology of Alzheimer Disease and Related Disorders. Basel, Karger, 2004, pp 79–83

Hydrogen Sulfide Is Severely Decreased in Alzheimer Disease Brains

Hideo Kimura

National Institute of Neuroscience, National Center of Neurology and Psychiatry, Ogawahigashi, Kodaira, Tokyo, Japan

Since the first description of H_2S toxicity in 1713 [1], most studies about H_2S have been devoted to its toxic effects, with little attention paid to its physiological function [2]. Warenycia et al. [3] found that the rat brain contains endogenous H_2S, and endogenous concentrations of H_2S have also been measured in human and bovine brains [4, 5]. The relatively high concentrations of H_2S in the brain (50–160 μM) suggest that it has a physiological function.

Endogenous H_2S in the brain is formed from *L*-cysteine by a pyridoxal-5′-phosphate-dependent enzyme, cystathionine β-synthase (CBS) [6–11]. CBS inhibitors, hydroxylamine and amino-oxyacetate, suppress H_2S production, while a CBS activator, S-adenosyl-*L*-methionine (SAM), enhances it. Observations with CBS knockout mice clearly show that CBS is the only enzyme that produces H_2S in the brain [11].

Two other gases, nitric oxide (NO) and carbon monoxide (CO), are endogenously produced by enzymes localized to the brain. NO is synthesized by NO synthase via the metabolism of arginine to citrulline [12, 13], and CO is produced by heme oxygenase via the metabolism of heme to biliverdin [14, 15]. Both NO and CO enhance the induction of hippocampal long-term potentiation (LTP), a synaptic model of learning and memory [16–22]. The activities of NO synthase are regulated by Ca^{2+}/calmodulin, and NO is released when N-methyl-*D*-aspartate (NMDA) receptors are activated by *L*-glutamate [23, 24]. In contrast, the regulation of CO production by neuronal excitation is not understood [22].

H_2S production in the brain is enhanced in response to neuronal excitation via the Ca^{2+} and calmodulin-mediated pathways [11]. In addition, physiological concentrations of H_2S specifically potentiate the activity of NMDA receptors, and hippocampal LTP is altered in CBS knockout mice [10, 11]. H_2S can also regulate the release of corticotropin-releasing hormone from the hypothalamus [25]. Based upon these observations, it has been proposed that H_2S may function as a neuromodulator or transmitter in the brain [10, 11].

Loss of CBS activity causes homocysteinurea, an autosomal recessive disease characterized, in part, by mental retardation [26]. CBS interacts with huntingtin, mutants of which cause Huntington's disease [27]. Polymorphisms of the CBS gene are significantly underrepresented in children with high IQs compared with those with average IQs, suggesting that CBS activity may be associated with overall cognitive function [28]. These observations suggest that the abnormal regulation of H_2S production may cause neuronal dysfunction.

The enzymatic activity of CBS has two metabolic outcomes [6, 26]. Most studies have been devoted to a pathway in which CBS catalyzes the reaction with substrate homocysteine to produce cystathionine [26], but little attention has been paid to another pathway in which CBS produces H_2S from *L*-cysteine as a substrate [6]. SAM, which enhances CBS activity in both metabolic pathways [29], is much reduced in AD brains [30]. This observation, in conjunction with the previous finding that homocysteine is accumulated in serum of AD patients [31], suggests that CBS activity must be reduced in AD. The present study shows that H_2S and SAM are decreased, but that homocysteine is increased in AD brains. These observations suggest that CBS activity must be reduced in AD brains and the decreased levels of H_2S may be involved in the cognitive decline observed in this disease.

Results and Discussion

Since the levels of SAM, an activator of CBS, are lower in AD brains than in the brains of normal individuals [30], the endogenous levels of H_2S in AD brains could be lower than in control brains. To examine this possibility, the endogenous H_2S levels in AD brains were measured and compared with the brains of age-matched normal individuals. The amounts of H_2S in the frontal cortex of each individual were then measured. Thirteen brains of AD patients contained 0.22 ± 0.05 nmol/mg protein of endogenous H_2S. Six brains of age-matched normal individuals contained 0.49 ± 0.07 nmol/mg protein of H_2S. Endogenous H_2S levels of AD brains are significantly lower than those of brains of age-matched normal individuals (ANOVA, $p < 0.01$).

Because CBS is the major enzyme that produces H_2S in the brain [11], there are three possibilities that may cause the changes in the endogenous H_2S levels. These are differences in the levels of the substrate for CBS, the amount of CBS, or the activity of CBS. To examine the first possibility, the amount of free cysteine in the brain was measured by HPLC. The endogenous cysteine level in AD brains was 7.4 ± 0.3 nmol/mg protein (n = 13) and 7.4 ± 0.4 nmol/mg protein in control brains (n = 6). Therefore there was no significant difference in the amount of endogenous free cysteine between AD and control brains.

To compare the amount of CBS in AD with controls, the levels of CBS in the brains were determined. 30 μg of brain homogenates was analyzed by Western blotting with an antibody against CBS and quantitated by an image analyzer. The CBS level of control brains was 100 ± 7.7 (n = 6), and the relative CBS level of AD brains was 96.9 ± 10.4 (n = 13). Therefore, there is no significant difference in the amount of CBS between control brains and AD.

Because the amounts of SAM, a CBS activator, in AD brains had been reported to be lower than in control brains [30], we attempted to confirm the previous data by measuring the amounts of SAM in our AD brain samples by HPLC. The endogenous level of SAM in AD brains is 0.16 ± 0.03 nmol/mg protein (n = 13) and that in control brains 0.53 ± 0.04 nmol/mg protein (n = 6). Therefore the endogenous levels of SAM in AD brains are significantly less than those of control brains ($p < 0.001$ by the ANOVA test). These observations are consistent with those obtained by Morrison et al. [30] and suggest that the low endogenous level of H_2S in AD brains may be caused by the decreased activity of CBS due to the lack of SAM.

The enzymatic activity of CBS has two metabolic outcomes [6, 26]. CBS catalyzes the reaction with cysteine as a substrate to produce H_2S [6]. In another pathway, CBS catalyzes the reaction with homocysteine as a substrate to produce cystathionine [26]. The activity of CBS in both pathways is regulated by SAM [29] and the serum homocysteine level is higher in AD patients than in normal individuals [31]. These observations suggest that the activity of CBS must be reduced in AD brains. To test this possibility, it may be best to measure CBS activity. However, because CBS activity in frozen rat brain samples is decreased to less than 10% of that in fresh samples [unpubl. obs.], it is impossible to accurately measure the H_2S-producing activity of CBS extracted from frozen AD brains. Alternatively, the amounts of homocysteine in the brains of AD and DAT patients were measured by HPLC and compared with those of control brains. The endogenous homocysteine level in AD brains was 5.6 ± 0.3 nmol/mg protein (n = 13) and 3.2 ± 0.2 nmol/mg protein in control (n = 6). The endogenous homocysteine levels in AD brains were indeed significantly greater than in control brains ($p < 0.001$ by the ANOVA test). These observations confirm that CBS activity is decreased in AD brains, and that this decline in CBS activity may be due to the lack of SAM.

Recent studies have suggested that abnormalities in the cerebral microvasculature may be relevant to the cause of AD [32, 33]. H_2S relaxes smooth muscle, and this effect is augmented by NO [34]. In addition to the synergic effect of H_2S and NO on smooth muscle relaxation, H_2S production in smooth muscle can also be regulated by NO [35]. These observations suggest that decreased H_2S may also cause the dysfunction of cerebral microvasculature that leads to AD.

In conclusion, H_2S, a neuromodulator, is decreased in AD brains. The activator of CBS, SAM, is also decreased, while one of the two CBS substrates, homocysteine, is increased. The reduction of H_2S and the CBS activity may be involved in some aspects of the cognitive decline associated with AD.

Acknowledgement

This work was supported by a grant from the National Institute of Neuroscience, National Center of Neurology and Psychiatry, Japan.

References

1 Ramazzin B: (1713) Diseases of Workers (1713)(translated from the Latin text De Morbis Artificum by WC Wright, 1940). Chicago, University of Chicago Press; reprinted 1964 in History Med 1964;23.
2 Reiffenstein RJ, Hulbert WC, Roth SH: Toxicology of hydrogen sulfide. Annu Rev Pharmacol Toxicol 1992;32:109–134.
3 Warenycia MW, Goodwin LR, Benishin CG, Reiffenstein RJ, Francom DM, Taylor JD, Dieken FP: Acute hydrogen sulfide poisoning: demonstration of selective uptake of sulfide by the brainstem by measurement of brain sulfide levels. Biochem Pharmacol 1989;38:973–981.
4 Goodwin LR, Francom D, Dieken FP, Taylor JD, Warenycia MW, Reiffenstein RJ, Dowling G: Determination of sulfide in brain tissue by gas dialysis/ion chromatography: postmortem studies and two case reports. J Anal Toxicol 1989;13:105–109.
5 Savage JC, Gould DH: Determination of sulfides in brain tissue and rumen fluid by ion-interaction reversed-phase high-performance liquid chromatography. J Chromatogr 1990;526:540–545.
6 Stipanuk MH, Beck PW: Characterization of the enzymic capacity for cysteine desulphydration in liver and kidney of the rat. Biochem J 1982;206:267–277.
7 Griffith OW: Mammalian sulfur amino acid metabolism: A overview. Methods Enzymol 1987; 143:366–376.
8 Erickson PF, Maxwell IH, Su LJ, Baumann M, Glode LM: Sequence of cDNA for rat cystathionine γ-lyase and comparison of deduced amino acid sequence with related *Escherichia coli* enzymes. Biochem J 1990;269:335–340.
9 Swaroop M, Bradley K, Ohura T, Tahara T, Roper MD, Rosenberg LE, Kraus JP: Rat cystathionine β-synthase. J Biol Chem 1992;267:11455–11461.
10 Abe K, Kimura H: The possible role of hydrogen sulfide as an endogenous neuromodulator. J Neurosci 1996;16:1066–1071.
11 Eto K, Ogasawara M, Umemura K, Nagai Y, Kimura H: Hydrogen sulfide is produced in response to neuronal excitation. J Neurosci, in press.
12 Palmer RMJ, Ashton DS, Moncada S: Vascular endothelial cells synthesize nitric oxide from *L*-arginine. Nature 1988;333:664–666.
13 Bredt DS, Snyder SH: Nitric oxide, a novel neuronal messenger. Neuron 1992;8:3–11.
14 Maines MD: Heme oxygenase: Function, multiplicity, regulatory mechanisms, and clinical applications. FASEB J 1988;2:2557–2568.
15 Verma A, Hirsch DJ, Glatt CE, Ronnett GV, Snyder SH: Carbon monoxide: A putative neural messenger. Science 1993;259:381–384.

16 O'Dell TJ, Hawkins RD, Kandel ER, Arancio O: Tests of the roles of two diffusible substances in long-term potentiation: Evidence for nitric oxide as a possible early retrograde messenger. Proc Natl Acad Sci USA 1991;88:11285–11289.
17 Schuman EM, Madison DV: A requirement for the intercellular messenger nitric oxide in long-term potentiation. Science 1991;254:1503–1506.
18 Haley JE, Wilcox GL, Chapman PF: The role of nitric oxide in hippocampal long-term potentiation. Neuron 1992;8:211–216.
19 Stevens CF, Wang Y: Reversal of long-term potentiation by inhibitors of haem oxygenase. Nature 1993;364:147–149.
20 Zhuo M, Small SA, Kandel ER, Hawkins RD: Nitric oxide and carbon monoxide produce activity-dependent long-term synaptic enhancement in hippocampus. Science 1993;260:1946–1950.
21 Bliss TVP, Collingridge GL: A synaptic model of memory: Long-term potentiation in the hippocampus. Nature 1993;361:31–39.
22 Snyder SH, Ferris CD: Novel neurotransmitters and their neuropsychiatric relevance. Am J Psychiatry 2000;157:1738–1751.
23 Garthwaite J, Charles SL, Chess-Williams R: Endothelium-derived relaxing factor release on activation of NMDA receptors suggests role as intercellular messenger in the brain. Nature 1988; 336:385–388.
24 Bredt DS, Snyder SH: Isolation of nitric oxide synthetase, a calmodulin-requiring enzyme. Proc Natl Acad Sci USA 1990;87:682–685.
25 Russo CD, Tringali G, Ragazzoni E, Maggiano N, Menini E, Vairano M, Preziosi P, Navarra P: Evidence that hydrogen sulphide can modulate hypothalamo-pituitary-adrenal axis function: In vitro and in vivo studies in the rat. J Neuroendocrinol 2000;12:225–233.
26 Mudd SH, Levy HL, Skovby F: Disorders of transsulfuration; in Scriver CR, Beaudet AL, Sly WS, Valle D (eds): The Metabolic Basis of Inherited Disease. New York, McGraw-Hill, 1989, pp 693–734.
27 Boutell JM, Wood JD, Harper PS, Jones AL: Huntingtin interacts with cystathionine β-synthase. Hum Mol Genet 1998;7:371–378.
28 Barbaux S, Plomin R, Whitehead AS: Polymorphisms of genes controlling homocysteine/folate metabolism and cognitive function. NeuroReport 2000;11:1133–1136.
29 Finkelstein JD, Kyle WE, Martin JJ, Pick AM: Activation of cystathionine synthase by adenosylmethionine and adenosylmethionine. Biochem Biophys Res Commun 1975;66:81–87.
30 Morrison LD, Smith DD, Kish SJ: Brain S-adenosylmethionine levels are severely decreased in Alzheimer's disease. J Neurochem 1996;67:1328–1331.
31 Clarke R, Smith D, Jobst KA, Fefsum H, Sutton L, Ueland PM: Folate, vitamin B_{12}, and serum total homocysteine levels in confirmed Alzheimer Disease. Arch Neurol 1998;55:1449–1455.
32 De la Torre JC, Mussivand T: Can disturbed brain microcirculation cause Alzheimer's disease? Neurol Res 1993;15:146–153.
33 Kalaria RN: The role of cerebral ischemia in Alzheimer's disease. Neurobiol Aging 2000;21: 321–330.
34 Hosoki R, Matsuki N, Kimura H: The possible role of hydrogen sulfide as an endogenous smooth muscle relaxant in synergy with nitric oxide. Biochem Biophys Res Commun 1997;237:527–531.
35 Zhao W, Zhang J, Lu Y, Wang R: The vasorelaxant effect of H_2S as a novel endogenous gaseous KATP channel opener. EMBO J 2001;20:6008–6016.

Dr. Hideo Kimura
National Institute of Neuroscience, NCNP
4–1–1 Ogawahigashi, Kodaira, Tokyo 187–8551 (Japan)
Tel. +81 42 346 1725, Fax +81 42 346 1755, E-Mail kimura@ncnp.go.jp

Takeda M, Tanaka T, Cacabelos R (eds): Molecular Neurobiology of Alzheimer Disease and Related Disorders. Basel, Karger, 2004, pp 84–93

Functional Analysis of the Presenilin Complex and γ-Secretase Activity

Taisuke Tomita, Nobumasa Takasugi, Makiko Tsuruoka, Manabu Niimura, Ikuo Hayashi, Yasuko Takahashi, Yuichi Morohashi, Noriko Isoo, Sayaka Tanaka, Chihiro Sato, Takeshi Iwatsubo

Department of Neuropathology and Neuroscience, Graduate School of Pharmaceutical Sciences, University of Tokyo, Tokyo, Japan

Mutations in presenilin (PS) genes cause early-onset familial Alzheimer's disease by altering γ-cleavage to increase the production of Aβ42 peptide [1]. Moreover, PS are required for the intramembranous cleavage of several type I transmembrane proteins, including Notch family proteins [2, 3]. Following an initial cleavage within the ectodomain, PS-dependent γ-secretase cleavage occurs within the transmembrane domain (TMD) and releases soluble peptides from the membrane (fig. 1). The liberated intracellular domain fragments are implicated in the transcriptional regulation, suggesting that γ-cleavage mediates not only the generation of Alzheimer-related Aβ peptides but also the novel signal transduction pathways [4, 5].

Two conserved critical aspartate residues within TMDs 6 and 7 of PS were shown to be required for γ-secretase activity [6]. Molecular and cell biological studies suggest that PS functions as a catalytic subunit of γ-secretase and belongs to an atypical GxGD-type polytopic aspartyl protease family [7]. We have previously generated a variety of PS mutants to clarify the relationships between the metabolism of PS polypeptides and γ-secretase activities [8–12]. Through these studies, we have reached the conclusion that stabilization and high-molecular-weight (HMW) protein complex formation of PS is most closely linked to the γ-secretase activities. Therefore, a search for PS cofactors, which are essential to stabilization and HMW complex formation, is a critical issue in PS research. Recently, biochemical and genetic studies have revealed that nicastrin, APH-1 and PEN-2 are required for Notch signaling, γ-secretase

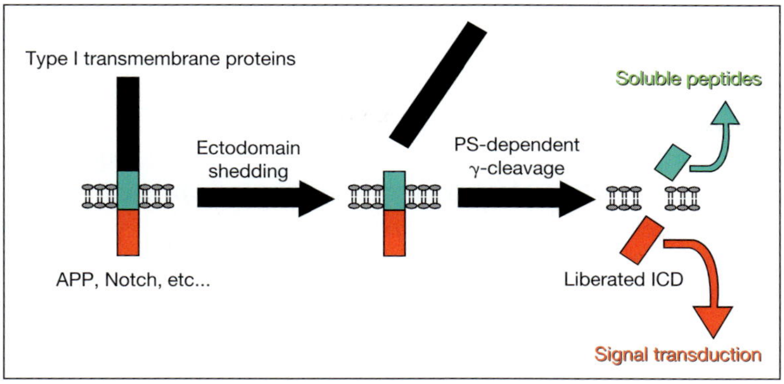

Fig. 1. Intramembrane proteolysis by PS-dependent γ-secretase. Type I transmembrane proteins are shed by ligand-dependent or constitutive endoproteolysis within the ectodomain. Membrane-tethered C-terminal fragments are then cleaved by PS-dependent γ-secretase activity within the TMD, and soluble peptides are released from the membrane.

activity and the accumulation of PS fragments [13–16]. However, the molecular and functional relationship between these cofactors and γ-secretase activity has remained unclear.

To study the molecular nature of PS and its cofactors, we employed *Drosophila* S2 cells, which are useful in overexpression of transgenes as well as inactivation of gene expression by the RNA interference (RNAi) technique. Moreover, the sequencing project for the *Drosophila* genome was completed and revealed that all γ-secretase-related genes are conserved [17].

We found that *Drosophila* APH-1 (dAPH-1) increases the stabilization of *Drosophila* PS (Psn) holoproteins that are incorporated into an HMW protein complex. Inactivation of *Drosophila* PEN-2(dPEN-2), another Psn cofactor, abrogates accumulation of Psn fragments, but promotes stabilization of Psn holoprotein. Coexpression of dPEN-2 with dAPH-1 and *Drosophila* nicastrin (dNCT) increases the formation of Psn fragments as well as γ-secretase activity to generate Aβ. These data illustrate the differential roles of PS cofactors in PS complex formation, APH-1 as the stabilizing cofactor of PS and PEN-2 as a factor conferring γ-secretase activities and facilitating the formation of PS fragments.

Materials and Methods

Materials and methods have been described in detail previously [18, 19].

Results and Discussion

Human APP Is Cleaved by Psn-Dependent γ-Secretase to Generate Aβ40 and Aβ42 in Drosophila *S2 Cells*

It has been reported that *Psn*-dependent γ-secretase cleavage is required for Notch signaling in *Drosophila*, although the molecular nature of the Psn complex has remained unclear [20]. To characterize the Psn complex and γ-secretase activity in *Drosophila* S2 cells, we raised antibodies against N-terminal and loop regions of Psn. Western blot analysis revealed that endogenous Psn proteins undergo endoproteolysis to generate NTF and CTF. Upon cycloheximide treatment that blocks total cellular protein synthesis, endogenous as well as exogenous Psn holoproteins were unstable, whereas the fragments were highly stabilized. To examine the capacity of Psn proteins to form HMW complexes, we solubilized the membrane fractions of S2 cells in 1% CHAPSO, and separated the extracted proteins on a linear glycerol velocity gradient. Western blot analysis revealed that Psn fragments were predominantly distributed in the 200- to over 400-kD HMW range, whereas Psn holoproteins were fractionated in the low-molecular-weight (LMW) range. Thus, these data suggest that Psn proteins are metabolized similarly to mammalian PS [21–24].

We transfected the double-stranded RNA (dsRNA) encoding various regions in Psn ORF, all of which resulted in a significant suppression of the expression of Psn holoprotein as well as fragments thereof. To evaluate the γ-secretase-like activity for proteolytic processing of the TMD sequence of human amyloid precursor protein (APP) in *Drosophila* S2 cells, we transiently transfected a cDNA encoding SC100 that corresponds to the C-terminal fragment of human APP starting at the 1st residue of Aβ preceded by a signal peptide, and analyzed the conditioned media by ELISA [25, 26]. S2 cells secrete Aβ40 as well as Aβ42. %Aβ42 was ~20%, which was increased by transfection of SC100 carrying the I716F mutation, known as the Aβ42-overproducing mutation in mammalian cells [27], suggesting that γ-secretase activity in S2 cells can process human APP similarly to that in mammalian cells. We generated stable S2 cell lines expressing SC100, and transfected them with Psn dsRNAs (fig. 2). Downregulation of the expression of *Psn* caused a significant reduction in Psn holoprotein and fragments as well as the secretion of Aβ peptides, which was accompanied by accumulation of SC100 and its derivatives. These data suggest that human APP undergoes endoproteolysis to generate Aβ peptides by a *Psn*-dependent γ-secretase activity.

dAPH-1 Functions as a Stabilizing Cofactor of the Psn Complex

We then transfected dsRNA for dNCT or dAPH-1 into S2 cells (fig. 2). In sharp contrast to Psn RNAi, they selectively inhibited the accumulation of

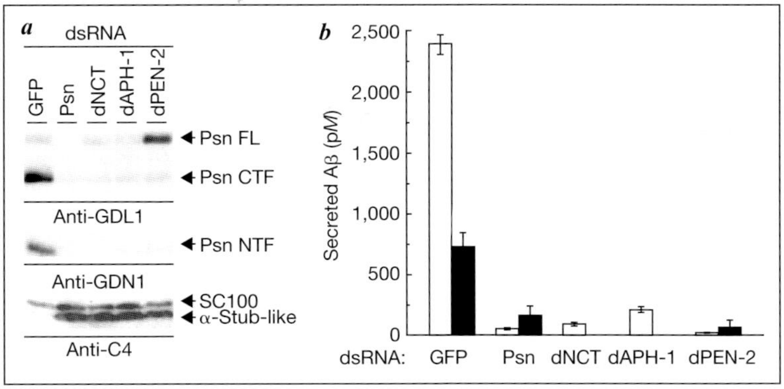

Fig. 2. Effects of RNAi for Psn and its cofactors on Psn metabolism and Aβ generation. ***a*** S2 cells were transfected with dsRNA for GFP or Psn cofactors (dsRNA), for 48 h and analyzed by immunoblotting. ***b*** Levels of Aβ40 (open column) and Aβ42 (closed column) (mean ± SE, n = 3) secreted from S2 cells stably transfected with SC100 and treated by RNAi for GFP, dNCT, dAPH-1 or dPEN-2 quantitated by sandwich ELISA.

Psn fragments, whereas the amount of Psn holoproteins was unchanged [14]. However, the secretion of Aβ peptides into culture media was also significantly decreased. These data suggest that dNCT as well as dAPH-1 is required for the accumulation of Psn fragments in S2 cells and γ-secretase cleavage of APP.

To gain insights into the function of Psn cofactors, we stably overexpressed dNCT and dAPH-1 in S2 cells (fig. 3). We found that the levels of Psn holoprotein were increased in S2 cells stably transfected with dAPH-1 with or without dNCT, but not in cells transfected singly with dNCT, whereas the levels of Psn fragments were not altered. This prompted us to examine the biochemical properties of increased Psn holoproteins in the presence of dAPH-1.

To examine the stability of Psn holoprotein, we treated S2 cells with or without stable cotransfection of dAPH-1 and dNCT with cycloheximide (fig. 3). In mock transfected S2 cells, holoprotein of Psn was rapidly degraded within 4 h of treatment, whereas fragments of Psn were highly stable. In sharp contrast, Psn holoprotein, that was significantly increased prior to CHX treatment, was highly stable and was not decreased during the 4-hour chase . We then separated CHAPSO-solubilized membrane fractions of double-stable S2 cells by glycerol velocity gradient centrifugation. Psn holoportein in cells overexpressing dAPH-1 was fractionated mostly in HMW ranges together with Psn fragments, in sharp contrast to the distribution of short-lived Psn holoproteins in WT cells. In addition, dAPH-1 and dNCT were also separated chiefly in HMW ranges. We performed

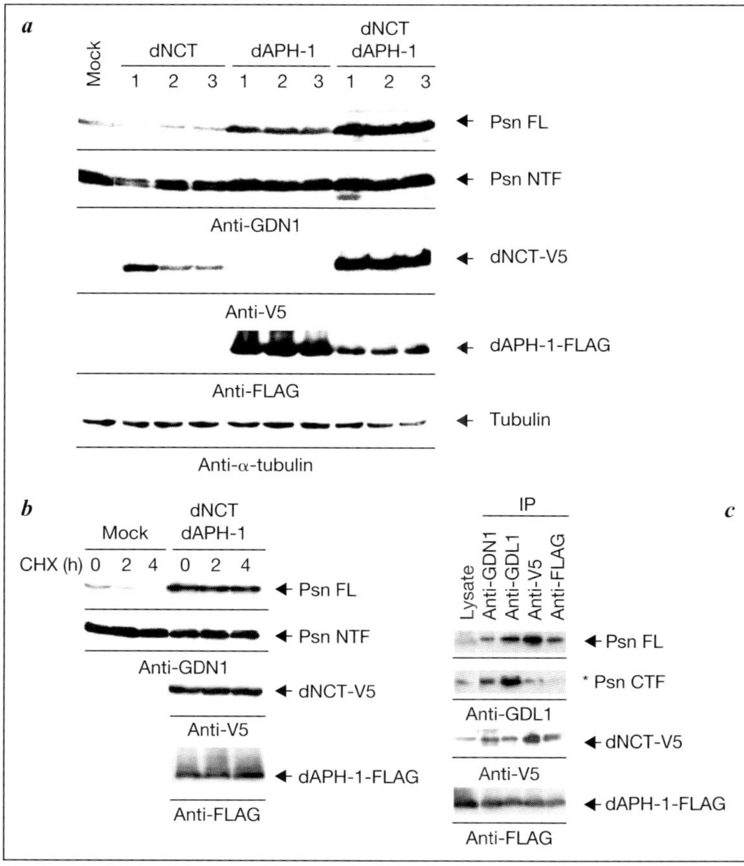

Fig. 3. Effects of overexpression of dAPH-1 on accumulation, stabilization and HMW complex formation of Psn holoprotein. ***a*** Immunoblot analysis of S2 cells stably expressing dNCT, dAPH-1 or doubly with dNCT and dAPH-1 for Psn NTF (upper panel), dNCT and dAPH-1 (middle two panels). Three independent stable cell lines for each cDNA are shown. ***b*** Analysis of stability of Psn, dNCT or dAPH-1 by cycloheximide (CHX) treatment and chase for 2–4 h. ***c*** Coimmunoprecipitation of Psn, dNCT and dAPH-1.

a coimmunoprecipitation study in double-stable cells, and found that Psn, dNCT and dAPH-1 interact with each other. These data suggest that the overexpressed dNCT, dAPH-1 and Psn holoproteins form the stabilized HMW protein complex that is reminiscent of the molecular characteristics of Psn fragments, which is the active form of γ-secretase.

We then compared the Aβ secretion from cells with or without coexpression of dNCT/dAPH-1, but found no significant changes (fig. 4), suggesting that

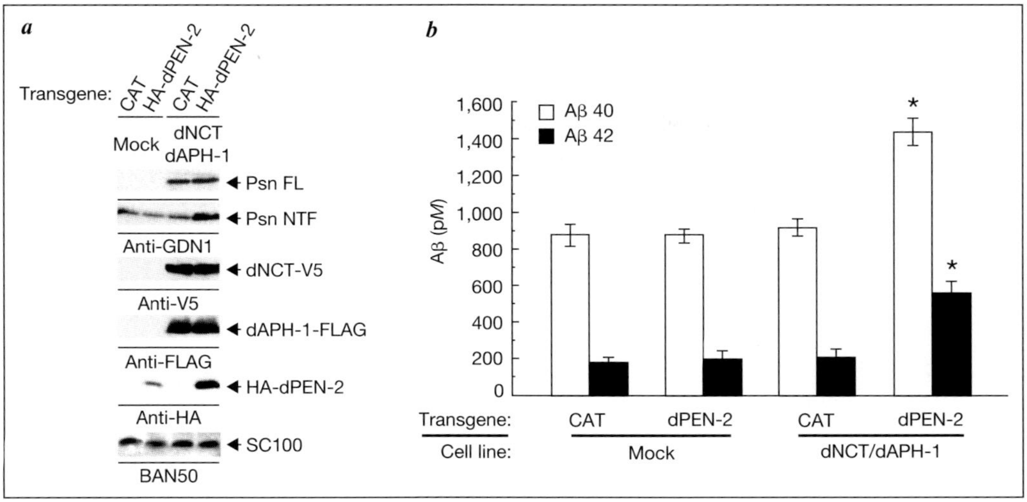

Fig. 4. Overexpression of dPEN-2 in addition to dNCT and dAPH-1 increases the accumulation of Psn fragments and γ-secretase activity. ***a*** S2 cells stably expressing dNCT and dAPH-1 or an empty vector were transiently transfected with CAT or dPEN-2, together with SC100, and analyzed by immunoblotting. ***b*** The levels of secreted Aβ40 (open column) and Aβ42 (closed column) (mean ± SE, n = 3) from S2 cells transfected with combinations of dNCT, dAPH-1 or dPEN-2 and SC100 as in ***a*** were quantitated by Aβ ELISA. *Statistically significant by ANOVA ($p < 0.01$).

some additional triggers or factors are needed for the upregulation of γ-secretase activity. Taken together, our data strongly suggest that dAPH-1 represents the 'stabilizing' cofactor of the PS complex and that some additional triggers or factors are needed for the activation of γ-secretase.

PEN-2 Is a PS Cofactor Conferring γ-Secretase Activities and Facilitating the Formation of PS Fragments

We then focused on dPEN-2, a novel protein that was identified as a PS enhancer gene in *Caenorhabditis elegans* [16]. dPEN-2 RNAi abolished the accumulation of Psn fragments as well as the Aβ secretion similarly to that with dNCT or dAPH-1 RNAi (fig. 2), which intriguingly was accompanied by an increase in the levels of Psn holoproteins. To clarify the difference of the knock-down effects of these genes, we transfected the mixed dsRNA into S2 cells, and found that the augmentation of the holoprotein level was observed only in dPEN-2 RNAi-treated cell lysates and was diminished by cotransfection of dNCT or dAPH-1 dsRNA. Psn holoproteins accumulated by dPEN-2

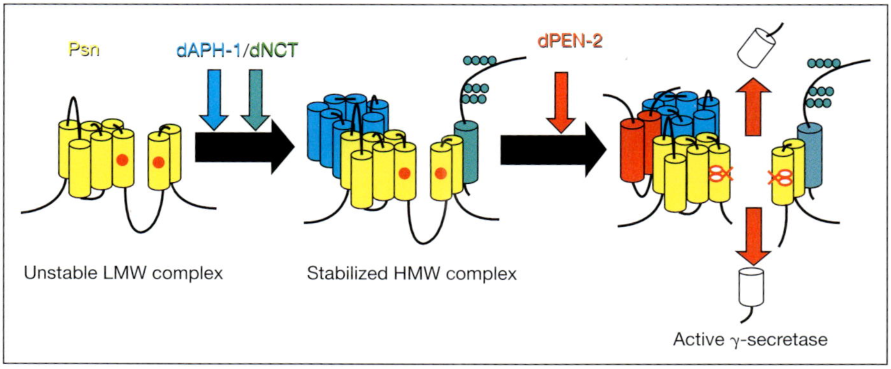

Fig. 5. Schematic depiction of assembly and activation of γ-secretase complex. Nascent Psn holoprotein forms an LMW protein complex and is rapidly degraded, while a fraction of Psn is stabilized by binding to dAPH-1 (and dNCT), although it still remains γ-secretase inactive. dPEN-2 elicits the final step of maturation of the γ-secretase complex, facilitating endoproteolysis of Psn and conferring γ-secretase activity.

RNAi were highly stabilized and fractionated in HMW ranges, suggesting that dPEN-2 RNAi abolished the accumulation of Psn fragments, which was accompanied by an increase in the levels of the stabilized HMW form of Psn holoproteins.

To rule out the possibility that RNAi inactivation of Psn and its cofactors reduces Aβ secretion by some indirect mechanism (e.g., altering trafficking or localization of γ-secretase or substrates), we examined the γ-secretase activities in membrane fractions of RNAi treated S2 cells by an in vitro γ-secretase assay, using C-terminally tagged recombinant APP-C100 as substrates [19, 28]. We found that the membrane fractions from S2 cells treated either with Psn, dNCT, dAPH-1 or dPEN-2 RNAi considerably diminished γ-secretase activities in these cells (2.7 ± 2.9, 9.7 ± 10.7, 0, 0% compared with GFP RNAi, respectively). These data strongly suggest that Psn and its cofactors are required for the bona fide γ-secretase activities, and the stabilized HMW Psn holoproteins accumulated by dPEN-2 RNAi are inactive in terms of γ-secretase activity similarly to Psn holoproteins accumulated by overexpression of dNCT and dAPH-1.

These results prompted us to examine the role of dPEN-2 on the assembly of PS complex and γ-secretase activity. We transfected dPEN-2 into S2 cells with or without co-expression of dNCT and dAPH-1 (fig. 4). The overexpression of dPEN-2 in dNCT and dAPH-1 double-stable S2 cells increased the accumulation of Psn fragments as well as Aβ secretion, whereas the overexpression in wild-type S2 cells caused no significant changes. It is strongly suggested that dPEN-2 is a cofactor that is required for the final step of Psn complex maturation after Psn is stabilized in a holoprotein form and incorporated into an HMW protein complex, conferring γ-secretase activities and leading to endoproteolysis of Psn.

In summary, using overexpression and the RNAi technique, we have been able to dissect the process of HMW protein complex formation and activation of γ-secretase function of PS (fig. 5). Overexpression of dAPH-1 was sufficient to elicit the stabilization and HMW complex formation of endogenous Psn in a holoprotein form, although this HMW protein complex harboring Psn holoprotein was still γ-secretase inactive. The final maturation step of the PS complex, which is comprised of Psn fragments and is γ-secretase active, may then be mediated by the additional function of dPEN-2, by directly interacting with PS/dAPH-1/dNCT complex. Further efforts to reconstitute γ-secretase activity from recombinant proteins of PS and the three cofactors in vitro will be needed to elucidate the configuration and function of PS complex and to develop therapeutic approaches to AD based around inhibition of γ-secretase.

References

1 Selkoe DJ: Alzheimer's disease: Genes, proteins, and therapy. Physiol Rev 2001;81:741–766.

2 De Strooper B, Saftig P, Craessaerts K, Vanderstichele H, Guhde G, Annaert W, Von Figura K, Van Leuven F: Deficiency of presenilin-1 inhibits the normal cleavage of amyloid precursor protein. Nature 1998;391:387–390.

3 De Strooper B, Annaert W, Cupers P, Saftig P, Craessaerts K, Mumm JS, Schroeter EH, Schrijvers V, Wolfe MS, Ray WJ, Goate A, Kopan R: A presenilin-1-dependent γ-secretase-like protease mediates release of Notch intracellular domain. Nature 1999;398:518–522.

4 Fortini ME: γ-Secretase-mediated proteolysis in cell-surface-receptor signalling. Nat Rev Mol Cell Biol 2002;3:673–684.

5 Urban S, Freeman M: Intramembrane proteolysis controls diverse signalling pathways throughout evolution. Curr Opin Genet Dev 2002;12:512–518.

6 Wolfe MS, Xia W, Ostaszewski BL, Diehl TS, Kimberly WT, Selkoe DJ: Two transmembrane aspartates in presenilin-1 required for presenilin endoproteolysis and γ-secretase activity. Nature 1999;398:513–517.

7 Haass C, Steiner H: Alzheimer disease γ-secretase: A complex story of GxGD-type presenilin proteases. Trends Cell Biol 2002;12:556–562.

8 Tomita T, Tokuhiro S, Hashimoto T, Aiba K, Saido TC, Maruyama K, Iwatsubo T: Molecular dissection of domains in mutant presenilin 2 that mediate overproduction of amyloidogenic forms of amyloid β peptides. Inability of truncated forms of PS2 with familial Alzheimer's disease mutation to increase secretion of Aβ42. J Biol Chem 1998;273:21153–21160.

9 Saura CA, Tomita T, Davenport F, Harris CL, Iwatsubo T, Thinakaran G: Evidence that intramolecular associations between presenilin domains are obligatory for endoproteolytic processing. J Biol Chem 1999;274:13818–13823.

10 Tomita T, Takikawa R, Koyama A, Morohashi Y, Takasugi N, Saido TC, Maruyama K, Iwatsubo T: C terminus of presenilin is required for overproduction of amyloidogenic Aβ42 through stabilization and endoproteolysis of presenilin. J Neurosci 1999;15:10627–10634.

11 Saura CA, Tomita T, Soriano S, Takahashi M, Leem JY, Honda T, Koo EH, Iwatsubo T, Thinakaran G: The nonconserved hydrophilic loop domain of presenilin (PS) is not required for PS endoproteolysis or enhanced Aβ42 production mediated by familial early onset Alzheimer's disease-linked PS variants. J Biol Chem 2000;275:17136–17142.
12 Tomita T, Watabiki T, Takikawa R, Morohashi Y, Takasugi N, Kopan R, De Strooper B, Iwatsubo T: The first proline of PALP motif at the C terminus of presenilins is obligatory for stabilization, complex formation and γ-secretase activities of presenilins. J Biol Chem 2001;276:33273–33281.
13 Yu G, Nishimura M, Arawaka S, Levitan D, Zhang L, Tandon A, Song YQ, Rogaeva E, Chen F, Kawarai T, Supala A, Levesque L, Yu H, Yang DS, Holmes E, Milman P, Liang Y, Zhang DM, Xu DH, Sato C, Rogaev E, Smith M, Janus C, Zhang Y, Aebersold R, Farrer LS, Sorbi S, Bruni A, Fraser P, St. George Hyslop P: Nicastrin modulates presenilin-mediated *notch/glp-1* signal transduction and βAPP processing. Nature 2000;407:48–54.
14 Hu Y, Ye Y, Fortini ME: Nicastrin is required for γ-secretase cleavage of the *Drosophila* Notch receptor. Dev Cell 2002;2:69–78.
15 Goutte C, Tsunozaki M, Hale VA, Priess JR: APH-1 is a multipass membrane protein essential for the *Notch* signaling pathway in *Caenorhabditis elegans* embryos. Proc Natl Acad Sci USA 2002; 99:775–779.
16 Francis R, McGrath G, Zhang J, Ruddy DA, Sym M, Apfeld J, Nicoll M, Maxwell M, Hai B, Ellis MC, Parks AL, Xu W, Li J, Gurney M, Myers RL, Himes CS, Hiebsch R, Ruble C, Nye JS, Curtis D: *aph-1* and *pen-2* are required for Notch pathway signaling, γ-secretase cleavage of βAPP, and presenilin protein accumulation. Dev Cell 2002;3:85–97.
17 Fortini ME, Skupski MP, Boguski MS, Hariharan IK: A survey of human disease gene counterparts in the *Drosophila* genome. J Cell Biol 2000;150:F23–F30.
18 Takasugi N, Takahashi Y, Morohashi Y, Tomita T, Iwatsubo T: The mechanism of γ-secretase activities through high molecular weight complex formation of presenilins is conserved in *Drosophila melanogaster* and mammals. J Biol Chem 2002;277:50198–50205.
19 Takasugi N, Tomita T, Hayashi I, Tsuruoka M, Niimura M, Takahashi Y, Thinakaran G, Iwatsubo T: The role of presenilin cofactors in the γ-secretase complex. Nature 2003;422:438–441.
20 Struhl G, Greenwald I: Presenilin is required for activity and nuclear access of Notch in *Drosophila*. Nature1999;398:522–525.
21 Seeger M, Nordstedt C, Petanceska S, Kovacs DM, Gouras GK, Hahne S, Fraser P, Levesque L, Czernik AJ, George Hyslop PS, Sisodia SS, Thinakaran G, Tanzi RE, Greengard P, Gandy S: Evidence for phosphorylation and oligomeric assembly of presenilin 1. Proc Natl Acad Sci USA 1997;94:5090–5094.
22 Capell A, Grunberg J, Pesold B, Diehlmann A, Citron M, Nixon R, Beyreuther K, Selkoe DJ, Haass C: The proteolytic fragments of the Alzheimer's disease-associated presenilin-1 form heterodimers and occur as a 100–150-kDa molecular mass complex. J Biol Chem 1998;273: 3205–3211.
23 Ratovitski T, Slunt HH, Thinakaran G, Price DL, Sisodia SS, Borchelt DR: Endoproteolytic processing and stabilization of wild-type and mutant presenilin. J Biol Chem 1997;272: 24536–24541.
24 Yu G, Chen F, Levesque G, Nishimura M, Zhang DM, Levesque L, Rogaeva E, Xu D, Liang Y, Duthie M, St George Hyslop PH, Fraser PE: The presenilin 1 protein is a component of a high molecular weight intracellular complex that contains β-catenin. J Biol Chem 1998;273:16470–16475.
25 Suzuki N, Cheung TT, Cai XD, Odaka A, Otvos L Jr, Eckman C, Golde TE, Younkin SG: An increased percentage of long amyloid β protein secreted by familial amyloid β protein precursor (βAPP717) mutants. Science 1994;264:1336–1340.
26 Iwata H, Tomita T, Maruyama K, Iwatsubo T: Subcellular compartment and molecular subdomain of β-amyloid precursor protein relevant to the Aβ42-promoting effects of Alzheimer mutant presenilin 2. J Biol Chem 2001;276:21678–21685.
27 Lichtenthaler SF, Wang R, Grimm H, Uljon SN, Masters CL, Beyreuther K: Mechanism of the cleavage specificity of Alzheimer's disease γ-secretase identified by phenylalanine-scanning mutagenesis of the transmembrane domain of the amyloid precursor protein. Proc Natl Acad Sci USA 1999;96:3053–3058.

28 Takahashi Y, Hayashi I, Tominari Y, Rikimaru K, Morohashi Y, Kan T, Natsugari H, Fukuyama T, Tomita T, Iwatsubo T: Sulindac sulfide is a non-competitive γ-secretase inhibitor that preferentially reduces Aβ42 generation. J Biol Chem 2003;278:18664–18670.

Taisuke Tomita
Department of Neuropathology and Neuroscience
Graduate School of Pharmaceutical Sciences, University of Tokyo
7–3–1 Hongo, Bunkyo, Tokyo 113–0033 (Japan)
Tel. +81 3 5841 4868, Fax +81 3 5841 4708, E-Mail taisuke@mol.f.u-tokyo.ac.jp

Takeda M, Tanaka T, Cacabelos R (eds): Molecular Neurobiology of Alzheimer Disease and Related Disorders. Basel, Karger, 2004, pp 94–107

Pharmacogenomic Studies with a Combination Therapy in Alzheimer's Disease

Ramón Cacabelos[a], *Lucía Fernández-Novoa*[b], *Victor Pichel*[a], *Valter Lombardi*[c], *Yasuhiko Kubota*[a], *Masatoshi Takeda*[d]

[a] EuroEspes Biomedical Research Center, Institute for CNS Disorders, Department of Clinical Neuroscience,
[b] Department of Molecular Genetics, EBIOTEC,
[c] Department of Biotechnology, EBIOTEC, Bergondo, Coruña, Spain;
[d] Department of Psychiatry and Behavioral Proteomics, Osaka University Graduate School of Medicine, Suita, Japan

Alzheimer's disease (AD) is a polygenic, multifactorial/complex disorder in which both mutational genetics and susceptibility genetics are implicated [1–3]. More than 30 different genes distributed across the human genome might be involved in the etiopathogenic cascades leading to neurodegeneration in AD [4]. In addition, genomic variation seems to be 2–3 times higher in the AD population as compared with the normal population with no family history of dementia [5]. Basic studies predict that biological factors present in genotyped sera of AD patients influence microglia activation, cytokine secretion, vascular endothelial changes, and the expression of major histocompatibility complex (MHC) class II antigens in a genotype-dependent fashion [6, 7]. Furthermore, different AD-related phenotypes including brain atrophy, cognitive deterioration, cerebrovascular changes, lymphocyte apoptosis, and brain bioelectrical activity also appear to be associated with specific genotypic profiles [8, 9]. All these data together suggest that the pharmacological treatment of AD needs to be remodeled on the basis of a pharmacogenomic approach to increase efficacy and safety and reduce side-effects and unnecessary costs [4, 10, 11]. In this regard, preliminary studies using a pharmacogenomic strategy clearly indicate that the therapeutic response in AD is genotype specific [4, 10–13]. These pharmacogenomic studies using genetic matrix models integrating monogenic, bigenic, trigenic or polygenic analysis allowed to conclude that different genetic clusters influence the therapeutic response in AD. For example, in the vast majority of the studies available, patients with the APOE-4/4 genotype show a differential phenotypic profile with the worst therapeutic response to conventional drugs (monotherapy) or to multifactorial treatments (combination therapy) [4, 10–13].

In the present study, we evaluated (a) the efficacy of a combination therapy in AD, (b) the influence of particular genotypes on the therapeutic response (monogenic-related response), and (c) the cognitive response associated with a trigenic cluster as a pharmacogenomic approach to AD therapeutics in an attempt to elucidate whether or not the genomic differentiation of AD patients can be of some benefit in obtaining better therapeutic outcomes in drug clinical trials.

Patients and Methods

In this study, 184 patients (mean age: 71.20 ± 8.69 years, range: 49–92 years; 121 females, age: 71.23 ± 8.19 years, range: 53–87 years; 63 males, age: 71.14 ± 9.65 years, range: 49–92 years) fulfilling the criteria for AD (DSM-IV, NINCDS-ADRDA) have been included. All the patients underwent the EuroEspes Dementia Diagnostic Protocol (EDDP) [14] including: (a) medical examination (general, psychiatric, neurologic); (b) neuropsychological assessment (MMSE, ADAS, FAST, GDS, FAST, BEHAVE-AD, HIS, HAM-A/D); (c) chest and neck X-ray; (d) ECG; (e) laboratory exams (blood biochemistry, hematology, urine, vitamin B_{12}, folic acid, iron, lues serology); (f) CT scan; (g) brain bioelectrical activity mapping (qEEG); (h) transcranial Doppler ultrasonography; and (i) genetic testing (APP, APOE, PS1, PS2) following informed consent [1].

The patients received CDP-choline (500 mg/day, p.o., an endogenous nucleotide) plus piracetam (2.4 g/day, p.o., a nootropic agent) for 1 year. On individual needs the patients were also given vitamin B_{12}, folic acid or iron when the plasma levels of these metabolic factors were below 130 pg/ml, 2.50 ng/ml, and 60 μg/dl, respectively, until full recovery of the normal levels: 170–1,000 pg/ml for vitamin B_{12}, >3.00 ng/ml for folic acid, and 60–160 μg/dl for iron. Approximately 10% of the patients received alprazolam (<1 mg/day) during short periods of time; 16% of the patients needed enalapril (5–10 mg/day) when blood pressure levels were higher than 160/90 mm Hg.

Psychometric assessment (MMSE score) was performed at baseline (BL) and at 1, 3, 6, 9, and 12 months during the treatment period. The cognitive performance of the patients was evaluated according to the following strategy: (a) global response with respect to BL MMSE scores (n = 184); (b) sex-related therapeutic response (F = 121, M = 63); (c) monogenic-related response (APOE, PS1, PS2) by differential genotypes: APOE-2/3 (age: 74.34 ± 9.26 years, range: 55–96), APOE-2/4 (age: 77.62 ± 5.52 years, range: 70–85), APOE-3/3 (age: 70.80 ± 9.73 years, range: 49–92), APOE-3/4 (age: 72.87 ± 6.41 years, range: 54–85), APOE-4/4 (age: 67.12 ± 8.28 years, range: 53–78), PS1–1/1 (age: 71.46 ± 8.70 years, range: 53–92), PS1–1/2 (age: 70.49 ± 8.91, range: 49–87), PS1–2/2 (age: 73.18 ± 7.80 years, range: 54–87), PS2+ (age: 69.48 ± 9.10 years, range: 49–85), PS2− (age: 71.82 ± 8.49 years, range: 52–92), and (d) trigenic (APOE+PS1+PS2)-related response including 36 potential genotypes resulting from the integration of all possible allelic combinations in a matrix model of which only the most frequent genotypes were analyzed [10]. APOE-2/2 was excluded from analysis because only a single case was available. PS2 genotypes were characterized according to the presence (PS2+) or absence (PS2−) of a genetic defect in the PS2 gene exon 5 [1]. APP, APOE, PS1 and PS2 genotyping was performed according to previously reported conventional procedures [1].

All values are expressed as mean ± SD for each point. Statistical analysis was performed by the paired t-test, ANOVA, and linear correlation analysis in standard computer programs (Sigma Plot, SPSS).

Results

Genotype Frequencies

No single-point mutations were found in the APP gene in our sample. The distribution of genotype frequencies for major APOE genotypes was the following: APOE-2/3 = 5.90%; APOE-2/4 = 1.82%; APOE-3/3 = 49.32%; APOE-3/4 = 36.13%; and APOE-4/4 = 6.83%. The frequencies of the PS1-related genotypes were 32.61% PS1–1/1, 52.72% PS1–1/2, and 14.67% PS1–2/2. Patients with a point mutation in the PS2 gene exon 5 (PS2+) accounted for 26.63% of the PS2-related genotypes, and 73.37% of the cases were devoid of this genetic defect (PS2−). The integration in a matrix model of all possible allelic combinations related to the APOE+PS1+PS2 cluster yields 36 different genotypes, as previously reported [4, 10, 11]. In our sample, we only analyzed the most frequent 18 genotypes of the genomic cluster with enough number of patients per genetic profile.

Efficacy of the Combination Therapy

Approximately 60% of the patients responded to this treatment (table 1, fig. 1). The therapeutic response was almost identical in males (r = −0.58, a coef.: 24.27, b coef.: −0.32) and females (r = −0.61, a coef.: 21.86, b coef.: −0.33) (table 1, fig. 1). A significant improvement in mental performance was observed during the first 6 months of treatment, with a progressive cognitive decline thereafter (table 1, fig. 1). In the AD population the best response to this combination therapy was seen in the 1st ($p < 0.00000000001$), 3rd ($p < 0.000001$), and 6th month of treatment ($p < 0.01$). After 9 months of treatment, there is a kind of inflection point after which approximately 80% of the patients deteriorate with MMSE scores below BL levels (table 1, fig. 1).

Monogenic-Related Therapeutic Response

When the therapeutic response is evaluated according to differential genotypes, patients with the APOE-2/3 genotype (r = +0.36, a coef.: 18.16, b coef.: +0.56) and APOE-3/4 genotype (r = +0.38, a coef.: 18.41, b coef.: +0.14) are

Table 1. Cognitive response to a combination therapy in AD

Parameters	Total	Females	Males
n	184	121	63
Age, years	71.20 ± 8.69	71.23 ± 8.19	71.14 ± 9.65
Range, years	49–92	53–87	49–92
Baseline MMSE	21.13 ± 6.63	20.31 ± 6.94	24.47 ± 5.72
1 month	22.84 ± 6.43[1]	21.99 ± 6.84[1]	24.47 ± 5.26[1]
3 months	22.48 ± 6.39[1]	21.65 ± 6.88[2]	24.09 ± 5.01[3]
6 months	21.61 ± 6.83[4]	20.69 ± 7.27	23.38 ± 5.51[5]
9 months	21.07 ± 7.02	20.33 ± 7.52	22.50 ± 5.74
12 months	20.05 ± 7.47[6]	19.15 ± 7.81[6]	21.77 ± 6.48[7]
r value	−0.60	−0.61	−0.58
a coefficient	22.68	21.86	24.27
b coefficient	−0.33	−0.33	−0.32

Values: mean ± SD.
[1]$p < 0.00000000001$ vs BL.
[2]$p < 0.000000002$ vs BL.
[3]$p < 0.0000001$ vs BL.
[4]$p < 0.01$ vs BL.
[5]$p < 0.02$ vs BL.
[6]$p < 0.0004$ vs BL.
[7]$p < 0.03$ vs BL.

the best responders, and patients with the APOE-4/4 genotype are the worst responders (r = −0.65, a coef.: 25.11, b coef.: −1.07) (fig. 2). The three different PS1-related genotypes behave in a similar manner, with a final MMSE score after 1 year of treatment below BL values (fig. 3), and the same applies for the two different PS2-related genotypes, though PS2+ carriers seem to respond worse (r = −0.96, a coef.: 20.84, b coef.: −0.55) than PS2− carriers (r = −0.70, a coef.: 21.63, b coef.: −0.45) (fig. 4).

Trigenic-Related Therapeutic Response

The integration of the APOE+PS1+PS2 genotypes in a trigenic matrix model yielded 18 major genotypes with a clear differential response to treatment (fig. 5–8). The best responders were the following genotype carriers: 331222− (r = +0.34, a coef.: 19.93, b coef.: +0.30), 341122− (r = +0.38, a coef.: 18.04, b coef.: +0.26), 341222− (r = +0.57, a coef.: 21.78, b coef.: +0.14), and 441112− (r = +0.40, a coef.: 25.92, b coef.: +0.20) (fig. 5–8). There is a tendency for better responders in those patients with APOE-3/3 and

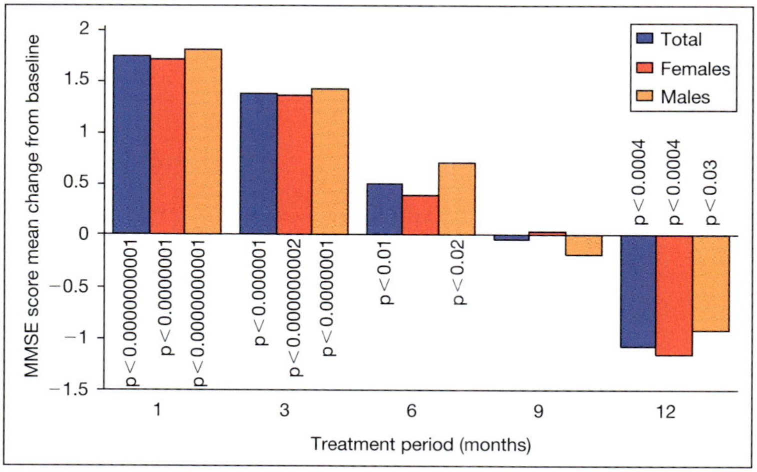

Fig. 1. Multifactorial therapy in AD. Sex-dependent effect of a combination therapy on cognition in AD.

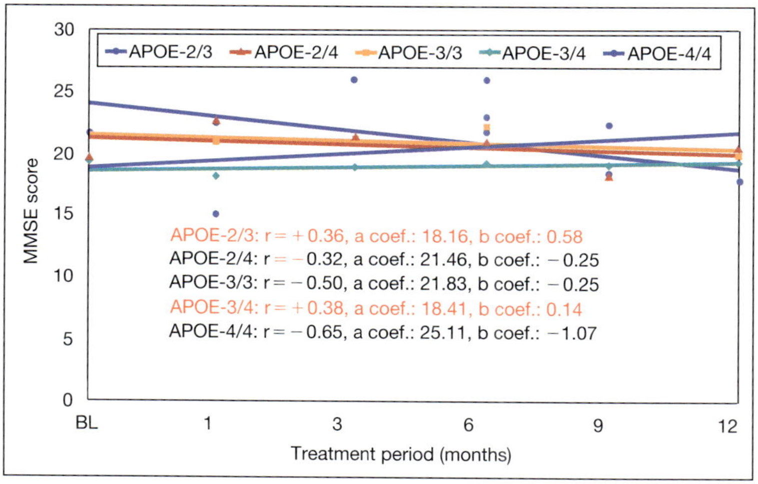

Fig. 2. APOE-related therapeutic response to a combination therapy in AD.

APOE-3/4 associated with either PS1–1/2, PS1–2/2 and PS2−. In contrast, except the 441112− cluster, practically all APOE-4/4 carriers associated with any other PS1 or PS2 genotypes are by far the worst responders (fig. 5–8). When we compare the trigenic-related responses with the global sample, we

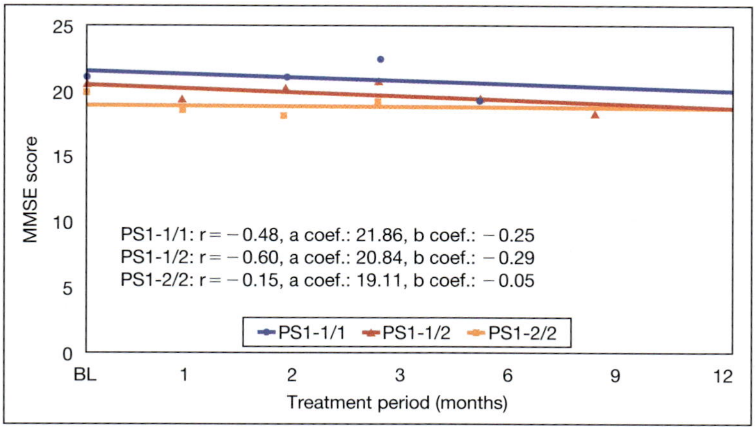

Fig. 3. Presenilin-1-related therapeutic response to a combination therapy in AD.

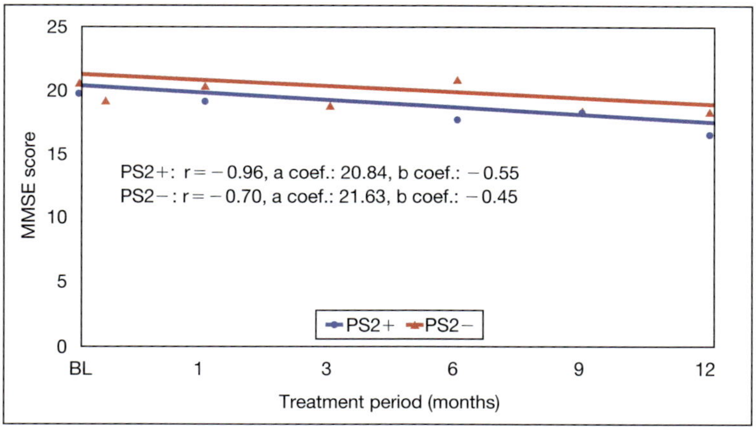

Fig. 4. Presenilin-2-related therapeutic response to a combination therapy in AD.

can distinguish the linear cognitive decline of the whole group (r = −0.60, a coef.: 22.68, b coef.: −0.33) (thick red line in fig. 6) in clear contrast with the wide dispersion in cognitive performance of the different genomic clusters represented by the trigenic genotypes characterized in this sample (fig. 6). When we analyze every individual cluster along the 1-year treatment, we can observe that APOE-4/4 carriers, especially those with the 441122+ genotype are the worst responders from the early beginning of the trial (fig. 7).

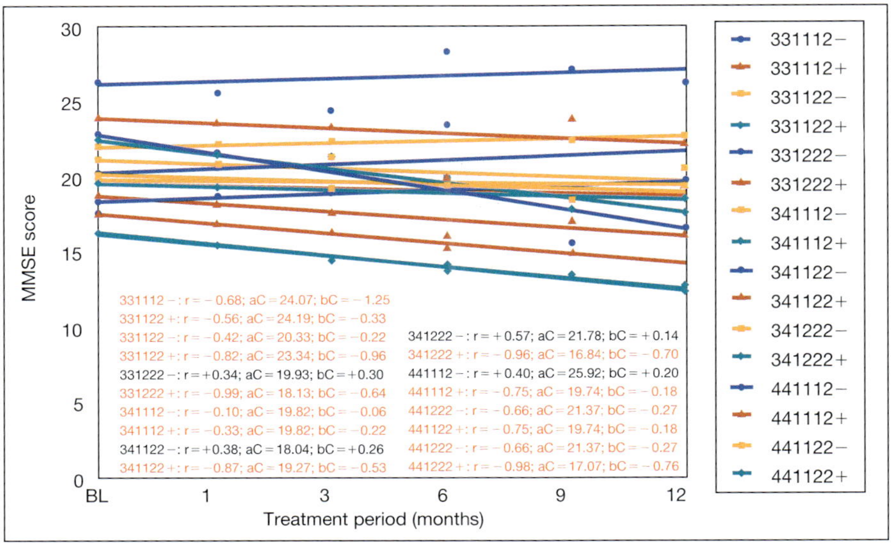

Fig. 5. Trigenic genotype-related therapeutic response to a combination therapy in AD.

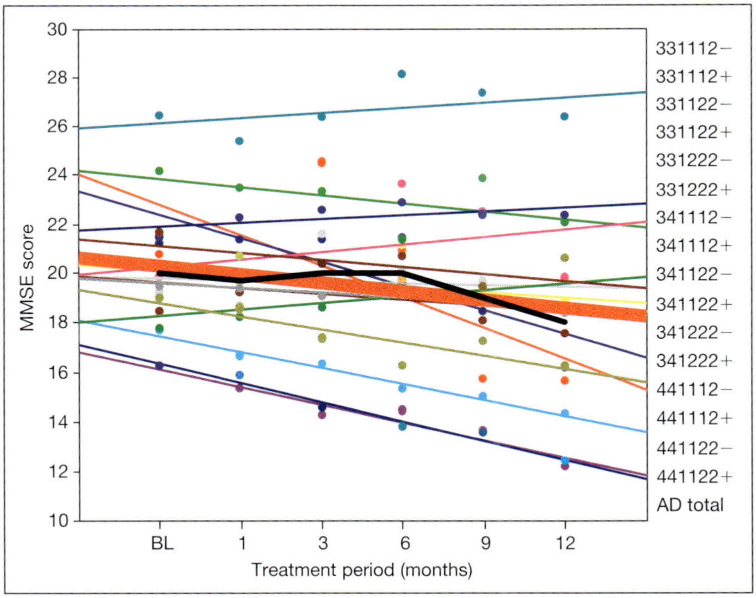

Fig. 6. Trigenic genotype-related therapeutic response to a combination therapy in AD differentiating independent genomic clusters (APOE+PS1+PS2) from the global AD population treated with CDP-choline (500 mg/day, p.o.) plus piracetam (2.4 g/day, p.o.) for one year.

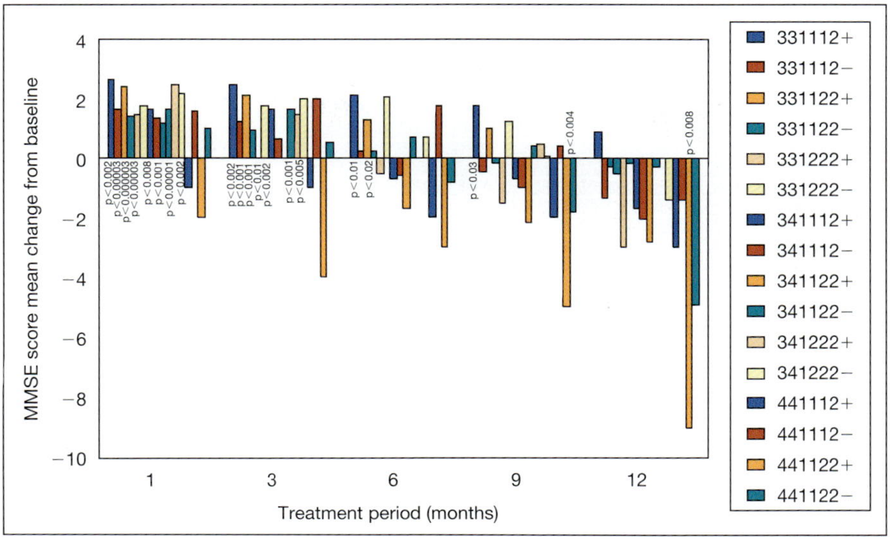

Fig. 7. Trigenic genotype-related pharmacogenomic response to a combination therapy in AD.

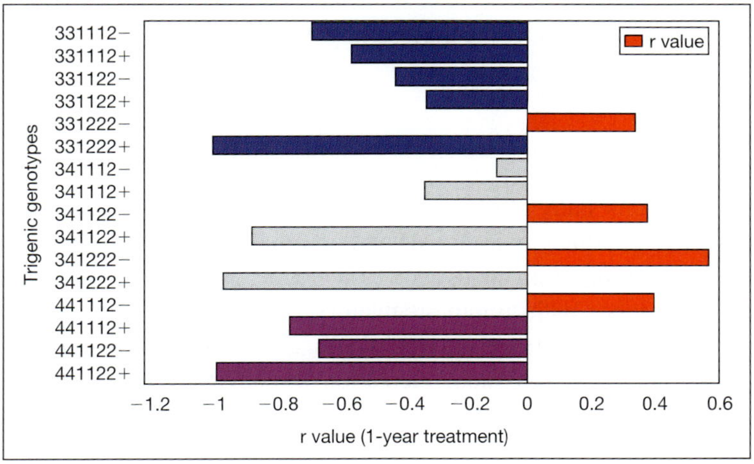

Fig. 8. Differentiation of genomic clusters integrated by trigenic genotypes (APOE+PS1+PS2) in a population of AD patients treated with a combination therapy for one year. Red bars represent the genomic clusters showing a positive response (r value) with MMSE scores higher than baseline values.

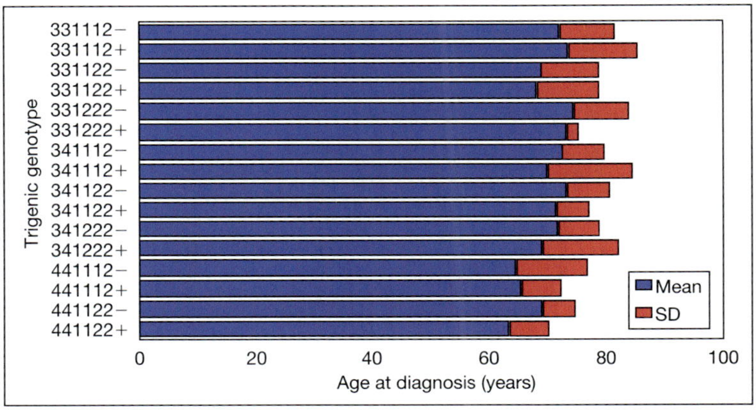

Fig. 9. Trigenic genotype-related age at diagnosis in AD-associated genomic clusters reflecting the influence of different genotypes on the age at onset in the Spanish population.

Genotype-Related Age at Onset

The average interval from the onset of the first symptoms of dementia until diagnosis in Spain ranges from 6 months to 3.8 years, with approximately 2.3 years on the average in most Spanish public hospitals [1]. In our study, it appears that the APOE-4/4 genotype also influences the age at onset anticipating by 5–10 years the onset of the disease (fig. 9).

Discussion

The first aim of this study was to verify whether or not a combination therapy can be a suitable treatment to improve cognition in AD. Our results demonstrate that the combination of an endogenous nucleotide, such as CDP-choline, acting as a choline donor with immunoregulatory properties [13, 15], with a nootropic agent (piracetam), at conventional doses, administered together with vitamin B_{12}, folic acid and iron supplementation in patients with AD and in patients with AD plus a partial deficit in folic acid and/or vitamine B_{12} and/or a ferropenic anemia, is a useful treatment to protect AD brains against neurodegeneration and cerebrovascular dysfunction [9]. These results are in agreement with data reported in previous studies carried out with CDP-choline alone [15, 16] or in combination with piracetam and anapsos in dementia [4, 10, 11, 13, 17, 18]. With this multifactorial therapy it appears feasible to delay cognitive deterioration in established cases of AD by about 9 months in approximately 60–70% of the patients (table 1, fig. 1, 6, 7). These data might indicate that when

AD neurons are stimulated with nonspecific compounds devoid of effects on the pathogenic cascades responsible for AD-related neurodegeneration they maintain an activated state during a certain period of time (<1 year) accompanied by an apparent improvement in cognition; however, since these compounds do not specifically inhibit the mechanisms inducing premature neuronal death, then overstimulation of AD neurons provokes an accelerated functional decline with consequent cognitive deterioration [13]. This phenomenon may also apply for the effects of cholinesterase inhibitors when given for long periods of time. This bimodal effect, with cognitive improvement in the early phases of treatment and a potentially accelerated decline in subsequent phases of therapeutic intervention should be taken into account when developing new compounds or when conventional drugs are given for long periods of time without apparent improvement in cognition or in daily life activities.

After demonstrating that this combination therapy is potentially useful in AD, the second aim of this study was to confirm that the therapeutic response in AD patients is genotype specific [10] and that different genomic clusters integrating AD-associated genes do influence the individual response of AD patients to a particular therapy, as established in pioneering pharmacogenomic studies [4, 10–13]. The genetic screening in AD reflects that mutational genetics probably accounts for less than 5% of AD cases [1, 2]. However, polymorphic variants of potential risk practically are present in more than 90% of AD patients [1, 4, 11]. Among the polymorphisms of risk, APOE-4/4 is likely to be one of the most devastating variants for AD patients carrying this genotype [1, 2]. According to our results, it seems clear that APOE-4/4 carriers are the worst responders in either monogenic-related analysis (APOE-related responses) (fig. 3) or in trigenic-related analysis integrating APOE+PS1+PS2 genomic variants (fig. 5–8). In contrast, the best responders are patients whose genome include the APOE-3/3 (331222−) and APOE-3/4 (341122−, 341222−) polymorphisms (fig. 5–8). Notwithstanding, it is becoming apparent that the association of genetic effects is determinant for the definition of phenotypic profiles, and that genetic interactions are probably more influential than isolated point mutations on the phenotypic expression of the disease. In this regard, we found that using genetic matrix models integrating several polymorphic variants associated with AD it is possible to identify genotype-phenotype correlations in AD [8]. For instance, we found such correlations in brain atrophy, brain mapping, cerebrovascular function, microglia activation, and lymphocyte apoptosis in AD patients [6–8, 19; Cacabelos et al., Pichel, in preparation]. Likewise, different AD-related polymorphic variants and genomic associations influence the response of microglia in culture when the genotyped serum of individual AD patients is added to the culture media, allowing a predictive therapeutic response in in vitro biochip models [9]. The influence of genomic associations on the therapeutic response to the combination therapy reported in this study is also evident when we observe that the best responders are patients with trigenic clusters in which the PS2+ genotype is absent, which suggests that the genetic defect present in the exon 5 of the PS2 gene (PS2+ genotype) exerts a negative effect on the responding capacity of PS2+ carriers to the combination therapy (fig. 5–8). To a lesser extent, it also appears that the PS1–1/2 and PS1–2/2 genotypes are more prone to facilitate a better therapeutic response since the best responders represented by the 331222−, 341122−, and 341222− clusters, with the exception of the 441112 cluster, all include the PS1–1/2 or PS1–2/2 genotypes (fig. 5–8). These findings might indicate that the responding capacity of AD patients to a particular treatment could be the result of gene-dependent effects associated with different polymorphic variants involved in AD pathogenesis.

Although this pharmacogenomic strategy is very primitive and time consuming due to the necessity of a large number of patients to perform a clinical trial and also many hours of computer work, it appears that the results obtained are clear-cut, indicating that pharmacological development and AD treatment can be optimized by using pharmacogenomic protocols. The application of pharmacogenomics to the preclinical level in drug development would help to find drugs with etiopathogenic activity operating on specific biochemical pathways (drug target) involved in amyloid deposition (i.e., secretases), neurofibrillary tangle formation (i.e., tau hyperphosphorylation), neuronal apoptosis (i.e., caspases), free radical formation (i.e., antioxidants), nitric oxide-related endothelial dysfunction (i.e., NO inhibitors, NO-releasing nonsteroidal anti-inflammatory drugs, NO-NSAIDs), neuroimmunological dysregulation (i.e., cytokine inhibitors) and the like [13, 20]. Pharmacogenetic studies would help to obtain drugs devoid of adverse events by understanding their metabolic pathways and the CYP family gene-related products involved in drug metabolism [21]; and finally, clinical pharmacogenomics would be essential to define efficacy optimizing the therapeutic outcome [4, 10–13, 22, 23]. As depicted in figures 5–8, the harmonization of patients in genomic clusters allows a precise differentiation of the therapeutic response easily identifying good responders, poor responders or nonresponders.

Functional genomics, proteomics, pharmacogenomics, high-throughput methods, combinational chemistry and bioinformatics will greatly contribute to accelerate drug development for AD and other complex disorders [4, 11]. Relying on the experience of several years developing and implementing preclinical and clinical pharmacogenomic studies in AD [1, 4, 5, 8–13], we recommend the regulatory authorities in the USA, European Union and Japan to stimulate the initiatives of the pharmaceutical industry to devise novel drug-developing strategies fostering pharmacogenomics in AD to optimize drug development and therapeutics, increasing efficacy and safety, and reducing unnecessary costs for the industry in R&D and for society.

Conclusions

From the results obtained in the present study and in previous pharmacogenomic studies reported by our group, we can conclude the following:

1. Multifactorial treatments combining neuroprotectants, endogenous nucleotides, nootropic agents, vasoactive substances, cholinesterase inhibitors, and NMDA antagonists associated with metabolic supplementation on an individual basis adapted to the phenotype of the patient may be useful to improve cognition and slow-down disease progression in AD [13].
2. In our personal experience the best results have been obtained combining (a) CDP-choline with piracetam and metabolic supplementation, (b) CDP-choline with piracetam and anapsos, (c) CDP-choline with piracetam and cholinesterase inhibitors (donepezil, rivastigmine), (d) CDP-choline with memantine, and (e) CDP-choline, piracetam and nicergoline.
3. In the present study the combination of CDP-choline and piracetam has proven to be effective, improving cognition during the first 9 months of treatment, and not showing apparent side-effects.
4. The therapeutic response in AD seems to be genotype-specific under different pharmacogenomic conditions [4, 10, 11, 13].
5. In monogenic-related studies, patients with the APOE-2/3 and APOE-3/4 genotypes are the best responders, and APOE-4/4 carriers are the worst responders.
6. PS1- and PS2-related genotypes do not appear to influence the therapeutic response in AD as independent genomic entities.
7. In trigenic-related studies, the best responders are those patients who carry the 331222−, 341122−, 341222−, and 441112− genomic clusters.
8. A genetic defect in exon 5 of the PS2 gene seems to exert a negative effect on cognition conferring on PS2+ carriers in trigenic clusters the condition of poor responders to combination therapy.
9. The worst responders in all genomic clusters are patients with the 441122+ genotype.
10. The APOE-4/4 genotype seems to accelerate neurodegeneration anticipating the onset of the disease by 5–10 years; and, in general, APOE-4/4 carriers show a faster disease progression and a poorer therapeutic response to all available treatments than any other polymorphic variant.
11. Pharmacogenomic studies using trigenic, tetragenic or polygenic clusters as a harmonization procedure to reduce genomic heterogeneity are very useful to widen the therapeutic scope of limited pharmacological resources (fig. 5, 6).
12. Pharmacogenomic strategies constitute powerful tools for preclinical and clinical drug development in AD.

13. Pharmacogenomic studies (efficacy) in combination with pharmacogenetic studies (safety) are of paramount importance for the future development of new compounds for the treatment of AD and other CNS disorders [24].
14. We recommend regulatory authorities to foster pharmacogenomic studies in AD with available compounds and with novel drugs to improve efficacy and safety, reduce adverse events, and cut-down unnecessary cost for the industry and the community.

Acknowledgements

We thank our coworkers at the Institute for CNS Disorders, EuroEspes Biomedical Research Center (CIBE), Coruña, Spain, for their valuable help in conducting the clinical, genetic and computational studies necessary for the implementation of pharmacogenomic protocols.

Most of the work we carry out at CIBE is financially supported by the EuroEspes Foundation and EBIOTEC.

References

1 Cacabelos R: Handbook of Neurogeriatrics. Alzheimer Disease and Other Dementias. Epidemiology and Genetics. Masson, Barcelona, 1999.
2 Saunders AM: Gene identification in Alzheimer's disease. Pharmacogenomics 2001;2:239–249.
3 Selkoe DJ: Alzheimer's disease: Genes, proteins, and therapy. Physiol Rev 2001;81:741–766.
4 Cacabelos R: Pharmacogenomics in Alzheimer's disease. Min Rev Med Chem 2002;2:59–84.
5 Cacabelos R, Fernández-Novoa L, Lombardi V, Takeda M: Genetic variation and pharmacogenomics in Alzheimer disease. Psychiat Neurol Jap 2003;105:47–67.
6 Lombardi VRM, García M, Cacabelos R: Microglial activation induced by factor(s) contained in sera from Alzheimer-related ApoE genotypes. J Neurosci Res 1998;54:539–553.
7 Lombardi VR, García M, Rey L, Cacabelos R: Characterization of cytokine production, screening of lymphocyte subset patterns and in vitro apoptosis in healthy and Alzheimer's disease individuals. J Neuroimmunol 1999;97:163–171.
8 Cacabelos R: Diagnosis of Alzheimer's disease: Defining genetic profiles (genotype vs phenotype). Acta Neurol Scand Suppl 1996;165:72–84.
9 Cacabelos R: Psychogeriatric research. A conceptual introduction to geriatric neuroscience. Psychogeriatrics 2001;1:158–188.
10 Cacabelos R, Alvarez A, Fernández-Novoa L, Lombardi VRM: A pharmacogenomic approach to Alzheimer's disease. Acta Neurol Scand Suppl 2000;176:12–19.
11 Cacabelos R: Pharmacogenomics for the treatment of dementia. Ann Med 2002;34:357–379.
12 Cacabelos R: Pharmacogenomics in Alzheimer's disease. Drug News Perspect 2000;13:252–254.
13 Cacabelos R, Alvarez XA, Lombardi V, Fernández-Novoa L, Corzo L, Pérez P, et al: Pharmacological treatment of Alzheimer disease: From psychotropic drugs and cholinesterase inhibitors to pharmacogenomics. Drugs Today 2000;36:415–499.
14 Cacabelos R: Dementia; in Jobe TH, Gaviria M, Kovilparambil A (eds): Clinical Psychiatry. Boston, Blackwell Science, 1997, pp 73–122.
15 Cacabelos R, Caamaño J, Gómez MJ, Fernández-Novoa L, Franco-Maside A, Alvarez XA: Therapeutic effects of CDP-choline in Alzheimer's disease. Cognition, brain mapping, cerebrovascular hemodynamics, and immune factors. Ann NY Acad Sci 1996;777:399–403.

16 Alvarez XA, Mouzo R, Pichel V, Pérez P, Laredo M, Fernández-Novoa L, et al: Double-blind placebo-controlled study with citicoline in APOE genotyped Alzheimer's disease patients. Effects on cognitive performance, brain bioelectrical activity, and cerebral perfusion. Meth Find Exp Clin Pharmacol 1999;21:633–644.
17 Alvarez XA, Pichel V, Pérez P, Laredo M, Corzo L, Zas R, et al: Double-blind, randomized, placebo-controlled pilot study with anapsos in senile dementia: Effects on cognition, brain bioelectrical activity and cerebral hemodynamics. Meth Find Exp Clin Pharmacol 2000;22:585–594.
18 Alvarez A, Miguel-Hidalgo JJ, Fernández-Novoa L, Díaz J, Sempere JM, Cacabelos R, Anapsos: Neuroimmunotrophic treatment in Alzheimer disease and neurodegenerative disorders. CNS Drug Rev 1997;3:181–206.
19 Fernández-Novoa L, Cacabelos R: Histamine function in brain disorders. Behav Brain Res 2001; 124:213–233.
20 Johnson JA: Drug target pharmacogenomics: An overview. Am J Pharmacogenomics 2001;1: 271–281.
21 Evans WE, Johnson JA: Pharmacogenomics: The inherited basis for interindividual differences in drug response. Annu Rev Genomics Hum Genet 2001;2:9–39.
22 McLeod HL, Evans WE: Pharmacogenomics: Unlocking the human genome for better drug therapy. Annu Rev Pharmacol Toxicol 2001;41:101–121.
23 Maimone D, Dominici R, Grimaldi LM: Pharmacogenomics of neurodegenerative disease. Eur J Pharmacol 2001;413:11–29.
24 Roses AD: Pharmacogenetics and future drug development and delivery. Lancet 2000;355: 1358–1361.

Prof. Dr. Ramón Cacabelos
EuroEspes Biomedical Research Center, Institute for CNS Disorders
15166-Bergondo, Coruña (Spain)
Tel. +34 981 780505, Fax +34 981 780511, E-Mail cacabelos@euroespes.com

Takeda M, Tanaka T, Cacabelos R (eds): Molecular Neurobiology of Alzheimer Disease and Related Disorders. Basel, Karger, 2004, pp 108–122

Nicotinic Receptor Stimulation Blocks Neurotoxicity Induced by Amyloid-β via the Phosphatidylinositol-3-Kinase Cascade

Takeshi Kihara, Shun Shimohama

Department of Neurology, Graduate School of Medicine, Kyoto University, Kyoto, Japan

Alzheimer's disease (AD) is one of the most common dementias in the elderly. This neurodegenerative disease is characterized by the presence of two types of abnormal deposits: senile plaques (SP) and neurofibrillary tangles (NFT), and extensive neuronal loss [1]. Fully developed SP are composed of a core of amyloid-β (Aβ) surrounded by dystrophic neurites and reactive glia [2]. Although these pathologic changes are found in the brains of AD patients, both can be found in the brains of normal elderly individuals [2, 3], possibly because early changes may predate development of symptoms. Aβ is one of the candidate causes of the neurodegeneration found in AD because a negative correlation was found between SP and neuron density [1]. Several mutations of the amyloid precursor protein (APP) are found in familial AD, and these mutations are involved in amyloidogenesis [4]. It has also been shown that familial AD mutations of presenilin-1 (PS-1) enhance the generation of Aβ (1–42) [5]. Therefore, Aβ plays a key role in the pathogenesis of AD.

Glutamate is an excitotoxic neurotransmitter in the CNS. It has been shown, however, that glutamate causes cell death, which is mediated via intracellular Ca^{2+} influx, activation of Ca^{2+}-dependent enzymes such as nitric oxide (NO) synthase, and the production of toxic oxygen radicals [6]. In addition, some reports have shown that Aβ causes a reduction in glutamate uptake in cultured astrocytes [7], indicating that Aβ-induced cytotoxicity might be mediated via glutamate cytotoxicity to some extent.

It has been reported that activated phosphatidylinositol-3-kinase (PI3K) and Akt kinase promote neuron survival [8]. Anti-apoptotic proteins such as Bcl-2, Bcl-x and Bad were thought to be involved in this survival system. Nicotinic receptors are ionotropic receptors, which allow Ca^{2+} to enter cells and function physiologically. It has been shown that the PI3K cascade is activated by tyrosine kinase or G-protein-mediated signals in neuronal cells [9]. Conversely, there is no evidence that nicotinic receptors contain a G-protein or tyrosine kinase. α-Amino-3-hydroxy-5-methylisoxazole-4-propionic acid (AMPA) receptors are also ionotropic receptors, and it was recently shown that a member of the Src family, Lyn, is physically associated with AMPA receptors and mediates signals to PI3K [10]. Thus, there is a possibility that ionotropic receptors such as nicotinic receptors could be associated with a tyrosine kinase such as Src.

In the present study we showed that Aβ-induced cell death is mediated via glutamate. The neuroprotective effect of nicotine was examined, focusing on the involvement of the PI3K cascade.

Experimental Procedures

Materials

The sources of drugs and materials used in this study were as follows: Eagle's Minimum Essential Medium (EMEM; Nissui Pharmaceutical Co.), Aβ protein fragments 25–35, 1–40, 1–42 (Bachem), (–)-nicotine, MK801 and monoclonal anti-α_7-nicotinic acetylcholine antibody (Research Biochemicals International), α-bungarotoxin (α-BTX; Wako), LY294002 (Biomol Research Laboratories, Inc.), PP2 (Calbiochem), PD98059, anti-phospho- and nonphospho-specific p44/p42 mitogen-activated protein (MAP) kinase and Akt antibodies (New England Biolabs), anti-microtubule-associated protein-2 (MAP-2) antibody (Sigma Chemical Co.), polyclonal anti-Fyn antibody (Upstate), polyclonal anti-PI3K p85 subunit antibody and anti-α_4-nicotinic acetylcholine antibody (SantaCruz), monoclonal anti-PI3K p85 subunit antibody and anti-Bcl-2 antibody (Transduction Laboratories).

Cell Cultures

Primary cultures were obtained from the cerebral cortex of fetal rats (17–19 days of gestation) by procedures described previously [11, 12]. Briefly, single cells dissociated from the cerebral cortex of fetal rats were plated out onto plastic coverslips and placed in Falcon dishes. Cultures were incubated in EMEM supplemented with 10% fetal calf serum (1–5 days after plating out) or 10% horse serum (6–12 days after plating out), glutamine (2 m*M*), glucose (total 11 m*M*), $NaHCO_3$ (24 m*M*), and HEPES (10 m*M*). Cultures were maintained at 37°C in a humidified atmosphere of 5% CO_2. Only mature cultures (10–14 days in vitro) were used for the experiments. The animals were treated in accordance with the guidelines published in the *NIH Guide for the Care and Use of Laboratory Animals*.

Treatment of the Cultures

All experiments were carried out in EMEM at 37°C. Cultured neurons were exposed to glutamate for 24 h followed by incubation with EMEM for a further 24 h. Aβ and nicotine were added to the medium before exposure to glutamate. Antagonists were simultaneously added with nicotine. When cells were exposed to glutamate, other drugs were removed.

Assessment of Neurotoxicity

Cell viability was assessed by trypan blue exclusion and immunostaining with anti-MAP-2 antibody. After the experiment, cell cultures were immediately stained with 1.5% trypan blue for 10 min at room temperature, fixed with isotonic formalin (pH 7.0, 2–4°C), and then rinsed with physiologic saline. Cells stained with trypan blue were considered nonviable. In the immunostaining evaluation, neurotoxicity in each experiment was defined as a reduction in the survival rate, which was expressed as percentage survival relative to the survival observed in control cultures. Immunostaining was performed by the methods described previously [13], using the primary antibody (anti-MAP-2 antibody, diluted 1:500) for 24 h. At least 200 neurons were counted in 10–20 randomly selected fields at ×100 total magnification in control cultures to determine the total number of neurons.

Preparation of Cell Extracts

Semi-confluent cultures were exposed to each treatment, and incubated at 37°C for various time intervals. Subsequently, cells were lysed in a buffer consisting of 20 m*M* Tris-HCl, pH 7.0, 2 m*M* EGTA, 25 m*M* 2-glycerophosphate, 1% Triton-X100, 2 m*M* dithiothreitol, 1 m*M* vanadate, 1 m*M* phenylmethylsulfonyl fluoride and 1% aprotinin and centrifuged at 15,000 *g* for 30 min at 4°C. The supernatants were used as the cell extracts.

Immunoblotting and Immunoprecipitation

Protein samples in sodium dodecyl sulfate (SDS) buffer were loaded onto SDS polyacrylamide gels. After electrophoresis, proteins were electrotransferred to a polyvinylidenedifluoride membrane (Immobilon, Millipore). Membranes were incubated with either antibody in 20 m*M* Tris-HCl, pH 7.6, 135 m*M* NaCl, 0.1% Tween-20 containing 5% nonfat dry milk. Subsequently, membranes were incubated with horseradish-peroxidase-conjugated anti-rabbit antibody. Immunoreactive bands were detected by LumiGLO (New England Biolabs, Inc., Beverly, Mass.). Protein samples were immunoprecipitated with antibodies (described above) and then incubated with protein-G-Sepharose, as described previously [14]. Samples were then subjected to SDS-PAGE followed by immunoblotting.

Statistics

Data are expressed as the percentage of surviving neurons relative to the number of neurons in control culture (vehicle only) and represent the mean ± SE. Statistical significance was determined using one-way analysis of variance (ANOVA) followed by Scheffe's multiple-comparisons test.

Results

Nicotine Protects Neurons from Aβ 25–35-Induced Cytotoxicity

Aβ 25–35 was previously reported to be neurotoxic [6]. In our experiment, a 48-hour exposure to 20 μ*M* Aβ caused a significant reduction in neuronal cells (fig. 1a). Incubating the cultures with nicotine significantly protected neurons

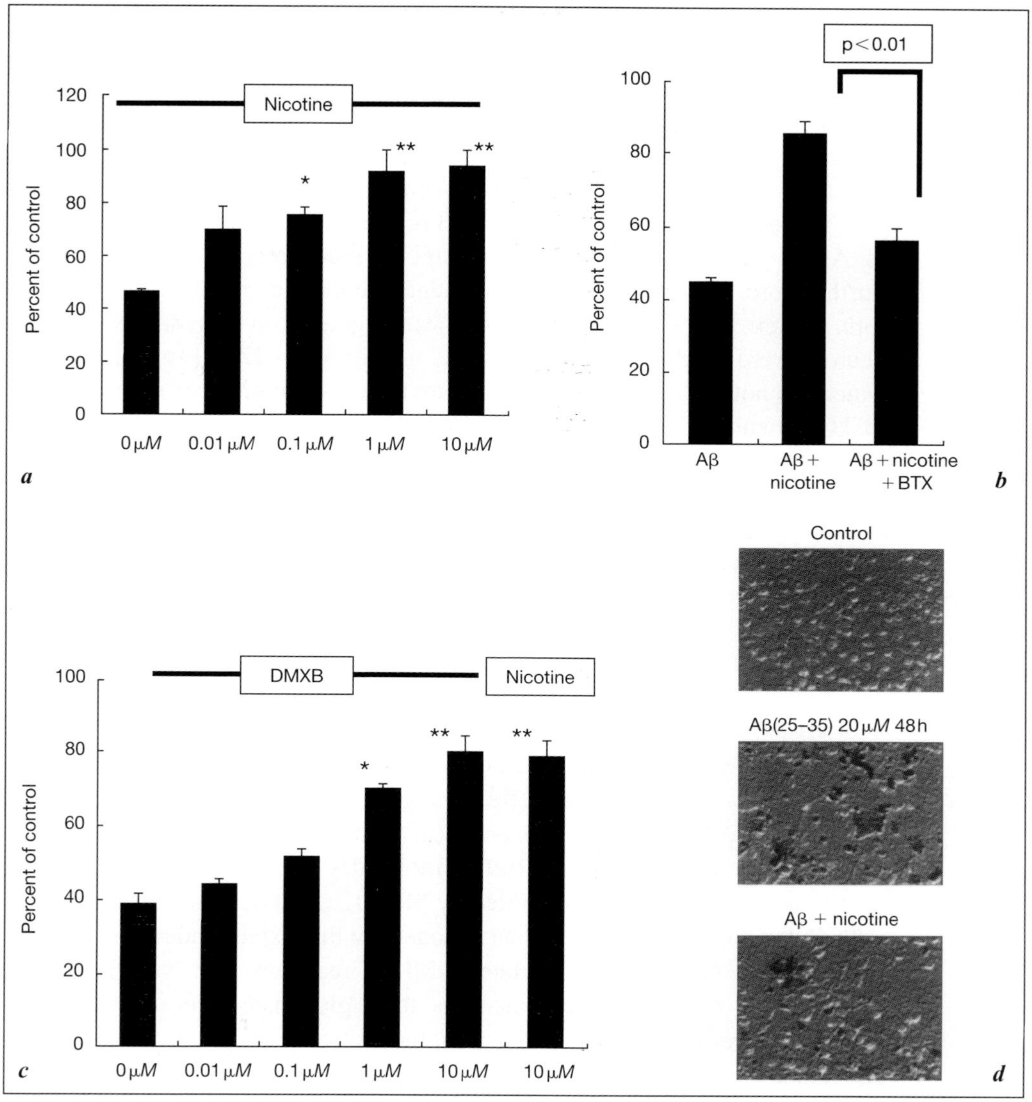

Fig. 1. ***a*** Concentration-dependent effects of nicotine on Aβ-induced neurotoxicity. Nicotine was added to medium containing 20 μ*M* Aβ and incubated for 48 h (n = 5). $*p < 0.05$, $**p < 0.01$, compared with Aβ alone (0 μ*M* nicotine). ***b*** Effects of the α_7-antagonist, α-BTX, on nicotine-induced protection against Aβ cytotoxicity. Nicotine (10 μ*M*) and α-BTX (1 n*M*) were added simultaneously to medium containing Aβ (20 μ*M*) (n = 5). ***c*** Neuroprotective effects of DMXB, a selective α_7-agonist, against Aβ cytotoxicity. DMXB was added to the medium containing Aβ (20 μ*M*) and incubated for 48 h (n = 5). $*p < 0.05$, $**p < 0.01$. ***d*** Trypan blue staining showing the protective effect of nicotine against Aβ(25–35) neurotoxicity.

from Aβ 25–35-induced cytotoxicity (fig. 1a, 1d). This effect was inhibited by α-BTX (fig. 1b). In addition, 3-(2,4)-dimethoxybenzylidene anabaseine (DMXB), a selective α_7-nicotinic receptor agonist, also protected cells from Aβ 25–35 toxicity (fig. 1c).

Nicotine Protects Neurons from Aβ-Enhanced Glutamate Cytotoxicity

There are controversies whether Aβ is directly toxic or not, because 20 μ*M* of Aβ is much higher than that found in the cerebrospinal fluid (CSF). Furthermore, Aβ 1–40 and 1–42 fragments are main component of SP. Therefore, we next investigated the effect of these fragments on survival rate. Cortical neurons were incubated with Aβ 1–40 (1 n*M*) or Aβ 1–42 (100 p*M*) for 7 days, which did not induce cell death. These are the concentrations of Aβ in the CSF of AD patients [15]. Treatment with a low dose of glutamate (20 μ*M*) alone did not significantly induce cell death (fig. 2a). Seven days of exposure to Aβ 1–40 and 1–42 followed by 24 h of exposure to glutamate (20 μ*M*) caused a significant reduction in the number of neuronal cells (fig. 2a). These findings suggested that Aβ 1–40 and 1–42 are not neurotoxic at physiological concentrations but make neurons vulnerable to excitatory amino acids. Coincubating the cultures with nicotine (0.5 μ*M*, 7 days) and Aβ 1–40 and 1–42 significantly reduced Aβ-enhanced glutamate cytotoxicity (fig. 2a).

Aβ-Induced Cytotoxicity Is Mediated via NMDA Receptors

Cortical neurons were incubated with Aβ 25–35 (20 μ*M*) for 48 h, which caused direct cytotoxicity. This effect was inhibited by MK801 when incubated with Aβ 25–35 (fig. 2b). When cortical neurons were incubated with both Aβ 1–40 (1 n*M*) and Aβ 1–42 (100 p*M*) for 7 days, which enhanced glutamate cytotoxicity, this effect was inhibited by MK801 when neurons were incubated with glutamate (fig. 2a). Those data indicate that the toxicity induced by Aβ was mediated via N-methyl-*D*-aspartate (NMDA) receptors. We have previously reported that nicotine protects neurons from glutamate-induced cytotoxicity [16–18]. Therefore, we hypothesized that the protective effect of nicotine against Aβ-induced cytotoxicity depends upon its effect on glutamate toxicity.

PI3K Contributes to the Protective Effect of Nicotine against Glutamate-Induced Cytotoxicity

To investigate the mechanism of the protective effect of nicotine, we focused on the PI3K cascade because PI3K has been reported to protect cells from apoptosis through Akt activation [8]. The glutamate toxicity model was adopted because a low concentration of Aβ alone was not toxic or only enhanced glutamate toxicity. Furthermore, we showed that nicotine did not directly affect the Aβ conformation, as described previously [19].

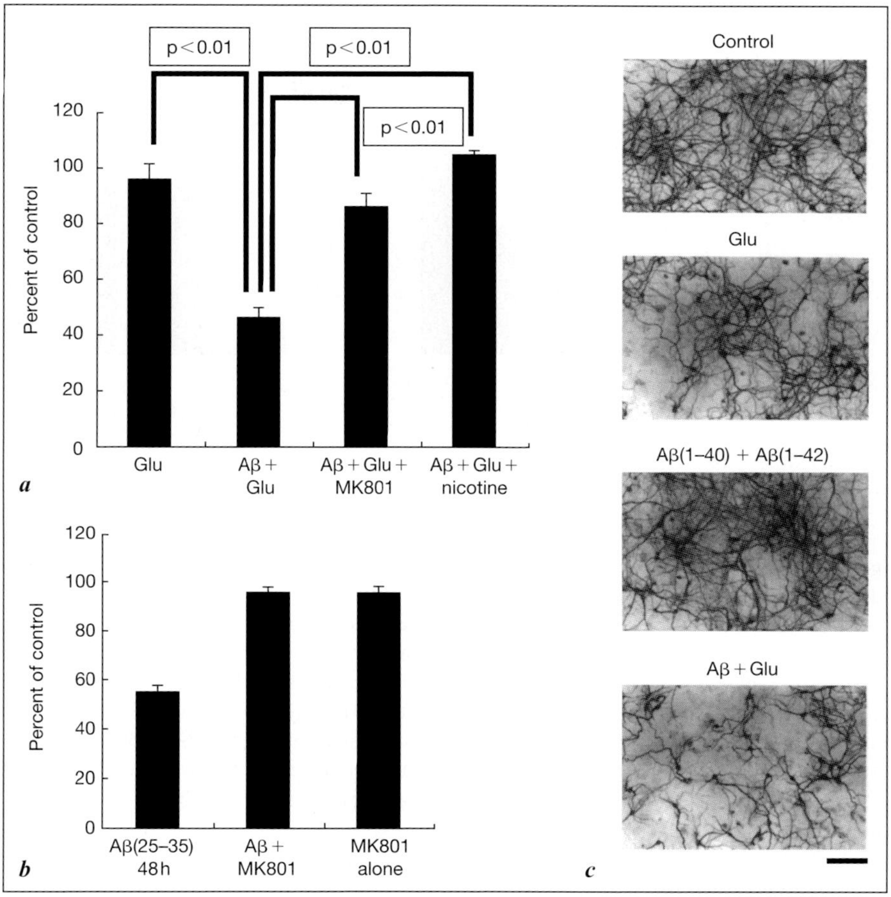

Fig. 2. ***a*** Effects of nicotine on Aβ-enhanced glutamate (Glu) neurotoxicity. Cultures were exposed to Aβ(1–40) (1 n*M*) and Aβ(1–42) (100 p*M*) for 7 days followed by a 24-hour incubation with 20 μ*M* glutamate. Toxicity caused by this treatment was inhibited by 1 μ*M* MK801. Nicotine protected cells from Aβ-enhanced glutamate neurotoxicity (n = 4). ***b*** Toxicity caused by Aβ(25–35) (20 μ*M* 48 h) was also inhibited by 10 μ*M* MK801. ***c*** Immunostained images showing the protective effect of nicotine against Aβ-enhanced glutamate (Glu) neurotoxicity.

Prolonged exposure to glutamate (50 μ*M*, 24 h) induced cytotoxicity. Incubating the cultures with nicotine (10 μ*M*) for 24 h prior to glutamate exposure significantly reduced glutamate cytotoxicity. Simultaneous application of either LY294002 (10 μ*M*) or wortmannin (100 n*M*), both PI3K inhibitors, with nicotine, reduced the protective effect of nicotine against glutamate cytotoxicity. α-BTX (1 n*M*), an α_7-selective nicotinic receptor antagonist, also blocked the

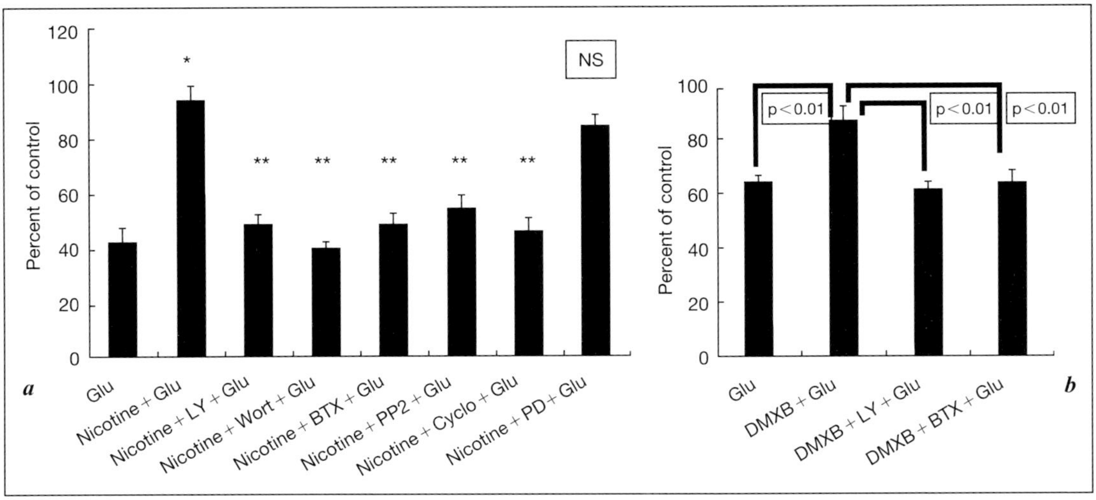

Fig. 3. ***a*** Effect of nicotine on glutamate-induced cytotoxicity. Cultures were exposed to glutamate (50 μ*M*) for 24 h followed by incubation in EMEM for 24 h, which induced cell death. Nicotine (1 μ*M*) was preincubated with the cultures for 24 h. If used, LY294002 (LY), wortmannin (Wort), α-BTX (BTX), PP2, cycloheximide (Cyclo) and PD98059 (PD) were added to the medium containing nicotine. After the preincubation, medium was replaced by glutamate-containing media for 24 h, and finally replaced with EMEM, as described above. LY294002, wortmannin and PI3K inhibitors, all significantly reduced the protective effect of nicotine. Furthermore, PP2, a Src family tyrosine kinase inhibitor, also reduced the protective effect of nicotine (n = 4). *$p < 0.01$ compared with glutamate alone, **$p < 0.01$ compared with glutamate + nicotine. NS = not significant. ***b*** Effect of DMXB on glutamate-induced cytotoxicity. DMXB (10 μ*M*) was added to the medium and incubated for 24 h. LY294002 significantly reduced the protective effect (n = 4).

protection, indicating that the α_7-subtype of nicotinic receptors is involved in this effect (fig. 3a).

DMXB (10 μ*M*), an α_7-selective nicotinic receptor agonist [20] also exerted a protective effect against glutamate-induced cytotoxicity. This effect was inhibited by 1 n*M* α-BTX, indicating that the effect of DMXB is mediated via α_7-receptors. Furthermore, the protection was also reduced by 10 μ*M* LY294002 (fig. 3b). From these findings, we concluded that α_7-nicotinic receptor stimulation exerts a neuroprotective effect against glutamate cytotoxicity, and that the PI3K cascade is involved in this effect. We have previously shown that α4β2 subunit nicotinic receptor stimulation also exerted a protective effect against Aβ- and glutamate-induced cytotoxicity [18, 21]. This effect, however, was not inhibited by LY294002 (data not shown).

Furthermore, a non-receptor tyrosine kinase inhibitor, PP2 (10 μ*M*), also reduced the protective effect. This implies that Src is involved in the mechanism of protection. Cycloheximide (1 μg/ml) inhibited the protection, implying that some protein synthesis is necessary. In contrast, PD98059 (50 μ*M*), a mitogen-activated protein kinase (MAPK) kinase (MAPKK, also known as MEK1) inhibitor, did not reduce the protective effect of nicotine (fig. 3a).

Nicotine Activates Akt through PI3K Activation and Upregulates Bcl-2

Akt is a serine/threonine protein kinase and a putative effector of PI3K. When PI3K is activated, it phosphorylates Akt. To investigate the activation of Akt by nicotine through PI3K, we examined the level of phosphorylated Akt detected by an antibody specific for phospho-Akt using Western blotting. In preliminary experiments, phosphorylation of Akt was detected just after the application of nicotine. The levels of phosphorylated Akt increased and reached a plateau after around 60 min of stimulation, and were maintained for 24 h (data not shown). Therefore, 60 min of stimulation was adopted for the following experiments.

Nicotine-induced Akt phosphorylation was blocked by the simultaneous application of LY294002, showing the involvement of PI3K (fig. 4a). PD98059 did not alter the phosphorylating effect of nicotine. Akt phosphorylation was blocked by α-BTX, indicating that the phosphorylating effect of nicotine is mediated by α_7-nicotinic receptors (fig. 4c). Conversely, dihydro-β-erythroidine (DHβE, 100 n*M*), a selective α4β2 nicotinic receptor antagonist, did not block nicotine-induced Akt phosphorylation (fig. 4c). PP2 blocked Akt phosphorylation, indicating the involvement of tyrosine kinase. MK801 (10 μ*M*) did not block Akt phosphorylation, showing that the secondary release of glutamate has no effect (fig. 4b).

Bcl-2 proteins are anti-apoptotic proteins which can prevent cell death induced by a variety of toxic insults [22]. It has been reported that Akt activation leads to the overexpression of Bcl-2 [23]. Because nicotine could activate Akt via PI3K, we examined the protein levels of Bcl-2. Treatment with nicotine for 24 h induced the augmented level of Bcl-2, and the nicotine-induced upregulation of Bcl-2 was reduced by LY294002, indicating the involvement of PI3K signal transduction (fig. 4d).

α_7-Nicotinic acetylcholine Receptors Physically Associate with PI3K and Fyn

The present results show that nicotine protects neurons from glutamate cytotoxicity by activating PI3K, which in turn activates Akt and upregulates Bcl-2. Nicotinic receptors are ionotropic, and there have been no reports of ionotropic receptors activating PI3K directly.

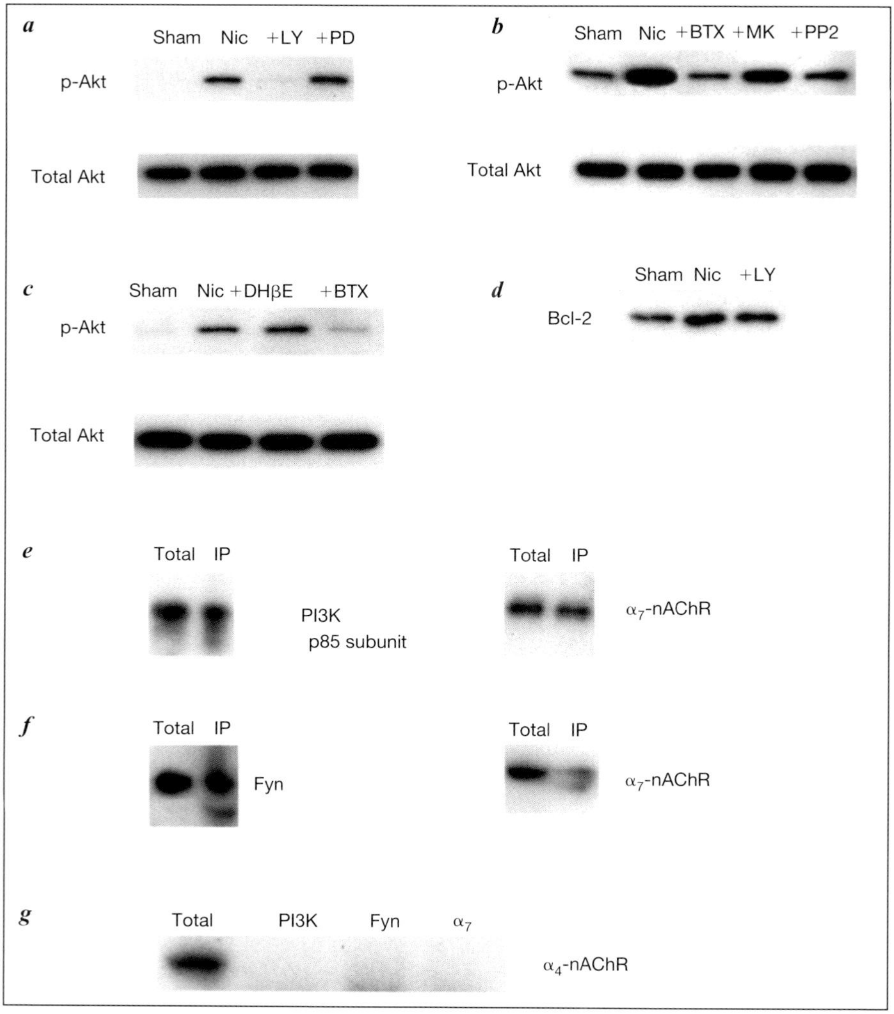

4

Fig. 4. Representative data of phosphorylation of Akt/PKB by nicotine detected by Western blot using antibody specific for phosphorylated Akt (p-Akt). LY = LY294002; PD = PD98059; Nic = nicotine; BTX = α-BTX; MK = MK801; DHβE = dihydro-β-erythroidine; IP = immunoprecipitated; nAChR = nicotinic acetylcholine receptor. ***a*** Nicotine (10 μ*M*) increased the levels of the phosphorylated Akt compared with the total Akt levels. This phosphorylation was significantly inhibited by LY294002 (10 μ*M*). PD98059 (50 μ*M*), a MEK-1 inhibitor, did not interfere with Akt phosphorylation. Each inhibitor was added simultaneously with nicotine (n = 6). ***b*** Akt phosphorylation was inhibited by α-BTX (1 n*M*) and PP2 (10 μ*M*), a Src inhibitor. MK801 (10 μ*M*) did not reduce the phosphorylation of Akt induced by nicotine. ***c*** Akt phosphorylation was not inhibited by dihydro-β-erythroidine (100 n*M*) in contrast to α-BTX (n = 6). ***d*** Bcl-2 was upregulated after nicotine treatment. LY294002 (10 μ*M*) inhibited the upregulation of Bcl-2, indicating that nicotine-induced

Therefore, we hypothesized that nicotinic receptors may act as metabotropic receptors, directly transmitting signals to PI3K. In other words, nicotinic receptors might associate with PI3K. In order to demonstrate this, lysates of cortical neurons were immunoprecipitated with a monoclonal anti-α_7-nicotinic receptor antibody. Immunoprecipitated samples were then separated by SDS-PAGE and stained with a polyclonal anti-PI3K p85 subunit antibody using the Western blot technique. The PI3K p85 subunit was detected in this immunoprecipitated sample, indicating that α_7-nicotinic receptors bind to the PI3K p85 subunit (fig. 4e). Conversely, lysate immunoprecipitated with the polyclonal anti-PI3K p85 subunit antibody contained protein detected by the monoclonal anti-α_7-nicotinic receptor antibody. The non-receptor-type tyrosine kinase, Fyn, was also coimmunoprecipitated with the α_7-nicotinic receptors (fig. 4f). The α_4-subunit of the nicotinic receptor was also investigated, and was not detected in lysates immunoprecipitated with the PI3K p85 subunit, Fyn, or the α_7-nicotinic receptor antibody (fig. 4g).

Discussion

Amyloid accumulation is one of the earliest changes in AD pathology and causes neuronal death in the central nervous system [24]. In the present study, Aβ was directly toxic to neurons at high concentrations, which might be mediated via NMDA receptors. Low concentrations of Aβ enhanced glutamate-induced neurotoxicity, which might also be mediated via NMDA receptors, although, NMDA receptors did not appear to be directly toxic. Previous reports have shown that Aβ stimulates NO production through Ca^{2+} entry triggered by activated NMDA-gated channels [25]. Other reports have suggested that Aβ inhibits glutamate uptake and increases extracellular glutamate [7]. There are also some reports that have proposed that Aβ enhances the toxicity induced by excitotoxins [26, 27]. These reports implied that Aβ-induced cytotoxicity might be, at least in part, mediated via glutamate toxicity. Oxidative stress, or free radical generation, might be another cause of Aβ-induced cytotoxicity. Glutamate also causes the generation of free radicals.

upregulation of Bcl-2 is mediated via the PI3K cascade (n = 6). ***e*** The PI3K was detected in immunoprecipitates produced using the anti-α_7-nicotinic receptor antibody (left). Conversely, the α_7-nicotinic receptor was detected in immunoprecipitates produced using the anti-p85 subunit PI3K antibody (right) (n = 6). ***f*** Fyn was detected in immunoprecipitates produced using the anti-α_7-nicotinic receptor antibody. Conversely, α_7-nicotinic receptors were detected in IP produced using the anti Fyn antibody (right) (n = 6). ***g*** The α_4-subunit of the nicotinic receptor was not detected in any of the IP samples produced using the anti-PI3K p85 subunit, Fyn, or the α_7-nicotinic receptor antibody.

Here, we showed that nicotinic receptor stimulation, especially α_7-receptor stimulation, inhibits glutamate toxicity, and that PI3K-Akt signal transduction contributes to this effect. In addition, the Bcl-2 family is stimulated downstream of the PI3K-Akt cascade and works as an anti-neuronal death factor. It has recently been reported that overexpression of α_7-receptor protects PC12 cells from hypoxic insults, and this protection is mediated via Akt phosphorylation [28]. PI3K-Akt activation promotes cell survival, and upregulation of Bcl-2 is one of the major reasons for cell survival [23, 29]. Nicotinic receptor stimulation transduces these survival signals besides its role as a transmitter. The beta sheet conformation of Aβ might influence its function, such as toxicity or modulation of survival signals. However, in our experiments, nicotine and nicotinic agonists did not influence beta sheet conformation [19]. Instead, signal transduction was shown to be important for the protective effect of nicotine.

It is not clear from our experiments whether other Src family members besides Fyn are associated with α_7-receptors. However, a relationship between nicotinic receptors and Fyn was suspected because the catecholamine release induced by nicotine is dependent upon the presence of Fyn and extracellular Ca^{2+}, and no other Src member was detected [30]. In our preliminary data, removal of extracellular Ca^{2+} suppressed Akt phosphorylation induced by nicotine (data not shown). We showed that an inhibitor of Src tyrosine kinase reduced Akt phosphorylation. In addition, PI3K and Fyn are physically associated with α_7-nicotinic receptors. Therefore, nicotinic receptor stimulation might phosphorylate Akt by signal transduction through Fyn to PI3K, and extracellular Ca^{2+} might contribute to this process. Recent data indicated that nicotine phosphorylates Akt through epidermal growth factor and Src in PC12h cells [31]. In this report, however, non-α_7-nicotinic receptors are involved in the phosphorylation. We used primary cortical neurons, and this cell difference might have contributed to this difference. Or several pathways exist, and the main pathway of primary neurons might be different from that of PC cells. Our hypothesis on a protective effect of nicotinic receptor stimulation is summarized in figure 5.

The brain contains nicotinic receptors of several subtypes with differing properties and functions. The abundant presence of α_7-receptors in the hippocampus, neocortex and basal ganglia [32], in conjunction with the memory-enhancing activity of selective α_7-nicotinic agonists, such as DMXB [33], suggests a significant role for α_7-receptors in learning and memory. In addition, the protective action of nicotine is mediated, at least partially, through α_7-receptors. It was reported that Aβ(1–42) binds to α_7-receptors [34], and this might inhibit α_7-nicotinic receptor-dependent learning and memory. The reduction of α_7-receptor activation might cause neurons vulnerable to various toxic insults such as glutamate. In our study, however, the lysate immunoprecipitated with anti-α_7 antibody did not contain Aβ(1–42) (data not shown). This might be because the antibody we used was

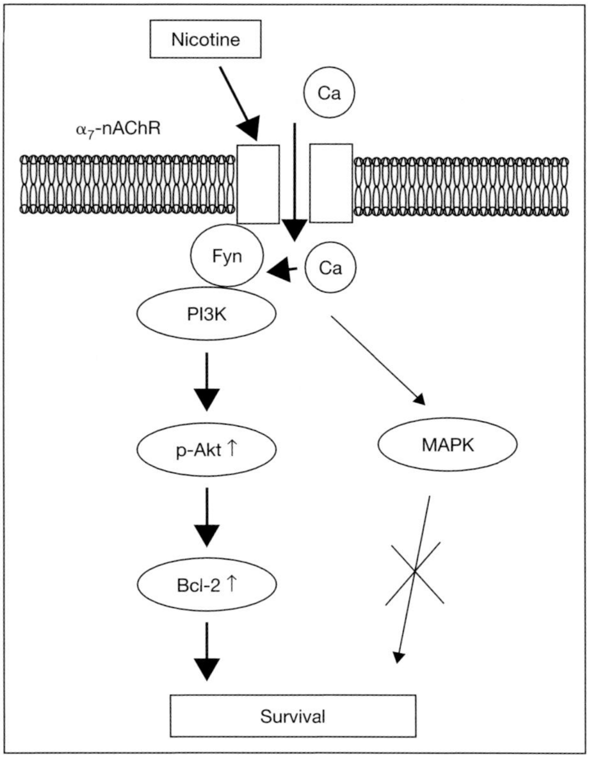

Fig. 5. A schematic model of the neuroprotection induced by α_7-nicotinic receptor stimulation.

different from that used in the report [34], but we could not prove that enhancement of glutamate toxicity depends upon the reduction of α_7-nicotinic receptors.

Recently, it was shown that ionotropic receptors have properties similar to metabotropic receptors. AMPA receptors are physically associated with a member of the Src family, Lyn [10]. The AMPA receptor activates Lyn, which then activates MAPK. Through the Lyn-MAPK pathway, AMPA receptors generate intracellular signals and transmit them from the cell surface to the nucleus. Nicotinic receptors are known to be ionotropic receptors. The present study indicated that nicotinic receptors also have metabotropic properties which contribute to neuronal survival. It is likely, however, that many unrecognized receptor functions still remain.

The cholinergic system is affected in dementia-causing diseases, Alzheimer's disease among others, and a reduction in the number of nicotinic receptors has been reported in these diseases [35, 36]. It is of interest that downregulation of nicotinic receptors can result in neuronal cell death or neurodegeneration [37]. Nicotine might function not only as a cholinergic agonist but also as a neuroprotective agent. Our present study suggests that nicotinic receptor stimulation could protect neurons from Aβ-enhanced glutamate toxicity. Thus, by an early diagnosis of Alzheimer's disease and protective therapy with nicotinic receptor stimulation, we could delay its progress.

Acknowledgements

This study was supported in part by Grants-in-Aid for Scientific Research from the Ministry of Education, Science, Culture, Sports, and Technology, and grants from the Ministry of Health, Labor, and Welfare of Japan, Japan Society for the Promotion of Science, and the Smoking Research Foundation. We also thank the Taiho Pharmaceutical Co., for providing us with DMXB.

References

1 Giannakopoulos P, Hof PR, Kovari E, Vallet PG, Herrmann FR, Bouras C: Distinct patterns of neuronal loss and Alzheimer's disease lesion distribution is elderly individuals older than 90 years. J Neuropathol Exp Neurol 1996;55:1210–1220.

2 Perry G, Cras P, Siedlak SL, Tabaton M, Kawai M: β protein immunoreactivity is found in the majority of neurofibrillary tangles of Alzheimer's disease. Am J Pathol 1992;140:283–290.

3 Nelson PT, Marton L, Saper CB: Alz-50 immunohistochemistry in the normal sheep striatum: A light and electron microscope study. Brain Res 1993;600:285–297.

4 Citron M, Oltersdorf T, Haass C, McConlogue L, Hung AY, Seubert P, Vigo PC, Lieberburg I, Selkoe DJ: Mutation of the β-amyloid precursor protein in familial Alzheimer's disease increases β-protein production. Nature 1992;360:672–674.

5 Tomita T, Maruyama K, Saido TC, Kume H, Shinozaki K, Tokuhiro S, Capell A, Walter J, Grunberg J, Haass C, Iwatsubo T, Obata K: The presenilin 2 mutation (N141I) linked to familial Alzheimer disease (Volga German families) increases the secretion of amyloid β protein ending at the 42nd (or 43rd) residue. Proc Natl Acad Sci USA 1997;94:2025–2030.

6 Tamura Y, Sato Y, Akaike A, Shiomi H: Mechanisms of cholecystokinin-induced protection of cultured cortical neurons against N-methyl-*D*-aspartate receptor-mediated glutamate cytotoxicity. Brain Res 1992;592:317–325.

7 Harris ME, Wang Y, Pedigo NW Jr, Hensley K, Butterfield DA, Carney JM: Amyloid β peptide (25–35) inhibits Na^+-dependent glutamate uptake in rat hippocampal astrocyte cultures. J Neurochem 1996;67:277–286.

8 del Peso L, Gonzalez-Garcia M, Page C, Herrera R, Nunez G: Interleukin-3-induced phosphorylation of BAD through the protein kinase Akt. Science 1997;278:687–689.

9 Perkinton MS, Sihra TS, Williams RJ: Ca^{2+}-permeable AMPA receptors induce phosphorylation of cAMP response element-binding protein through a phosphatidylinositol 3-kinase-dependent stimulation of the mitogen-activated protein kinase signaling cascade in neurons. J Neurosci 1999; 19:5861–5874.

10 Hayashi T, Umemori H, Mishina M, Yamamoto T: The AMPA receptor interacts with and signals through the protein tyrosine kinase Lyn. Nature 1999;397:72–76.

11 Kihara T, Shimohama S, Sawada H, Kimura J, Kume T, Kochiyama H, Maeda T, Akaike A: Nicotinic receptor stimulation protects neurons against β-amyloid toxicity. Ann Neurol 1997; 42:159–163.

12 Shimohama S, Ogawa N, Tamura Y, Akaike A, Tsukahara T, Iwata H, Kimura J: Protective effect of nerve growth factor against glutamate-induced neurotoxicity in cultured cortical neurons. Brain Res 1993;632:296–302.

13 Sawada H, Ibi M, Kihara T, Urushitani M, Akaike A, Kimura J, Shimohama S: Dopamine D_2-type agonists protect mesencephalic neurons from glutamate neurotoxicity: Mechanisms of neuroprotective treatment against oxidative stress. Ann Neurol 1998;44:110–119.
14 Kume T, Kouchiyama H, Kaneko S, Maeda T, Kaneko S, Akaike A, Shimohama S, Kihara T, Kimura J, Wada K, Koizumi S: BDNF prevents NO mediated glutamate cytotoxicity in cultured cortical neurons. Brain Res 1997;756:200–204.
15 Jensen M, Schroder J, Blomberg M, Engvall B, Pantel J, Ida N, Basun H, Wahlund LO, Werle E, Jauss M, Beyreuther K, Lannfelt L, Hartmann T: Cerebrospinal fluid Aβ42 is increased early in sporadic Alzheimer's disease, declines with disease progression. Ann Neurol 1999;45:504–511.
16 Akaike A, Tamura Y, Yokota T, Shimohama S, Kimura J: Nicotine-induced protection of cultured cortical neurons against N-methyl-*D*-aspartate receptor-mediated glutamate cytotoxicity. Brain Res 1994;644:181–187.
17 Shimohama S, Akaike A, Kimura J: Nicotine-induced protection against glutamate cytotoxicity. Nicotinic cholinergic receptor-mediated inhibition of nitric oxide formation. Ann NY Acad Sci 1996;777:356–361.
18 Kaneko S, Maeda T, Kume T, Kochiyama H, Akaike A, Shimohama S, Kimura J: Nicotine protects cultured cortical neurons against glutamate-induced cytotoxicity via α_7-neuronal receptors and neuronal CNS receptors. Brain Res 1997;765:135–140.
19 Kihara T, Shimohama S, Akaike A: Effects of nicotinic receptor agonists on β-amyloid β-sheet formation. Jpn J Pharmacol 1999;79:393–396.
20 Hunter BE, de Fiebre CM, Papke RL, Kem WR, Meyer EM: A novel nicotinic agonist facilitates induction of long-term potentiation in the rat hippocampus. Neurosci Lett 1994;168:130–134.
21 Kihara T, Shimohama S, Urushitani M, Sawada H, Kimura J, Kume T, Maeda T, Akaike A: Stimulation of α4β2 nicotinic acetylcholine receptors inhibits β-amyloid toxicity. Brain Res 1998; 792:331–334.
22 Zhong LT, Kane DJ, Bredesen DE: BCL-2 blocks glutamate toxicity in neural cell lines. Mol Brain Res 1993;19:353–355.
23 Matsuzaki H, Tamatani M, Mitsuda N, Namikawa K, Kiyama H, Miyake S, Tohyama M: Activation of Akt kinase inhibits apoptosis and changes in Bcl-2 and Bax expression induced by nitric oxide in primary hippocampal neurons. J Neurochem 1999;73:2037–2046.
24 Yankner BA, Duffy LK, Kirschner DA: Neurotrophic and neurotoxic effects of amyloid-β protein: Reversal by tachykinin neuropeptides. Science 1990;250:279–282.
25 O'Mahony S, Harkany T, Rensink AA, Abraham I, De Jong GI, Varga JL, Zarandi M, Penke B, Nyakas C, Luiten PG, Leonard BE: β-Amyloid-induced cholinergic denervation correlates with enhanced nitric oxide synthase activity in rat cerebral cortex: Reversal by NMDA receptor blockade. Brain Res Bull 1998;45:405–411.
26 Dornan WA, Kang DE, McCampbell A, Kang EE: Bilateral injections of βA(25–35) + IBO into the hippocampus disrupt acquisition of spatial learning in the rat. Neuroreport 1993;5:165–168.
27 Morimoto K, Yoshimi K, Tonohiro T, Yamada N, Oda T, Kaneko I: Co-injection of β-amyloid with ibotenic acid induces synergistic loss of rat hippocampal neurons. Neuroscience 1998;84: 479–487.
28 Utsugisawa K, Nagane Y, Obara D, Tohgi H: Overexpression of α_7-nicotinic acetylcholine receptor prevents G1-arrest and DNA fragmentation in PC12 cells after hypoxia. J Neurochem 2002;81: 497–505.
29 Eves EM, Xiong W, Bellacosa A, Kennedy SG, Tsichlis PN, Rosner MR, Haay N: Akt, a target of phosphatidylinositol 3-kinase, inhibits apoptosis in a differentiating neuronal cell line. Mol Cell Biol 1998;18:2143–2152.
30 Allen CM, Ely CM, Juaneza MA, Parsons SJ: Activation of Fyn tyrosine kinase upon secretagogue stimulation of bovine chromaffin cells. J Neurosci Res 1996;44:421–429.
31 Nakayama H, Numakawa T, Ikeuchi T: Nicotine-induced phosphorylation of Akt through epidermal growth factor receptor and Src in PC12h cells. J Neurochem 2002;83:1372–1379.
32 Clarke PB, Schwartz RD, Paul SM, Pert CB, Pert A: Nicotinic binding in rat brain: Autoradiographic comparison of [^{3}H]acetylcholine, [^{3}H]nicotine, and [^{125}I]-α-bungarotoxin. J Neurosci 1985;5: 1307–1315.

33 Meyer EM, Tay ET, Papke RL, Meyers C, Huang GL, de Fiebre CM: 3-[2,4-Dimethoxybenzylidene]anabaseine (DMXB) selectively activates rat α_7-receptors and improves memory-related behaviors in a mecamylamine-sensitive manner. Brain Res 1997;768:49–56.
34 Wang HY, Lee DH, D'Andrea MR, Peterson PA, Shank RP, Reitz AB: β-Amyloid(1–42) binds to α_7-nicotinic acetylcholine receptor with high affinity. J Biol Chem 2000;275:5626–5632.
35 Shimohama S, Taniguchi T, Fujiwara M, Kameyama M: Changes in nicotinic and muscarinic cholinergic receptors in Alzheimer-type dementia. J Neurochem 1986;46:288–293.
36 Whitehouse PJ, Kalaria RN: Nicotinic receptors and neurodegenerative dementing diseases: Basic research and clinical implications. Alzheimer Dis Assoc Disord 1995;9:3–5.
37 Zoli M, Picciotto MR, Ferrari R, Cocchi D, Changeux JP: Increased neurodegeneration during ageing in mice lacking high-affinity nicotine receptors. EMBO J 1999;18:1235–1244.

Dr. Shun Shimohama
Department of Neurology, Graduate School of Medicine, Kyoto University
54 Shogoin-Kawaharacho, Sakyoku, Kyoto 606–8507 (Japan)
Tel. +81 75 751 3771, Fax +81 75 751 3265, E-Mail i53367@sakura.kudpc.kyoto-u.ac.jp

Takeda M, Tanaka T, Cacabelos R (eds): Molecular Neurobiology of Alzheimer Disease and Related Disorders. Basel, Karger, 2004, pp 123–133

Involvement of Unfolded Protein Responses in Alzheimer's Disease

Takashi Kudo[a], *Taiichi Katayama*[b], *Kazunori Imaizumi*[c], *Daisuke Kanayama*[a], *Mikiko Sowa*[a], *Masayasu Okochi*[a], *Masaya Tohyama*[b], *Masatoshi Takeda*[a]

Departments of [a]Psychiatry and Behavioral Science, and [b]Anatomy and Neuroscience, Osaka University Graduate School of Medicine, Osaka, [c]Division of Structural Cell Biology, Nara Institute of Science and Technology, Takayama, Japan

Alzheimer's disease (AD) is a progressive neurodegenerative disorder characterized pathologically by cerebral neuritic plaques of amyloid-β peptide (Aβ) and neurofibrillary tangles of phosphorylated tau. Some early-onset cases of autosomal-dominant familial Alzheimer's disease (FAD) are caused by mutations in the amyloid precursor protein (APP) gene located on chromosome 21, presenilin-1 (PS1) located on chromosome 14, and presenilin-2 (PS2) located on chromosome 1. Of these three genes, mutations in PS1 are the most prevalent in cases of FAD. However, the mechanism by which mutations in PS1 predispose individuals to FAD has not yet been determined. FAD-linked PS1 variants alter proteolytic processing of APP [1, 2], and mutations in PS1 increase cellular susceptibility to apoptosis induced by various insults, including withdrawal of trophic factors and exposure to Aβ [3, 4]. The mechanism by which PS1 mutations promote cell death is not known, but cell culture studies have revealed perturbed calcium homeostasis and increased production of free radicals in affected cells. It has been reported that PS1 is located in subcellular compartments and appears to be present at particularly high levels in the endoplasmic reticulum (ER), the intermediate compartment, and the cis-Golgi region [5–7]. This has led us to study the relationship between ER function and PS1.

The ER performs the proper folding of several secretory proteins and membrane proteins. A variety of conditions, for instance the disturbance of calcium homeostasis, nutrient deprivation, elevated secretory protein synthesis, or altered glycosylation, disturb the proper folding and cause the accumulation of unfolding or misfolding proteins in the ER lumen. This is termed ER stress. Because the incompletely folded molecules threaten living cells, efficient quality control systems of the ER are absolutely essential for cell stability. Upon accumulation of unfolded proteins in the ER lumen, the unfolded protein response (UPR) is activated, thereby reducing the amount of new protein translocated into the ER lumen, increasing retrotranslocation and degradation of ER-located protein, and activating the protein-folding capacity of the ER.

This article focuses on the relationship between UPR and the pathogenesis of AD by reviewing our recent data.

Three Types of UPR

To date, three cellular responses have been referred to as UPR. Firstly, in response to the accumulation of unfolded proteins in the ER, eukaryotic cells activate an intracellular-signaling pathway from the ER to the nucleus, resulting in transcriptional upregulation of molecular chaperones, such as GRP78/BiP and GRP94. The chaperones assist or facilitate normal folding of unfolded or misfolded proteins [8]. Continuous delivery of newly synthesized proteins is a burden on the ER when proper folding is prevented under ER stress conditions. Therefore, the second strategy of cells against ER stress is a generalized suppression of translation mediated by the serine/threonine kinase, pancreatic ER kinase or PKR-like ER kinase (PERK). PERK phosphorylates the translational initiation factor eIF2α, causing translational attenuation [9, 10]. Thirdly, ER stress activates the ER-associated degradation (ERAD) system, by which misfolded proteins are transported out of the ER to the cytoplasm and are then ubiquitinated and degraded by the 26S proteasome [11]. By these three protective responses, cells can overcome ER stress, potentially leading to apoptosis. It has been revealed that cells deficient in the UPR are more vulnerable to ER stress-induced apoptosis [12–14].

ER-Resident Molecules for UPR

To date, three ER-resident transmembrane molecules, IRE1, PERK, and ATF6 , which sense the accumulation of unfolded proteins to induce UPR, have been identified in eukaryotes (fig. 1).

IRE1 is a type I transmembrane Ser/Thr protein kinase that also has site-specific RNase activity. Under nonstressed conditions, the most abundant ER

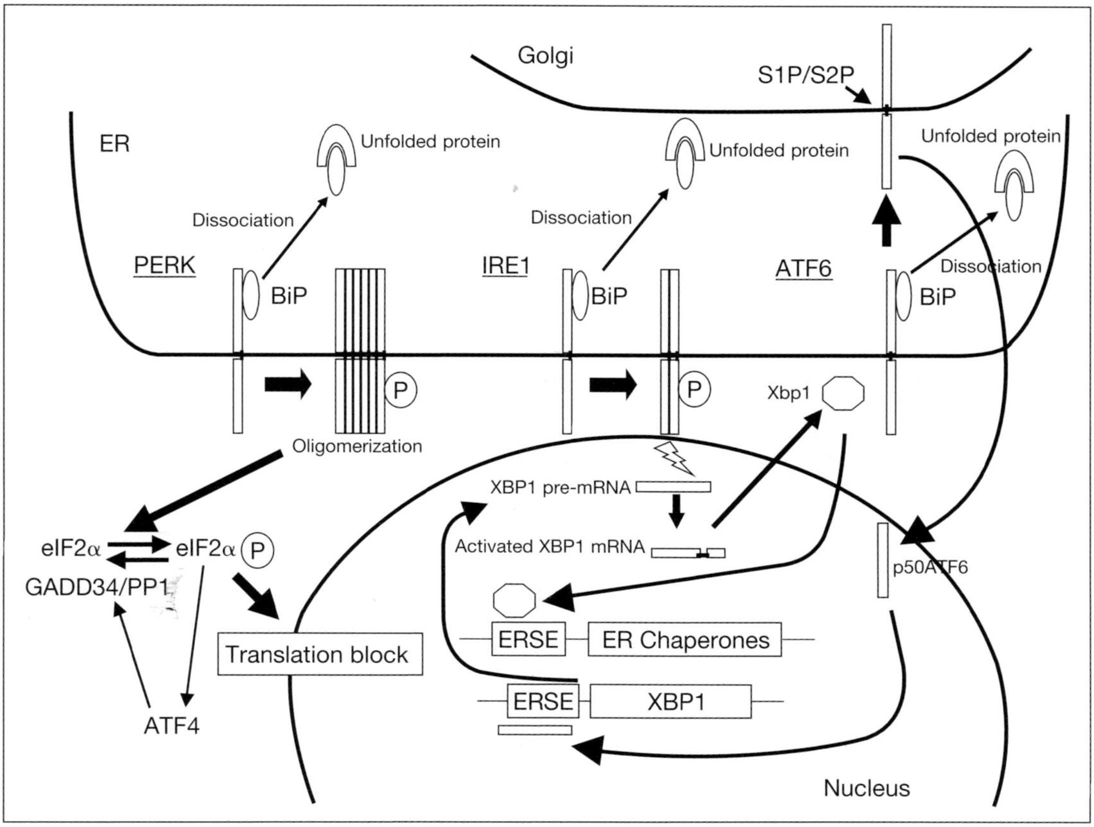

Fig. 1. Mechanisms of ER stress responses. Activation of the signaling-mediated ER stress transducers is triggered by dissociation of GRP78/BiP from stress transducers. The dissociation leads to oligomerization of stress transducers inducing their autophosphorylation (P) and the resultant activation of downstream signaling. Activation of PERK results in phosphorylation of eIF2α, and leads to inhibition of translation initiation. Simultaneously, GADD34 is expressed via ATF4 induction to set the stage for the resumption of mRNA translation after ER stress is resolved. Autophosphorylation and dimerization of IRE1 causes activation of endonuclease domains that have the potential to cleave XBP1 mRNA, and generate an activated form of XBP1. It can bind and activate the UPREs upstream of ER chaperone genes, such as GRP78/BiP and GRP94. In the Golgi, ATF6 is cleaved by S1P/S2P at or close to the cytosolic face of the membrane in response to ER stress. The N-terminal cytoplasmic domain (p50ATF6), which contains the DNA-binding, dimerization and transactivation domains, is translocated into the nucleus and activates transcription of XBP1 gene containing the ER stress response element (ERSE).

chaperone, GRP78/BiP, binds to IRE1. The presence of unfolded proteins in the ER lumen deprives IRE1 of GRP78/BiP to promote its dimerization and autophosphorylation to activate it as an RNase. In mammalian cells, XBP1 mRNA is cleaved by IRE1 to remove a 26-nucleotide intron and cause a frameshift to encode a protein that acts as a potent transcriptional activator [15, 16]. It can bind and activate the UPR elements (UPREs) upstream of ER chaperone genes, such as GRP78/BiP and GRP94 [17, 18].

PERK is type I transmembrane protein whose N-terminal luminal domain is sensitive to the upstream ER stress signal and whose C-terminal cytoplasmic domain directly phosphorylates the eukaryotic translation initiation factor-2α (eIF2α) [19, 20]. Because the stress-sensing domain of PERK is similar to that of IRE1, it is also maintained in an inactive state by the binding of the ER chaperone GRP78/BiP to their related luminal domains. Under conditions of ER stress, the increased unfolded proteins pull GRP78/BiP apart from PERK. Loss of GRP78/BiP binding correlates with oligomerization, autophosphorylation, and activation of downstream signaling by PERK [21]. The phosphorylation of PERK results in phosphorylation of the subunit of eIF2α at serine 51. The phosphorylation of eIF2α leads to interference with the formation of a 43S initiation complex, resulting in inhibition of translation initiation [10]. Although its phosphorylation inhibits translation of most mRNAs, the translation of the transcription factor ATF4 is paradoxically increased, increasing the transcription of specific target genes under conditions of ER stress [22]. Expression of one of these genes, GADD34, dephosphorylates eIF2α to set the stage for the resumption of mRNA translation after ER stress is resolved [23].

ATF6 is a type II transmembrane protein localized in the ER and is activated by ER-stress-induced proteolysis. On ER stress, processing of ATF6 by site-1 protease (S1P) and site-2 protease (S2P) occurs within the transmembrane segment and the N-terminal fragment facing the cytoplasm is separated from the ER membrane to translocate into the nucleus. It results in activation of XBP1 gene transcription, which facilitates protein folding [24]. Originally, S1P and S2P are the processing enzymes that cleave the ER-associated transmembrane sterol-response element–binding protein (SREBP) upon cholesterol deprivation [25]. The SREBP-cleavage-activating protein (SCAP) confers specificity for SREBP transport to the Golgi compartment, and consequently cleavage in response to sterol deprivation [26]. GRP78/BiP also interacts with ATF6 to prevent trafficking of ATF6 to the Golgi compartment. Therefore GRP78/BiP release permits ATF6 transport to the Golgi compartment, where it gains access to S1P and S2P proteases [27]. The regulation of signaling through the interaction of GRP78/BiP is an attractive hypothesis providing a direct mechanism by which all three ER stress sensors could be activated by the same stimulus.

ER Stress and Cell Death

Excess or prolonged ER stress leads to apoptotic cell death. Caspase-12 is an ER-associated effector in the caspase activation cascade [28]. ER stress induces tumor-necrosis-factor-receptor-associated factor 2 (TRAF2) release from procaspase 12, allowing it to bind activated IRE1. Release of TRAF2 permits clustering of procaspase-12 at the ER membrane to cause its activation [29]. Caspase-12 can activate caspase-9, which in turn activates caspase-3, the executioner for cell death. Procaspase-12 can also be activated by calpain in response to calcium release from the ER [30].

Activated IRE1 recruits the cytosolic adaptor TRAF2 to the ER membrane. TRAF2 activates the apoptosis-signaling kinase 1 (ASK1), a mitogen-activated protein kinase kinase kinase (MAPKKK) [31]. Activated ASK1 leads to activation of the JNK protein kinase and caspase activations through mitochondria [32].

CHOP is a transcription factor that is markedly induced by ER stress. It promotes cell cycle arrest and/or programmed cell death [33].

PS1 Mutants Are Susceptible to ER Stress

SK-N-SH neuroblastoma cells were stably transfected with complementary DNA constructs encoding wild-type PS1, PS1 with an alanine to glutamate mutation (A246E), or PS1 with a deletion of exon 9 (ΔE9). The A246E and ΔE9 mutations are FAD-linked mutations. We then added tunicamycin, which induces ER stress by preventing protein glycosylation, to the medium. Tunicamycin induced increased cell death in cells expressing mutant PS1 as compared with cells expressing the wild-type protein. The PS1 mutations also increased susceptibility to other ER stresses, such as a calcium iono-phore (A23187), which depletes intracellular calcium stores.

FAD Mutants of PS1 Attenuate UPR Initiated by IRE1

Under tunicamycin treatment (3 mg/ml, for 6 h), which prevents protein glycosylation and causes ER stress, the induction of GRP78 mRNA expression was significantly inhibited in permanent cell lines expressing mutant PS1 (ΔE9) compared with those transfected with wild-type PS1 or empty vector (fig. 2a). To confirm that FAD-linked mutants generally cause inhibition of GRP78 mRNA induction, other FAD-linked PS1 mutants, such as A246E, M146V and I213T, were studied with transiently transfected cells. These mutations also suppressed the induction of GRP78 mRNA. To exclude the possibility

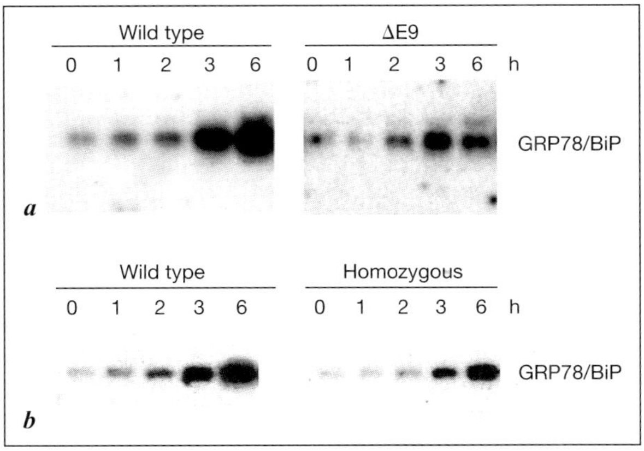

Fig. 2. Induction of GRP78/BiP mRNA in the cell with mutant PS1 and wild PS1. ***a*** SK-N-SH cells were stably transfected with each PS1 construct. Induction of GRP78/BiP and/or CHOP mRNA was attenuated in homozygous fibroblasts and ΔE9-expressing SK-N-SH cells when cells were treated with 3 μg/ml tunicamycin. These results were obtained for doses from 0.5 to 3 μg/ml tunicamycin. ***b*** Primary cultured fibroblasts from PS1 mutation (*I213T*) knock-in mice (wild and homozygous). Time courses of the levels of GRP78/BiP with tunicamycin are shown.

that the effects were caused by overexpression of PS1 mutants, we studied primary cultured neurons from embryos of PS1 mutant 'knock-in' mice, which express mutant PS1 (I213T) protein at normal physiological levels. Such 'knock-in' mice are believed to provide an animal model that closely imitates FAD in humans [34]. The induction of GRP78 mRNA by ER stress was slightly decreased in neurons from mice heterozygous for the knock-in PS1 mutation, and was significantly decreased in those from mice homozygous for the knock-in mutation (fig. 2b). In these knock-in mice, the efficiencies of transcription and translation from the mutated PS1 allele were equivalent to each other, indicating that the reduction of the expression of GRP78 mRNA was dependent on the expression level of mutated PS1 proteins. Thus, the FAD-linked PS1 mutations appeared to affect the system of GRP78 induction.

Immunoprecipitation and immunohistochemical studies showed that PS1 interacts with IRE1 on the ER membrane. As mentioned above, IRE1 leads to downstream signaling by a process that depends on oligomerization and autophosphorylation of its kinase domain. Therefore, we studied the effects of mutations in PS1 on autophosphorylation of IRE1. SK-N-SH cells stably transfected with wild-type PS1, mutant PS1, or empty vector were stimulated with thapsigargin (1 μ*M*) and Western blotting of IRE1 was performed. When cells were treated with 1 μ*M* thapsigargin, within 15 min, the bands of IRE1 were

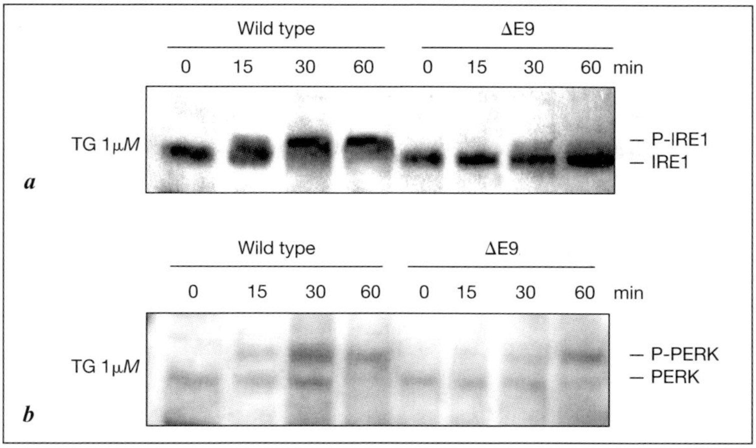

Fig. 3. Phosphorylation of IRE1 and PERK in the cell with mutant PS1 and wild PS1. ***a*** SK-N-SH cells stably transfected with each PS1 construct were stimulated with tapsigargin (TG; 1 and 5 μ*M*) for the indicated times, and Western blotting of IRE1 was performed. When cells were treated with 1 μ*M* tapsigargin, the bands of IRE1 were completely shifted (phosphorylated-IRE1: P-IRE1) within 15 min in cells expressing wild-type PS1 (*PS1W*). In contrast, no shift of Ire1-immunoreactive bands was seen in PS1ΔE9 cells treated with the same dose of tapsigargin. ***b*** Phosphorylation of PERK (P-PERK) in N2a stable cell lines expressing PS1 constructs. The changes in the levels of phosphorylated PERK in PS1 mutant cells are shown. PS1 mutant inhibited or delayed the phosphorylation of PERK after ER stress when cells were treated with 1 μ*M* tapsigargin.

completely shifted (phosphorylated-IRE1) in cells expressing wild PS1. In contrast, no shifts of IRE1-immunoreactive bands were seen in mutant PS1 expressing cells that were treated with the same dose of thapsigargin within 30 min (fig. 3a). These results indicate that FAD-linked PS1 mutants attenuate the autophosphorylation of IRE1 to impair the signaling for GRP78 induction.

To confirm that vulnerability to ER stress in cells expressing PS1 mutants is based on attenuated induction of GRP78 mRNA by inhibition of IRE1 function, we infected SK-N-SH cells bearing mutant PS1 with recombinant GRP78 using Semliki forest virus (SFV-GRP78 fusion). Sensitivity to ER stress caused by treatment with tunicamycin in neuroblastoma lines expressing PS1 mutants was reversed by infection with recombinant GRP78. These results indicated that the expression of GRP78 protects against neuronal death caused by ER stress, and that the reduction in GRP78 gene expression causes vulnerability to ER stress in cells expressing PS1 mutants.

We examined whether GRP78 protein levels were changed in the brains of AD patients. Western blotting with anti-KDEL antibody for both GRP78 and

GRP94, showed that levels of these proteins were significantly decreased in AD patients compared with age-matched controls. The levels in the brains of patients with FAD-linked PS1 mutations were decreased to a greater extent. Immunohistochemical analysis revealed that KDEL immunoreactivity was reduced in hippocampal and cortical neurons of both SAD and FAD patients compared with controls. Thus, amounts of ER-resident molecular chaperones such as GRP78 and GRP94 were downregulated in brains of AD patients, and these findings are likely to be associated with the pathology of AD.

FAD Mutants of PS1 Attenuate UPR Initiated by PERK

Activation of PERK during ER stress correlates with the autophosphorylation of those cytoplasmic kinase domains. At first, to examine the effects of PS1 mutants on the activation of PERK, Western blotting was perfomed using lysate from N2a cells expressing wild-type PS1 or mutants PS1 stimulated by 1 μM thapsigargin for 5–60 min. Phosphorylation of PERK retards their mobility on SDS polyacrylamide gels, and serves as a convenient marker for their activation status. In N2a cells expressing mock or wild-type PS1, the bands of PERK were completely shifted within 15 min after treatment with thapsigargin. In contrast, in cells expressing PS1 mutants, the mobility shifts were not observed 15–30 min after the treatment (fig. 3b). These results indicate that FAD-linked PS1 mutants disturb the autophosphorylation of PERK during ER stress.

Because disturbed function of PERK is known to cause the downregulation of phosphorylation of eIF2α, we examined the levels of phosphorylated eIF2α after ER stress in N2a cells expressing PS1 mutations. Western blotting using anti-phosphorylated eIF2α antibody showed that phosphorylation of eIF2α was inhibited in cells expressing mutant PS1. Thus, activation of PERK and resultant phosphorylation of eIF2α were disturbed by the expression of PS1 mutation.

FAD Mutants of PS1 Attenuate UPR Initiated by ATF6

Under ER stress, the translocation of ATF6 fragments into the nucleus was investigated in cell with wild-type PS1 or mutant PS1 by immunohistochemistry. In wild-type fibroblasts, the N-terminal fragments of ATF6 were quickly translocated into nucleus by ER stress. In contrast, in homozygous PS1 knock-in fibroblasts, the translocation was delayed.

The disturbance of the ATF6 pathway in homozygous PS1 knock-in fibroblasts was also confirmed by Western blotting. In wild-type fibroblasts, the appearance of a 50-kD ATF6 fragment was detected quickly by ER stress, whereas it was delayed in homozygous fibroblasts. Therefore the FAD-linked PS1 mutation also attenuates the signaling pathway though ATF6.

Discussion

To date, three ER-resident transmembrane molecules, IRE1, PERK, and ATF6, which sense the accumulation of unfolded proteins to induce UPR, have been identified. Our study showed that FAD-linked mutant PS1 disturbs all these molecules. However, it remains unclear why mutant PS1 inhibits the activation of ER stress transducers. As mentioned above, it is proposed that the activation of the signaling-mediated ER stress transducers could be triggered by dissociation of GRP78/BiP from stress transducers. The dissociation causes the oligomerization of stress transducers to induce its autophosphorylation and the resultant activation of downstream signaling. If PS1 mutants form malfolded structures, GRP78/BiP may constitutively bind to PS1 molecules to promote its folding. The complex formation of PS1, PERK (or IRE1) and GRP78/BiP may inhibit the dissociation of GRP78/BiP from PERK or IRE1 under ER stress conditions.

PS1 mutations alter the processing of APP and cause increased production of the more amyloidogenic Aβ peptide, Aβ(1–42). The ER and ER–Golgi intermediate compartment may be important sites for generation of Aβ(1–42) [35]. Interestingly, under ER stress, unfolded proteins are retrieved to the ER by the retrograde transport to prevent them from moving to the cell surface [36]. In view of the alteration in UPR systems in cells expressing PS1 mutants, it is possible that increased levels of Aβ(1–42) in these cells might be the result of retention of the unfolded APP in the ER due to the impaired protein-folding system. It was reported that amounts of secreted Aβ have been shown to be reduced by transfection with GRP78 [37]. This result is consistent with our speculation that Aβ production may be associated with the UPR system.

In summary, our results indicate a new mechanism by which PS1 mutations may affect the sensing of ER stress. Experimental manipulation of IRE1, PERK, or eIF2α phosphorylation or GRP78 expression might allow the development of therapeutic strategies for FAD.

References

1 Borchelt DR, et al: Familial Alzheimer's disease-linked presenilin 1 variants elevate Aβ1–42/1–40 ratio in vitro and in vivo. Neuron 1996;17:1005–1013.

2 Duff K, et al: Increased amyloid-β42(43) in brains of mice expressing mutant presenilin 1. Nature 1996;383:710–713.

3 Guo Q, et al: Alzheimer's presenilin mutation sensitizes neural cells to apoptosis induced by trophic factor withdrawal and amyloid β-peptide: Involvement of calcium and oxyradicals. J Neurosci 1997;17:4212–4222.
4 Guo Q, et al: Increased vulnerability of hippocampal neurons to excitotoxic necrosis in presenilin 1 mutant knock-in mice. Nat Med 1999;5:101–106.
5 Walter J, et al: The Alzheimer's disease-associated presenilins are differentially phosphorylated proteins located predominantly with the endoplasmic reticulum. Mol Med 1996;2:673–691.
6 Culvenor JG, et al: Alzheimer's disease-associated presenilin 1 in neuronal cells: Evidence for localization to the endoplasmic reticulum-Golgi intermediate compartment. J Neurosci Res 1997; 49:719–731.
7 Annaert WG, et al: Presenilin 1 controls γ-secretase processing of amyloid precursor protein in pre-Golgi compartments of hippocampal neurons. J Cell Biol 1999;147:277–294.
8 Sidrauski C, et al: The unfolded protein response: An intracellular signalling pathway with many surprising features. Trends Cell Biol 1998;8:245–249.
9 Shi Y, et al: Identification and characterization of pancreatic eukaryotic initiation factor 2α-subunit kinase, PEK, involved in translational control. Mol Cell Biol 1998;18:7499–7509.
10 Harding HP, et al: Protein translation and folding are coupled by an endoplasmic-reticulum-resident kinase. Nature 1999;397:271–274.
11 Bonifacino JS, Weissman AM: Ubiquitin and the control of protein fate in the secretory and endocytic pathway. Annu Rev Cell Dev Biol 1998;14:19–57.
12 Liu H, et al: Endoplasmic reticulum chaperones GRP78 and calreticulin prevent oxidative stress, Ca^{2+} disturbances, and cell death in renal epithelial cells. J Biol Chem 1997;272:21751–21759.
13 Harding HP, et al: Perk is essential for translational regulation and cell survival during the unfolded protein response. Mol Cell 2000;5:897–904.
14 Imai Y, et al: Parkin suppresses unfolded protein stress-induced cell death through its E3 ubiquitin-protein ligase activity. J Biol Chem 2000;275:35661–35664.
15 Calfon M, et al: IRE1 couples endoplasmic reticulum load to secretory capacity by processing the XBP-1 mRNA. Nature 2002;415:92–96.
16 Yoshida H, et al: XBP1 mRNA is induced by ATF6 and spliced by IRE1 in response to ER stress to produce a highly active transcription factor. Cell 2001;107:881–891.
17 Tirasophon W, et al: A stress response pathway from the endoplasmic reticulum to the nucleus requires a novel bifunctional protein kinase/endoribonuclease (Ire1p) in mammalian cells. Genes Dev 1998;12:1812–1824.
18 Wang X-Z, et al: Cloning of mammalian Ire1 reveals diversity in the ER stress responses. EMBO J 1998;17:5708–5717.
19 Harding H, et al: Translation and protein folding are coupled by an endoplasmic reticulum resident kinase. Nature 1999;397:271–274.
20 Shi Y, et al: Identification and characterization of pancreatic eukaryotic initiation factor 2 alpha-subunit kinase, PEK, involved in translational control. Mol Cell Biol 1998;18:7499–7509.
21 Bertolotti A, et al: Dynamic interaction of GRP78/BiP and the ER stress transducers in the unfolded protein response. Nat Cell Biol 2000;2:326–332.
22 Harding H et al: Regulated translation initiation controls stress-induced gene expression in mammalian cells. Mol Cell 2000;6:1099–1108.
23 Novoa I, et al: Feedback inhibition of the unfolded protein response by GADD34-mediated dephosphorylation of eIF2. J Cell Biol 2001;153:1011–1022.
24 Haze K, et al: Mammalian transcription factor ATF6 is synthesized as a transmembrane protein and activated by proteolysis in response to endoplasmic reticulum stress. Mol Biol Cell 1999;10: 3787–3799.
25 Ye J, et al: ER stress induces cleavage of membrane-bound ATF6 by the same proteases that process SREBPs. Mol Cell 2000;6:1355–1364.
26 Nohturfft A, et al: Regulated step in cholesterol feedback localized to budding of SCAP from ER membranes. Cell 2000;102:315–323.
27 Shen J, et al: ER stress regulation of ATF6 localization by dissociation of GRP78/BiP binding and unmasking of Golgi localization signals. Dev Cell 2002;3:99–111.

28 Nakagawa T, et al: Caspase-12 mediates endoplasmic-reticulum-specific apoptosis and cytotoxicity by amyloid-beta. Nature 2000;403:98–103.
29 Macejak DG, Sarnow P: Translational regulation of the immunoglobulin heavy-chain binding protein mRNA. Enzyme 1990;44:310–319.
30 Nakagawa T, Yuan J: Cross-talk between two cysteine protease families. Activation of caspase-12 by calpain in apoptosis. J Cell Biol 2000;150:887–894.
31 Nishitoh H, et al: ASK1 is essential for endoplasmic reticulum stress-induced neuronal cell death triggered by expanded polyglutamine repeats. Genes Dev 2002;16:1345–1355.
32 Yoneda T, et al: Activation of caspase-12, an endoplastic reticulum (ER) resident caspase, through tumor necrosis factor receptor-associated factor 2-dependent mechanism in response to the ER stress. J Biol Chem 2001;276:13935–13940.
33 Friedman AD: GADD153/CHOP, a DNA damage-inducible protein, reduced CAAT/enhancer binding protein activities and increased apoptosis in 32D c13 myeloid cells. Cancer Res 1996;56: 3250–3256.
34 Nakano Y, et al: Accumulation of murine amyloidβ42 in a gene-dosage dependent manner in PS1 'knock-in' mice. Eur J Neurosci 1999;11:2577–2581.
35 Cook DG, et al: Alzheimer's Aβ(1–42) is generated in the endoplasmic reticulum/intermediate compartment of NT2N cells. Nat Med 1997;3:1021–1023.
36 Hammond C, Helenius A: Quality control in the secretory pathway: Retention of a misfolded viral membrane glycoprotein involves cycling between the ER, intermediate compartment, and Golgi apparatus. J Cell Biol 1994;126:41–52.
37 Yang Y, et al: The chaperone BiP/GRP78 binds to amyloid precursor protein and decreases A40 and A42 secretion. J Biol Chem 1998;273:25552–25555.

Takashi Kudo
Department of Psychiatry and Behavioral Science
Osaka University Graduate School of Medicine
2–2, Yamadaoka, Suita, Osaka 565–0871 (Japan)
Tel. +81 6 6879 3053, Fax +81 6 6879 3059, E-Mail kudo@psy.med.osaka-u.ac.jp

Takeda M, Tanaka T, Cacabelos R (eds): Molecular Neurobiology of Alzheimer Disease and Related Disorders. Basel, Karger, 2004, pp 134–156

Advances in the Development of Biomarkers for Alzheimer's Disease – From CSF Total Tau and Amyloid-β(1–42) Proteins to Phosphorylated Tau and Amyloid-β-Antibodies

Harald Hampel, Stefan Teipel, Frank Faltraco, Sylvia Brettschneider, Alexander Goernitz, Katharina Buerger, Hans-Juergen Moeller

Alzheimer Memorial Center and Geriatric Psychiatry Branch, Department of Psychiatry, Ludwig-Maximilian University, Munich, Germany

Alzheimer's disease (AD) is the most common form of dementia in the elderly. Due to the increasing proportion of elderly people in the societies of the industrialized countries, the prevalence and incidence of AD will rise considerably within the next few decades, with an estimated 20–30 million subjects in the US suffering from AD in 2030 [1–3]. Therefore, AD has an ever-growing epidemiological, sociodemographic, and economic relevance worldwide. In clinical practice, the diagnosis of AD is still largely based on the exclusion of secondary causes and other dementias [4]. Based on a growing body of evidence on the neuropathological, morphological and functional phenotype of AD and hypothesized pathophysiological pathways, a large number of positive diagnostic criteria have been evaluated. Candidate biological markers of disease can now be applied in three important clinical areas: (1) to achieve early clinical or even preclinical diagnosis of AD so that treatment can be started earlier in the course of the disease; (2) to differentiate the disease from other relevant forms of dementia and from various psychiatric disorders with similar psychopathology (e.g. geriatric major depression), and (3) to map intraindividual disease progression and evaluate potential effects of therapy.

Criteria for a useful biomarker have been proposed by an international consensus group on molecular and biochemical markers of AD in 1998 [5]. According to these guidelines, a biomarker for AD should detect a fundamental feature of neuropathology and be validated in neuropathologically confirmed cases. Its sensitivity should exceed 80% and its specificity should be higher than 80%. A biomarker should be reliable, reproducible, noninvasive, simple to perform and inexpensive. Since cerebrospinal fluid (CSF) is in direct contact with the extracellular space of the central nervous system, biochemical changes in the brain could potentially be reflected in CSF. Therefore, CSF constitutes a potential source for clinically useful biomarker candidates of AD.

Based on pathophysiological models of AD, three main types of CSF candidate markers will be reviewed: (1) Based on the neurofibrillary pathology in AD, total tau protein (t-tau) has been studied in CSF as a marker of neuronal degeneration. In addition, assays have been developed to detect characteristic hyperphosphorylation sites of tau (p-tau). (2) Following the amyloid hypothesis of AD, CSF amyloid-β(1–42) [Aβ(1–42)] has been investigated. (3) Following evidence for clearance of amyloid plaques through passive immunization against human Aβ in a transgenic mouse model of amyloid pathology, antibodies against Aβ have been studied as potential biomarkers of AD.

Several studies have investigated CSF inflammatory markers, immunological mediators, neurotrophins, metalloproteinases or isoprostenes. These studies yielded inconclusive results, were based on small samples and/or their results were disappointing as these markers seem to mirror less specific pathophysiological aspects of AD than tau- and amyloid-related biomarkers. These efforts are not reviewed here.

One has to bear in mind that none of the aforementioned CSF markers fulfils all of the proposed guidelines for a useful biomarker of AD [5]. Some come closer than others to these criteria. Additionally, the evaluation of any new diagnostic marker should follow three subsequent steps. First, the technical feasibility of the new marker has to be established, including the availability of a validated assay with high precision and reliability of measurement and well-described reagents and standards. A large range of potential markers has successfully passed this first step. Second, the marker has to be evaluated under controlled clinical conditions on selected samples of diseased and comparison groups. The goal of these studies is to evaluate the clinical feasibility of a marker and to provide a first assessment of sensitivity and specificity. Only few potential markers have passed this step so far, among them the tau- and amyloid-related CSF proteins. Finally, a new marker has to be studied in a population-based sample to allow assessment of true diagnostic properties, providing evidence for the clinical usefulness of this marker. This step also serves as a basis for cost-effectiveness analyses, which are important, because every new marker has to be evaluated in the context of the already available set of diagnostic and therapeutic options and the socioeconomic resources of the health system. No marker so far has passed this last step. At present, several multicenter studies are being performed with the scope of evaluating new biomarkers in a population-based design, one important example is the Working Group on Biological Measures as part of the National Institute on Aging Alzheimer's Imaging Initiative or the German Competency Network on Dementias. These initiatives focus on CSF tau proteins and Aβ as core markers, since these markers are most advanced in the process of evaluation and are associated with key mechanisms of pathology implicated in AD. Therefore, these markers will also be the focus of this chapter, which concentrates on recent advances in measuring t-tau, tau protein phosphorylated at specific epitopes (p-tau), Aβ and Aβ antibodies.

Tau Protein in CSF

One of the major neuropathological hallmarks of AD are neurofibrillary tangles composed of paired helical filaments (PHF). The principal protein subunit of PHF is abnormally hyperphosphorylated tau (p-tau) [7]. Physiologically, tau protein is located in neuronal axons, in components of the cytoskeleton and in the intracellular transport systems. t-tau and truncated forms of monomeric and p-tau can be traced in the CSF. Using antibodies that detect all isoforms of tau proteins independent of phosphorylation, or specific phosphorylation sites, ELISAs have been developed to measure t-tau and CSF p-tau concentrations [8–10].

CSF t-Tau Protein in the Differentiation between AD and Normal Aging

t-Tau protein, a general marker of neuronal destruction, has been intensely studied in more than 2,000 AD patients and 1,000 age-matched elderly controls over the last 5–10 years [8–40]. The most consistent finding is a statistically significant increase in CSF t-tau protein in AD. The mean level of CSF t-tau protein concentration is about 3 times higher in AD compared to elderly controls. Across the reviewed studies, sensitivity and specificity levels varied due to the different control groups and statistical methods used. Specificity levels were between 65 and 86% and sensitivity between 40 and 86% [41]. In several studies, a significant elevation was also found in patients with early dementia [21, 38, 42]. In mild dementia, the potential of CSF t-tau protein to discriminate between AD and normal aging is high, with an average 75% sensitivity and 85% specificity.

An age-associated increase in t-tau protein has been shown in nondemented subjects [43, 44]. Therefore, the effect of age should be considered when t-tau protein levels are employed for diagnosis. Age-dependent reference values for t-tau protein have already been established: <300 pg/ml, for subjects between 21 and 50 years of age $<$450 pg/ml between 51 and 70 years and $<$500 pg/ml between 70 and 93 years [45].

CSF t-Tau Protein in the Differentiation between AD and Geriatric Major Depression

Geriatric major depression is an important psychiatric differential diagnosis of AD, as psychopathological symptoms considerably overlap and often only a follow-up assessment will allow clear clinical differentiation between both underlying entities. Subgrouping according to age of a sample of AD patients, healthy controls and patients with major depression resulted in a correct classification rate of 94.5% in the 'young old' subjects ($<$70 years) compared to only 68.4% in the 'old old' ($\geq$70 years). This report supports the notion that elevated CSF t-tau protein, particularly in subjects $<$70 years of age, is highly indicative of a neurodegenerative process [25].

CSF t-Tau Protein in the Differential Diagnosis of AD and Other Neurodegenerative Diseases

The potential of CSF t-tau protein, however, is limited in its ability to discriminate AD from other relevant dementias. With a sensitivity of 81%, CSF t-tau protein reached only 57% specificity in distinguishing AD from other dementias [28, 46].

Several groups reported an increase in t-tau protein in vascular dementia, as well as in frontotemporal dementia and Lewy body dementia [8, 14, 17, 18, 22, 26, 30, 33, 47–54]. A moderate increase has been reported in semantic dementia [55]. Other studies, however, found normal levels compared to controls in these entities [8, 10, 22, 28, 30, 34, 36, 46, 51, 53, 56].

The value of t-tau protein in differential diagnosis has been uncertain because the studies either were incomplete or lacked control and comparison groups. In addition, t-tau protein has low specificity in the differentiation of AD from other primary dementias. Therefore, t-tau protein has not been suggested as a marker for the differential diagnosis of AD. t-Tau protein reflects unspecific processes of axonal damage and neuronal degeneration. This notion is further supported by the increase in CSF t-tau protein in disorders with extensive and/or rapid neuronal degeneration such as Creutzfeldt-Jakob disease [57, 58]. A highly significant increase of 580% was documented in Creutzfeldt-Jakob disease compared to AD patients. At a cut-off level of 2,130 pg/ml t-tau protein yielded a sensitivity of 93% and a specificity of 100% between AD and Creutzfeldt-Jakob disease [59]. An elevation of CSF t-tau protein, correlating with clinical severity, has been shown in normal-pressure hydrocephalus [60]. Moreover, a marked transient increase in CSF t-tau protein has been demonstrated after acute stroke. The transient increase in CSF t-tau protein correlated to the infarct size measured by cranial computed tomography [61]. Elevated levels of CSF t-tau protein in patients with diffuse axonal damage after traumatic brain injury have been found to decrease with clinical improvement [62]. In contrast, CSF t-tau protein concentrations are only slightly elevated [8, 17, 30, 34, 63, 64] or normal [31] in neurological disorders that involve a more locally restricted neuron loss, such as alcoholic dementia, Parkinson's disease, progressive supranuclear palsy, and corticobasal degeneration.

Value of CSF t-Tau Protein for the Prediction of AD in Mild Cognitive Impairment

Mild cognitive impairment (MCI) is a major risk factor for AD. Ten to fifteen percent of patients with MCI have been reported to convert to AD within 1 year [65]. In patients suffering from MCI who converted to AD during follow-up, elevated t-tau protein levels at baseline were found in relatively small samples [15, 19]. Memory-impaired subjects who later progressed to manifest AD could be discriminated by high CSF t-tau protein from those who did not progress, with 90% sensitivity and 100% specificity [15]. Longitudinally, elevated CSF levels of t-tau protein in MCI subjects were found and remained elevated after conversion to clinical AD [66]. Another study showed that 88% of patients with MCI had elevated t-tau protein concentrations and/or low CSF Aβ(1–42) levels at baseline [67]. Thus, elevated CSF t-tau protein in MCI may have the potential to predict AD.

CSF t-Tau Protein during AD Progression

Cross-sectional studies correlating CSF t-tau protein concentrations with the cognitive status in AD were not conclusive. Some studies showed a correlation between elevation of t-tau protein and cognitive decline [20, 22, 68], whereas others found no systematic effect [10, 16, 18, 28, 49, 69]. Longitudinal studies of t-tau protein in mild to moderately demented AD patients showed no statistically significant correlation with progression during follow-up [18, 22, 51, 70, 71]. CSF t-tau protein remained stable at elevated levels for up to 2 years in mild to moderate AD. Initial and follow-up levels of t-tau protein correlated strongly, suggesting a stable rate of neurodegeneration during this time period. It was hypothesized that CSF t-tau protein would decrease over time if treatment of AD achieved disease modification and neuroprotection [72].

βA-Protein in CSF

According to the amyloid hypothesis the increased production of Aβ(1–42) leads to accumulation and oligomerization of Aβ(1–42) and consecutively to deposition as diffuse plaques. This process is followed by activation of inflammation and progressive injury of synapsis and neurons. Altered neuronal ionic homeostasis together with oxidative stress leads to alternation of kinase and phosphatase activity followed by formation of tangles [74].

Extracellular senile plaques consisting of Aβ are a histopathological hallmark for the diagnosis of AD [73]. They are the source of a pathogenic protein with 42 amino acids [Aβ(1–42)] [74]. Several groups have developed and studied different bioassays specifically designed for Aβ(1–42) protein [for a review, see ref. 15].

CSF Aβ(1–42) Protein in the Differentiation between Normal Aging and AD

To date, at least 900 patients with clinical AD and 500 healthy individuals have been enrolled in independent research studies. The most consistent finding is a marked decrease in Aβ(1–42) protein in AD (≈50% of control levels). Aβ(1–42) protein alone showed a sensitivity that varied from 78 to 100% and a specificity ranging between 47 and 81% in distinguishing AD from elderly controls. There is a pronounced overlap between the different groups [6, 20, 23, 33, 34, 36, 39, 67, 68, 75–79].

One study, however, reported a significant increase in Aβ(1–42) protein in AD [6], which may well be due to methodological bias (e.g., assay specificity for mono- versus oligomers) or differences in patient and control sampling. In this individual study, increased CSF Aβ(1–42) protein was also found in MD [6], while two other studies found normal levels in these patients [34, 77].

In contrast to t-tau protein, CSF levels for Aβ(1–42) protein seem to be unrelated to age. Based on recent data, a cut-off level of >500 pg/ml has been suggested to discriminate AD best from normal aging [45].

CSF Aβ(1–42) Protein in the Differential Diagnosis between AD and Other Neurodegenerative Diseases

Data on the potential of CSF Aβ(1–42) protein to distinguish AD from other dementias and neurological disorders were obtained in a number of independent studies. Compared to neurological controls, a slight decrease was found in non-Alzheimer's dementia [39]. Normal levels [36] or decreased levels [75] were found in Parkinson's disease. In Lewy body dementia, a disorder also characterized by the presence of senile plaques, low levels of Aβ(1–42) protein have been detected as well. The range of Aβ(1–42) protein levels overlaps that in AD patients [19, 33, 46, 52]. In addition, low CSF Aβ(1–42) protein is found in a relatively large percentage of patients with frontotemporal dementia and vascular dementia [28, 34]. In summary, CSF Aβ(1–42) protein does not seem to significantly support the differential diagnosis of AD.

The reduction in CSF Aβ(1–42) protein in AD has been hypothesized to indirectly reflect amyloid deposition in senile plaques, resulting in lower CSF levels in AD. A marked reduction in CSF Aβ(1–42) protein, however, is also found in Creutzfeldt-Jacob disease, even in cases without Aβ-positive plaques [59, 78]. These findings question the notion of a direct relationship between SP formation and CSF Aβ(1–42) protein levels.

Value of CSF Aβ(1–42) Protein for the Prediction of AD in MCI

It has been hypothesized that a decrease in CSF Aβ(1–42) protein might indicate an early stage of AD and be detectable before clinical symptoms of dementia become overt.

One study found a significant decrease of CSF Aβ(1–42) protein in MCI subjects compared to controls [19]. In another study on MCI patients who eventually developed AD, however, Aβ(1–42) protein levels did not differ significantly from age-matched normal controls [66].

We found Aβ(1–42) protein to be an early indicator of AD in MCI subjects taking potential confounding factors into account such as age, severity of cognitive decline, time of observation, apolipoprotein E-ε4 (APOE-ε4) carrier status, and gender [80].

CSF Aβ(1–42) Protein during AD Progression

Studies correlating CSF Aβ(1–42) protein concentrations with cognitive performance in AD were partly contradictory. Cross-sectionally, the concentration of Aβ(1–42) protein and cognitive measures were either inversely correlated [20, 81, 82] or no significant correlation was found [28, 39, 77, 83]. In a longitudinal study, a decrease in CSF Aβ(1–42) protein was found in a 3-year follow-up [70]. A highly significant correlation between low CSF concentrations at baseline and 1-year follow-up was found. There was no correlation between CSF levels and duration or severity [77]. The potential value of Aβ(1–42) protein in the course of AD progression should be further evaluated.

Combination of CSF Aβ(1–42) and t-Tau Proteins in the Differential Diagnosis of AD

Since the diagnostic accuracy of CSF t-tau protein and Aβ(1–42) protein alone is not sufficient to differentiate AD from other clinically relevant dementias, a combination of both markers has been suggested.

In a large multicenter sample, 236 subjects were divided into groups of 93 AD patients, 33 patients with non-AD dementias, 56 patients with other neurological disorders, and 54 healthy controls. Combination of both markers provided a sensitivity of 71% and a specificity of 83% for AD versus all other demented and nondemented groups [20].

In another study, classification tree analysis comparing AD and control subjects yielded 90% sensitivity and 80% specificity [39]. This analysis classified 26 out of 74 members of a group with various neurological disorders including

56 non-AD dementias (such as frontotemporal dementia, vascular dementia, Lewy body dementia and Parkinson disease) as having AD. This finding suggests that the AD Aβ(1–42) t-tau protein CSF profile is generally not specific for AD and may occur in a variety of conditions that cause brain damage, or may be present in dementia but clinically not detectable.

In a community-based sample of 241 subjects, specificity to differentiate between AD and Lewy body dementia was 67%, mainly due to low Aβ(1–42) protein in with Lewy body dementia patients, and only 48% in AD versus vascular dementia because of high CSF-t-tau protein levels in the group with vascular dementia [67].

Combined t-tau and Aβ(1–42) protein yielded 85% sensitivity and 85% specificity in discriminating 150 AD from 100 control subjects and 84 medical disorders, including inflammatory, neoplastic and other neurological disorders in a large multicenter study [28]. At the same level of sensitivity, however, the combined assays yielded 58% specificity in distinguishing AD from 79 non-AD dementias, including vascular dementia, frontotemporal dementia, normal-pressure hydrocephalus and other neurodegenerative diseases with dementia.

These studies indicate that the combination of both markers may not sufficiently improve differential diagnosis between AD and other clinically relevant dementias.

p-Tau Protein in CSF

To increase the diagnostic accuracy of CSF biomarkers, immunoassays to specifically detect p-tau protein in CSF have recently been developed and have become available for research within the last 2 years. These assays use monoclonal antibodies specific for phosphorylated epitopes of tau protein. In first clinical studies, CSF p-tau protein assessments have been described that focussed on different phosphoepitopes. Tau protein phosphorylated at serine 199 [p-tau(199)] [29]; threonine 231 [p-tau(231)] [84]; threonine 231 and serine 235 [p-tau(231–235)] [29]; threonine 181 [p-tau(181)] [85, 86], and serine 396 and serine 404 [p-tau(396/404)] [87].

The first studies applying these new biomarkers have recently been published. A comparative study for the different epitopes of tau protein biomarkers is now under way. For clarity of results, the data of the different p-tau protein epitopes are reviewed separately in the following paragraphs.

CSF p-Tau(231)

Immunohistochemical studies indicate that p-tau(231) is specific for AD and appears early in the pathogenesis, even before PHF formation [88, 89].

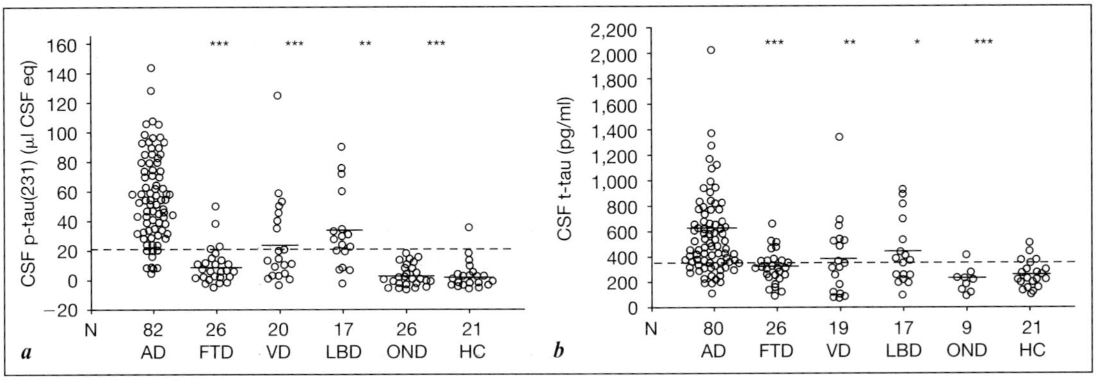

Fig. 1. Levels of p-tau(231) (***a***) and t-tau (***b***) in the CSF of patients and control subjects. Dashed lines represent the cutoff levels yielding the best sensitivity and specificity for discriminating patients with AD from those with non-AD and healthy control subjects (by means of ROC curves). Horizontal solid lines indicate means. FTD = Frontotemporal dementia; VD = vascular dementia; LBD = Lewy body dementia; OND = other neurological disorder. *p < 0.05; **p < 0.005; ***p < 0.001. All comparisons are with the AD group.

Therefore, the value of p-tau(231) as a specific CSF AD biomarker is currently being studied.

CSF p-Tau(231) in the Differential Diagnosis of AD

In a first study, we found that p-tau(231) discriminates with a sensitivity of 85% and a specificity of 97% (overall accuracy of 91%) between AD patients and patients with other neurological disorders [84]. In an independent sample of 192 subjects, we showed that p-tau(231) is superior to t-tau protein in the differential diagnosis of AD. CSF levels of p-tau(231) discriminated with a sensitivity of 90.2% and a specificity of 80% between AD and all other non-AD subjects. In particular, at a specificity level of 92.3% for both markers, p-tau(231) raised sensitivity levels of discrimination between AD and FTD, in comparison to t-tau protein, from 57.7 to 90.2% [90] (fig. 1, table 1).

Potential of CSF p-Tau(231) to Predict AD in MCI

Another promising value of p-tau(231) may be its ability to predict a cognitive decline in MCI patients. In a longitudinal study, 77 MCI patients showed elevated levels for p-tau(231) in comparison to healthy controls at baseline [91]. High CSF p-tau(231) levels at baseline significantly correlated with subsequent cognitive decline (fig. 2). In addition to p-tau(231), old age and apolipoprotein E-ε4 carrier status independently predicted cognitive decline in this large sample of 77 MCI subjects. The entire model explained 27% of variance in rates of

Table 1. Sensitivity, specificity and correctly allocated cases using ROC analysis for CSF p-tau(231) and t-tau given in relative numbers [90]

	CSF p-tau(231)			CSF t-tau		
	sensitivity	specificity	correctly allocated cases	sensitivity	specificity	correctly allocated cases
AD vs. non-AD	90.2	80.0	84.4	81.3	68.5	74.4
AD vs. OND	92.7	100.0	94.4	90.0	88.9	89.9
AD vs. HC	100.0	90.5	98.1	81.3	90.5	83.2
AD vs. other dementias	92.7	65.1	80.7	81.2	58.1	71.1
AD vs. FTD	90.2	92.3	90.7	57.5	92.3	66.0
AD vs. VD	92.7	65.0	87.3	93.8	47.4	84.8
AD vs. LBD	76.8	76.5	76.8	87.5	52.9	81.4

cognitive decline. This study suggests that high p-tau(231) may be a predictor of progressive cognitive decline in subjects with MCI.

CSF p-Tau(231) during the Course of AD Progression

Investigating intraindividual changes over time in 17 AD patients, we showed that CSF p-tau(231) declined during the course of AD (6-year follow-up; up to 7 serial CSF measurements) [92]. Concentrations of CSF p-tau(231), but not t-tau protein decreased over time in AD patients. The rate of change in p-tau(231) concentrations was inversely correlated with dementia severity at baseline. The decrease in CSF p-tau(231) with AD progression might reflect increasing sequestration of p-tau protein into the tangle, suggesting that with disease progression, p-tau(231) becomes more insoluble rather than penetrating into the CSF, whereas the solubility of t-tau protein remains unaffected (fig. 3).

There is inconsistency in longitudinal data, which report either changing tau levels in AD or no changes. However, brain-derived proteins, including tau, predominantly exhibit higher ventricular than lumbar levels [93]. This led to the hypothesis, that adjusting for ventricular enlargement of AD would correct for the dilution of tau and improve the detection of changes. In a recent 1-year longitudinal study [94], ventricular volumes and lumbar CSF p-tau(231) levels were determined in 7 MCI and 1 case with very mild AD and compared to 10 healthy controls. Ventricle volumes in MCI were approximately 40% greater at baseline and follow-up compared to controls. With ventricle volume correction, an increase in p-tau(231) levels in the MCI group relative to controls was demonstrated.

In summary, these data indicate that CSF p-tau(231) might be a valuable tool to improve the early and differential diagnosis of AD and for monitoring disease

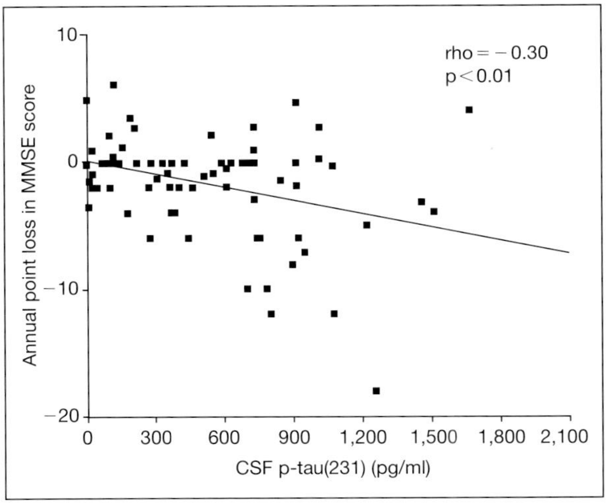

Fig. 2. Correlation between CSF p-tau(231) at baseline and rate of change in Mini Mental State Examination (MMSE) score in 77 subjects with MCI investigated longitudinally; rho indicates Spearman's rho.

progression. Further studies are warranted to replicate these findings in independent and autopsy-confirmed samples. These efforts are currently under way.

CSF p-Tau(231–235) in MCI

One study focused on p-tau(231–235) in MCI subjects who converted to AD compared to individuals complaining of their memory without cognitive impairment. Results showed significantly higher t-tau protein and detectable p-tau(231–235) protein levels in the MCI group. p-Tau(231–235) yielded a specificity of 100% and a sensitivity of 65% in differentiating MCI subjects who eventually developed AD compared to memory complainers [95].

CSF p-Tau(181)

The discriminative power of CSF p-tau(181) has been investigated in a number of studies with different diagnostic entities. Results for p-tau(181) showed a significant increase in AD compared to frontotemporal dementia and controls [86]. Focusing on the differentiation between AD and Lewy body dementia, specificity at a given sensitivity was improved by p-tau(181) compared with t-tau protein [46, 85]. Comparison of receiver operator characteristics (ROC) curves led to a correct classification for cases with AD and Lewy body dementia of more than 80%.

In a study on 101 subjects comparing p-tau(181) and t-tau protein in different diagnostic subgroups, p-tau(181) was increased in patients with probable and

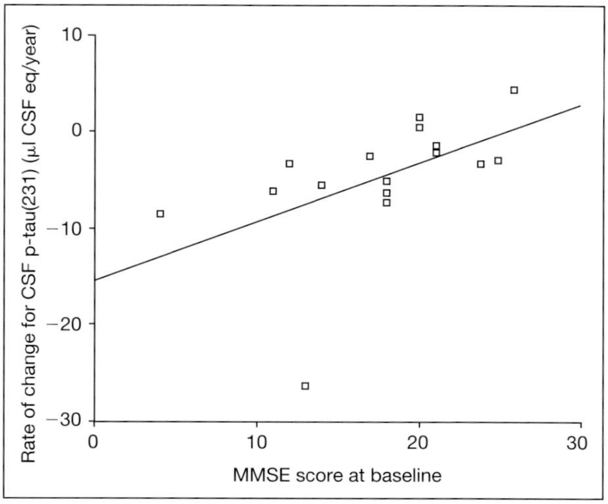

Fig. 3. Correlation between Mini Mental State Examination (MMSE) score at baseline and rate of change in CSF p-tau(231) of AD patients investigated longitudinally. Mean rate of cognitive decline (MMSE score) was −3.63 points per year (SD ± 3.9).

possible AD compared with vascular dementia and dementia in Parkinson disease [44]. Compared with frontotemporal dementia, Parkinson disease and vascular dementia and normal aging, both p-tau(181) and t-tau proteins were increased in probable AD. In possible AD, p-tau(181) was increased compared to frontotemporal dementia and vascular dementia. Data on a group of 51 AD patients (25 probable, 18 possible and 8 incipient AD cases) compared to 16 probable VD cases and 10 healthy controls have recently become available [96]. All AD cases were drug naive. Therefore, differences between groups cannot be attributed to confounding effects of medication. Patients with AD and vascular dementia, as well as healthy controls were distinguished using p-tau(181) whereas vascular dementia compared to healthy controls showed no statistically significant differences in concentration. Among the whole group of AD patients and controls, p-tau(181) demonstrated 71% sensitivity and 94% specificity compared to t-tau which had a sensitivity of 63% and 100% specificity. Taken together, diagnostic accuracy was better for p-tau(181) (78%) than for t-tau (71%). In summary, p-tau(181) may be a valuable biomarker, especially in the differential diagnosis between AD and Lewy body dementia and vascular dementia.

The notion that p-tau proteins reflect tau pathophysiology more distinctively is also supported by a marked transient increase in CSF t-tau protein after acute stroke, while CSF p-tau(181) did not change [61].

CSF p-Tau(199) in the Differential Diagnosis of AD and Other Neurodegenerative Diseases

In one study applying p-tau(199), this biomarker was superior to t-tau protein in separating AD from a patient group of non-AD subjects [29].

In a large multicenter sample of 570 subjects [53], p-tau(199) protein levels were elevated in the AD group, independently of age, gender, cognitive status and APOE-ε4 carrier status. For the AD group versus the combined groups of demented and nondemented subjects in this study, ROC analysis showed 85% sensitivity and 85% specificity for p-tau(199) [29]. Group separation between AD and distinct differential diagnoses was not examined.

CSF p-Tau(396/404)

A recently developed ultrasensitive bienzyme-substrate-recycle enzyme-linked immunosorbent assay for abnormally hyperphosphorylated tau protein in CSF recognizes attomolar amounts of tau protein. The assay is about 400 times (for Ser 396) and about 1,300 times (for Ser 404) more sensitive than conventional ELISAs in determining the hyperphosphorylated tau protein and total tau protein, respectively [87].

In CSF of 52 AD patients, 56 normal controls, 46 patients with vascular dementia, and 37 non-AD patients significantly elevated levels of p-tau(396/404) were only found in AD. Using the ratio of p-tau protein to t-tau protein of ≥0.33 as a cutoff for AD diagnosis, the clinical diagnosis could be confirmed in 96% of the clinically diagnosed patients. The results of this study suggest that p-tau (396/404) is a promising marker, especially in the differential diagnosis between AD and vascular dementia.

β-Amyloid Antibodies in CSF

Naturally occurring antibodies against Aβ peptide have been detected in the CSF and blood of patients with AD, neurological diseases and healthy controls [97–100, 119]. Investigating Aβ antibody in CSF by ELISA, Du et al. [97] found a statistically significant decrease in antibody levels in patients with AD compared to age-matched healthy controls. So far, only little is known about Aβ antibodies with respect to their function, induction, specificity and role in disease processes. Aβ antibodies may possibly mediate pathways for peripheral and central degradation of Aβ, thereby contributing to plaque clearance, as hypnotized by some groups [for Review see 120]. Further on, a study demonstrated that purified Aβ antibodies isolated from immunoglobulin block fibril formation, disrupt preformed fibrils and prevent neurotoxicity in neuronal cell cultures [121]. However, considering data from recent investigations in transgenic AD mouse models, there is evidence for a therapeutic impact of Aβ antibodies. A reduction in cerebral plaque load and cognitive impairment was observed after active and passive immunization resulting in increased production or administration of Aβ antibodies in these animal models [101–107]. Unfortunately, the phase II clinical trial of the Aβ vaccination approach had to be withdrawn because of signs consistent with meningoencephalitis in an increasing number of patients [108]. Setbacks in active immunization are shifting the focus on passive administration of antibodies. Dodel et al. [109] detected naturally occurring Aβ antibodies in commercially available immunoglobulin G products. In a clinical approach, they investigated patients with different neurological diseases treated with intravenous immunoglobulin preparations (IVIG). A significant decrease in total Aβ and Aβ(1–42) in CSF compared to baseline values was observed after treatment. In serum, a significant increase in total Aβ without change in the Aβ(1–42) level was observed [109]. These data are in agreement with observations in transgenic mouse models [103, 110]. Soon, we are expecting the results of a clinical trial on administrating IVIG to patients with AD to investigate safety and tolerability of the treatment and to measure the effect on central and peripheral Aβ levels. Further studies are warranted to determine the role of Aβ antibodies in Aβ clearance, their individual variability during different stages of AD and their predictive value for identifying responders to immunization and IVIG.

Discussion

Several reports demonstrated that CSF t-tau protein and Aβ(1–42) are highly specific and sensitive in differentiating AD from normal aging and from geriatric major depression. They seem valuable as well in detecting early neurodegenerative changes in MCI patients. Their clinical use, however, is limited, because they are not specific enough to accurately separate AD from other forms of dementia. Even the combined testing of both markers does not amend this situation.

On this basis, different phosphorylated epitopes of tau protein were examined in independent international multicenter studies. The first results are extremely promising, showing that different p-tau protein epitopes may substantially contribute to improved diagnostic accuracy. p-Tau(231) may be particularly useful in distinguishing AD from FTD. p-Tau(181) may improve the distinction between AD and Lewy body dementia. In addition, CSF p-tau(231) may be a prognostic parameter to predict cognitive decline and AD in MCI subjects. Further studies are needed to decide whether detection of multiple phosphoepitopes may allow a distinct representation of AD-related pathology at different stages of the disease, based on an 'evolutionary' model of a sequential phosphorylation pattern of tau protein [89].

Naturally occurring antibodies directed against $A\beta$ were detected in intravenous immunoglobulin preparations and exist in human CSF and plasma. Little is known about their role in the disease process; their CSF titers may be helpful in improving our understanding of the effects of emerging future immunologic therapies for AD.

In assessing the clinical significance of these findings, several confounding factors have to be taken into account. An important factor is the uncertainty of the clinical diagnostic criteria against which these markers have been tested. Neuropathological studies suggest that high proportions (40–80%) of clinically diagnosed patients with vascular dementia have notable concomitant AD pathology [111, 112]. For example, in the dementia registry of a health maintenance organization [113], only 36% of the patients had pathologically definite AD and no other findings, while 45% had pathologically definite AD plus coexistent vascular pathological features and 22% had pathological findings of AD plus pathological features of Lewy body dementia. Thus, in clinically diagnosed patients it is difficult to achieve high specificity for CSF biomarkers. The only way to resolve this question is to study the markers in neuropathologically confirmed subjects. Additionally, even if they are asymptomatic, age-matched controls may still harbor presymptomatic AD brain lesions [114, 115]. This may additionally reduce the specificity for CSF biomarkers of AD.

Furthermore, the true clinical value of a marker can only be assessed in a population-based sample. So far all markers have been evaluated in highly selected subgroups, representing relatively pure forms of the disease. It is very likely that age-related comorbidity in demented subjects from the population will considerably influence the diagnostic value of the CSF biomarkers. Therefore, it is the scope of several multicenter initiatives to assess the true diagnostic behavior of amyloid and tau-related CSF proteins in population-based samples. Finally, confounding conditions, such as age, gender, disease stage, genetic factors, storage time, and freeze-thaw cycles of CSF aliquots have to be taken into account.

The accuracy of the clinical or preclinical diagnosis of a multifactorial disorder such as AD might be increased by the cumulative information from clinical and neuropsychological examinations and brain imaging (e.g. magnetic resonance imaging or positron emission tomography) [116]. In the near future, novel methodological approaches in the characterization and quantification of proteins in biofluids might reveal additional information and potentially new candidate biomarkers. The peptide pattern of a sample can, for instance, be depicted as a multidimensional peptide mass fingerprint with each peptide's position being characterized by its molecular mass and chromatographic behavior. Such a fingerprint of a CNS sample consists of more than 6,000 different signals. First data on this promising new approach to analyze CNS diseases at the peptide level have been reported recently [117]. Besides the clinical diagnostic application, biomarkers may be used to gain insight into pathophysiological processes of the disease. These studies can complement information from postmortem neuropathological studies, and will be particularly useful in combination with genotypical markers and phenotypical morphological and functional markers derived from neuroimaging, as well as in the study of the intraindividual course of disease over time. In conclusion, the newly established immunoassays detecting phosphorylated tau, βA proteins and βA antibodies, and the rapidly developing modern brain imaging methods open up exciting avenues for research programs on the true diagnostic value of biological markers of AD using large-scale multicenter studies. International initiatives are on the way and may yield positive predictive markers of AD in the near future. With these markers we will be able to monitor, effectively and inexpensively, whether new candidate treatments of AD are working. For these reasons, the National Institutes on Aging (NIA) commissioned a working group on biomarkers as part of its Initiative on Neuroimaging in Alzheimer's Disease. A recently published review reports this group's deliberations [118]. The mission of the working group on biological markers was to provide the NIA with a list of biological measures suitable for a multicentric, longitudinal study of Alzheimer's Disease, with special consideration given to MCI. These measures, partly reviewed in this article, are considered to have potential value in diagnosis, prognosis, or assessing the beneficial effects of treatment. In addition, the NIA also established other working groups, one each for magnetic resonance imaging (volumetric), positron emission tomography and single photon emission computed tomography, and subjects and protocol design. The proceedings of those working groups will greatly help to establish and finally integrate specific diagnostic information from the most meaningful modalities aimed at enhancing diagnostic and prognostic data on AD.

References

1 Fratiglioni L, De Ronchi D, Aguero-Torres H: Worldwide prevalence and incidence of dementia. Drugs Aging 1999;15:365–375.

2 Letenneur L, Commenges D, Dartigues JF, Barberger-Gateau P: Incidence of dementia and Alzheimer's disease in elderly community residents of south-western France. Int J Epidemiol 1994; 23:1256–1261.

3 Buhl L, Bojsen-Moller M: Frequency of Alzheimer's disease in a postmortem study of psychiatric patients. Dan Med Bull 1988;35:288–290.

4 McKhann G, Drachman D, Folstein M, Katzman R, Price D, Stadlan EM: Clinical diagnosis of Alzheimer's disease: Report of the NINCDS-ADRDA Work Group under the auspices of the Department of Health and Human Services Task Force on Alzheimer's disease. Neurology 1984; 34:939–944.

5 Consensus report of the Working Group on: Molecular and Biochemical Markers of Alzheimer's Disease. The Ronald and Nancy Reagan Research Institute of the Alzheimer's Association and the National Institute on Aging Working Group. Neurobiol Aging 1998;19:109–116.
6 Jensen M, Schroeder J, Blomberg M, Engvall B, Pantel J, Ida N, Basun H, Wahlund LO, Werle E, Jauss M, Beyreuther K, Lannfelt L, Hartmann T: Cerebrospinal fluid A beta42 is increased early in sporadic Alzheimer's disease and declines with disease progression. Ann Neurol 1999;45: 504–511.
7 Iqbal K, Alonso AC, Gong CX, Khatoon S, Pei JJ, Wang JZ, Grundke-Iqbal I: Mechanisms of neurofibrillary degeneration and the formation of neurofibrillary tangles. J Neural Transm Suppl 1998;53:169–180.
8 Blennow K, Wallin A, Agren H, Spenger C, Siegfried J, Vanmechelen E: Tau protein in cerebrospinal fluid: A biochemical marker for axonal degeneration in Alzheimer disease? Mol Chem Neuropathol 1995;26:231–245.
9 Vandermeeren M, Mercken M, Vanmechelen E, Six J, van de Voorde A, Martin JJ, Cras P: Detection of tau proteins in normal and Alzheimer's disease cerebrospinal fluid with a sensitive sandwich enzyme-linked immunosorbent assay. J Neurochem 1993;61:1828–1834.
10 Vigo-Pelfrey C, Seubert P, Barbour R, Blomquist C, Lee M, Lee D, Coria F, Chang L, Miller B, Lieberburg I, et al: Elevation of microtubule-associated protein tau in the cerebrospinal fluid of patients with Alzheimer's disease. Neurology 1995;45:788–793.
11 Jensen M, Basun H, Lannfeldt L: Increased cerebrospinal fluid tau in patients with Alzheimer's disease. Neurosci Lett 1995;186:189–191.
12 Riemenschneider M, Buch K, Schmolke M, Kurz A, Guder WG: Cerebrospinal protein tau is elevated in early Alzheimer's disease. Neurosci Lett 1996;212:209–211.
13 Rosler N, Wichart I, Jellinger KA: Total tau protein immunoreactivity in lumbar cerebrospinal fluid of patients with Alzheimer's disease. J Neurol Neruosurg Psychiatry 1996;60:237–238.
14 Arai H, Morikawa Y, Higuchi M, Matsui T, Clark CM, Mura M, Machida N: Cerebrospinal fluid tau levels in neurodegenerative diseases with distinct tau-related pathology. Biochem Biophys Res Commun 1997;236:261–264.
15 Arai H, Nakagawa T, Kosaka Y, Higuchi M, Matsui T, Okamura N, Tashiro M, Sasaki H: Elevated cerebrospinal fluid tau protein level as a predictor of dementia in memory-impaired individuals. Alzheimer's Res 1997;3:211–213.
16 Arai H, Terajima M, Miura M, Higuchi S, Muramatsu T, Machida N, Seiki H, Takase S, Clark CM, Lee VMY, Trojanowski JQ, Sasaki H: Tau in cerebrospinal fluid: A potential diagnostic marker in Alzheimer's disease. Ann Neurol 1995;38:649–652.
17 Arai H, Higuchi S, Sasaki H: Apolipoprotein E genotyping and cerebrospinal fluid tau protein: Implications for the clinical diagnosis of Alzheimer's disease. Gerontology 1997;43:2–10.
18 Andreasen N, Vanmechelen E, van de Voorde A, Davidsson P, Hesse C, Tarvonen S, Raiha I, Sourander L, Winblad B, Blennow K: Cerebrospinal fluid tau protein as a biochemical marker for Alzheimer's disease: A community based follow-up study. J Neurol Neurosurg Psychiatry 1998; 64:298–305.
19 Andreasen N, Minthon L, Vanmechelen E, Vanderstichele H, Davidsson P, Winblad B, Blennow K: Cerebrospinal fluid tau and A-beta42 as predictors of development of Alzheimner's disease in patients with mild cognitive impairment. Neurosci Lett 1999;273:5–8.
20 Kanai M, Matsubara E, Isoe K, et al: Longitudinal study of cerebrospinal fluid levels of tau, A-beta 1–40, and A-beta 1–42(43) in Alzheimer's disease: A Study in Japan. Ann Neurol 1998;44:17–26.
21 Kurz A, Riemenschneider M, Buch K, Willoch F, Bartenstein P, Muller U, Guder W: Tau protein in cerebrospinal fluid is significantly increased at the earliest clinical stage of Alzheimer disease. Alzheimer Dis Assoc Disord 1998;12:372–377.
22 Nishimura T, Takeda M, Nakamura Y, Yosbida Y, Arai H, Sasaki H, Shouji M, Hirai S, Khise K, Tanaka K, Hamamoto M, Yamamoto H, Matsubayashi T, Urakami K, Adachi Y, Nakashima K, Toji H, Nakamura S, Yoshida H: Basic and clinical studies on the measurement of tau protein in cerebrospinal fluid as a biological marker for Alzheimer's disease and related disorders: multicenter study in Japan. Methods Find Exp Clin Pharmacol 1998;20:227–235.
23 Shoji M, Matsubara E, Kanai M, Watanabe M, Nakamura T, Tomidokoro Y, Shizuka M, Wakabayashi K, Igeta Y, Ikeda Y, Mizushima K, Amari M, Ishiguro K, Kawarabayashi T, Harigaya V,

Okamoto K, Hirai S: Combination assay of CSF tau, A-beta 1–40, A-beta 1–42(43) as a biochemical marker of Alzheimer's disease. J Neurol Sci 1998;158:134–140.
24 Tapiola T, Lehtovirta M, Ramberg J, Helisalmi S, Linnaranta K, Riekkinen P Sr, Soininen H: CSF tau is related to apolipoprotein E genotype in early Alzheimer's disease. Neurology 1998;50:169–174.
25 Buerger K, Padberg F, Nolde T, Stübner S, Teipel SJ, Haslinger A, Boetsch T, Sunderland T, Arai H, Rapoport SI, Möller HJ, Hampel H: CSF tau protein shows a better discrimination in young old (<70 years) than in old patients with Alzheimer's disease. Neurosci Lett 1999;277:21–24.
26 Green AJ, Harvey RJ, Thompson EJ, Rossor MN: Increased tau in the cerebrospinal fluid of patients with frontotemporal dementia and Alzheimer's disease. Neurosci Lett 1999;259:133–135.
27 Hampel H, Teipel SJ, Padberg F, Schwarz MJ, Haslinger A, Kötter HU, Buch K, Dukoff R, Lasser R, Stübner S, Müller N, Sunderland T, Rapoport SI, Möller HJ: Discriminant power of combined CSF tau and soluble gp130 in the diagnosis of Alzheimer's disease. Brain Res 1999;823:104–112.
28 Hulstaert F, Blennow K, Ivanoiu A, Schoonderwaldt HC, Riemenschneider M, De Deyn PP, Bancher C, Cras P, Wiltfang J, Mehta PD, Iqbal K, Pottel H, Vanmechelen E, Vanderstichele H: Improved discrimination of AD patients using beta-amyloid (1–42) and tau levels in CSF. Neurology 1999;52:1555–1562.
29 Ishiguro K, Ohno H, Arai H, Yamaguchi H, Urakami K, Park JM, Sato K, Kohno H, Imahori K: Phosphorylated tau in human cerebrospinal fluid is a diagnostic marker for Alzheimer's disease. Neurosci Lett 1999;270:91–94.
30 Molina L, Touchon J, Herpé M, Lefranc D, Duplan L, Cristol JP, Sabatier R, Vermersch P, Pau B, Mourton-Gilles C: Tau and apo E in CSF: Potential aid for discriminating Alzheimer's disease from other dementias. Neuro report 1999;10:3491–3495.
31 Morikawa Y, Arai H, Matsushita S, Kato M, Higuchi S, Miura M, Kawakami H, Higuchi M, Okamura N, Tashiro M, Matsui T, Sasaki H: Cerebrospinal fluid tau protein levels in demented and nondemented alcoholics. Alcohol Clin Exp Res 1999;23:575–577.
32 Kahle PJ, Jakowec M, Teipel SJ, Hampel H, Petzinger GM, Di Monte DA, Silverberg GD, Moller HJ, Yesavage JA, Tinklenberg JR, Shooter EM, Murphy GM: Combined assessment of tau and neuronal thread protein in Alzheimer's disease CSF. Neurology 2000;54:1498–1504.
33 Kanemaru K, Kameda N, Yamanouchi H: Decreased CSF amyloid beta42 and normal tau levels in dementia with Lewy bodies. Neurology 2000;54:1875–1876.
34 Sjogren M, Minthon L, Davidsson P, Granerus AK, Clarberg A, Vanderstichele H, Vanmechelen E, Wallin A, Blennow K: CSF levels of tau, beta-amyloid (1–42) and GAP-43 in frontotemporal dementia, other types of dementia and normal aging. J Neural Transm 2000;107:563–579.
35 Sjogren M, Rosengren L, Minthon L, Davidsson P, Blennow K, Wallin A: Cytoskeleton proteins in CSF distinguish frontotemporal dementia from AD. Neurology 2000;54:1960–1964.
36 Motter R, Vigo-Pelfrey C, Kholodenko D, Barbour R, Johnson-Wood K, Galasko D, Chang L, Miller B, Clark C, Green R: Reduction of beta-amyloid peptide42 in the cerebrospinal fluid of patients with Alzheimer's disease. Ann Neurol 1995;38:643–648.
37 Munroe WA, Southwick PC, Chang L, Scharre DW, Echols CL Jr, Fu PC, Whaley JM, Wolfert RL: Tau protein in cerebrospinal fluid as an aid in the diagnosis of Alzheimer's disease. Ann Clin Lab Sci 1995;25:207–217.
38 Galasko D, Clark C, Chang L, Miller B, Green RC, Rotter R, Seubert O: Assessment of CSF levels of tau protein in mildly demented patients with Alzheimer's disease. Neurology 1997;48:632–635.
39 Galasko D, Chang L, Motter R, Clark CM, Kaye J, Knopman D, Thomas R, Kholodenko D, Schenk D, Lieberburg I, Miller B, Green R, Basherad R, Kertiles L, Boss A, Seubert P: High cerebrospinal fluid tau and low amyloid-beta-42 levels in the clinical diagnosis of Alzheimer disease and relation to apolipoprotein E genotype. Arch Neurol 1998;55:937–945.
40 Mori H, Hosoda K, Matsubara E, Nakamoto T, Furiya Y, Endoh R, Usami M, Shoji M, Maruyama S, Hirai S: Tau in cerebrospinal fluids: Establishment of the sandwich ELISA with antibody specific to the repeat sequence in tau. Neurosci Lett 1995;186:181–183.
41 Blennow K, Vanmechelen E, Hampel H: CSF total tau, A-β42 and phosphorylated tau protein as biomarkers for Alzheimer's disease. Mol Neurobiol 2001;24:87–97.
42 Riemenschneider M, Buch K, Schmolke M, Kurz A, Guder WG: Diagnosis of Alzheimer's disease with cerebrospinal fluid tau protein and aspartate aminotransferase (letter). Lancet 1997; 350:784.

43 Bürger K, Zinkowski R, Teipel SJ, Arai H, DeBernardis J, Kerkman D, McCulloch C, Padberg F, Faltraco F, Goernitz A, Tapiola T, Rapoport SI, Pirttilä T, Möller HJ, Hampel H: Differentiation between geriatric major depression and Alzheimer's disease with CSF tau protein phosphorylated at threonine 231. Am J Psychiatry 2003;160:376–379.
44 Sjoegren M, Davidsson P, Tullberg M, Minthon L, Wallin A, Wikkelso C, Granerus AK, Vanderstichele E., Blennow K: Both total and phosphorylated tau are increased in Alzheimer's disease. J Neurol Neurosurg Psychiatry 2001;70:624–630.
45 Sjogren M, Vanderstichele H, Agren H, Zachrisson O, Edsbagge M, Wikkelso C, Skoog I, Wallin A, Wahlund LO, Marcusson J, Nagga K, Andreasen N, Davidsson P, Vanmechelen E, Blennow K: Tau and Abeta42 in cerebrospinal fluid from healthy adults 21–93 years of age: Establishment of reference values. Clin Chem 2001;47:1776–1781.
46 Parnetti L, Lanari A, Amici S, Gallai V, Vanmechelen E, Hulstaert F: CSF phosphorylated tau is a possible marker for discriminating Alzheimer's disease from dementia with Lewy bodies. Neurol Sci 2001;22:77–78.
47 Skoog I, Vanmechelen E, Andreasson LA, Palmertz B, Davidsson P, Hesse C, Blennow K: A population-based study of tau protein and ubiquitin in cerebrospinal fluid in 85-year-olds: Relation to severity of dementia and cerebral atrophy, but not to the apolipoprotein E4 allele. Neurodegeneration 1995;4:433–442.
48 Tato RE, Frank A, Hernanz A: Tau protein concentrations in cerebrospinal fluid of patients with dementia of the Alzheimer type. J Neurol Neurosurg Psychiatry 1995;59:280–283.
49 Andreasen N, Minthon L, Clarberg A, Davidsson P, Gottfries J, Vanmechelen E, Vanderstichele H, Winblad B, Blennow K: Sensitivity, specificity, and stability of CSF-tau in AD in a community-based patient sample. Neurology 1999;53:1488–1494.
50 Higuchi M, Tashiro M, Arai H, Okamura N, Hara S, Higuchi S, Itoh M, Shin RW, Trojanowski JQ, Sasaki H: Glucose hypometabolism and neuropathological correlates in brains of dementia with Lewy bodies. Exp Neurol 2000;162:247–256.
51 Mecocci P, Cherubini A, Bregnocchi M, Chionne F, Cecchetti R, Lowenthal DT, Senin U: Tau protein in cerebrospinal fluid: A new diagnostic and prognostic marker in Alzheimer disease? Alzheimer Dis Assoc Disord 1998;12:211–214.
52 Vanmechelen E, Vanderstichele H, Hulstaert F, Andreasen N, Minthon L, Winblad B, Davidsson P, Blennow K: Cerebrospinal fluid tau and beta-amyloid (1–42) in dementia disorders. Mech Ageing Dev 2001;122:2005–2011.
53 Itoh N, Arai H, Urakami K, Ishiguro K, Ohno H, Hampel H, Buerger K, Wiltfang J, Otto M, Kretzschmar H, Moeller HJ, Imagawa M, Kohno H, Nakashima K, Kuzuhara S, Sasaki H, Imahori K: Large-scale, multicenter study of cerebrospinal fluid tau protein phosphorylated at serine 199 for the antemortem diagnosis of Alzheimer's disease. Ann Neurol 2001;50:150–156.
54 Fabre SF, Forsell C, Viitanen M, Sjogren M, Wallin A, Blennow K, Blomberg M, Andersen C, Wahlund LO, Lannfelt L: Clinic-based cases with frontotemporal dementia show increased cerebrospinal fluid tau and high apolipoprotein E epsilon4 frequency, but no tau gene mutations. Exp Neurol 2001;168:413–418.
55 Andersen C, Fabre FS, Ostberg P, Lannfelt L, Wahlund L: Tau protein in cerebrospinal fluid from semantic dementia patients. Neurosci Lett 2000;294:155–158.
56 Arai H, Satoh-Nakagawa T, Higuchi M, Morikawa Y, Miura M, Kawakami H, Seki H, Takase S, Sasaki H: No increase in cerebrospinal fluid tau protein levels in patients with vascular dementia. Neurosci Lett 1998;256:174–176.
57 Otto M, Wiltfang J, Tumani H, Zerr I, Lantsch M, Kornhuber J, Weber T, Kretzschmar HA, Poser S: Elevated levels of tau-protein in cerebrospinal fluid of patients with Creutzfeldt-Jakob disease. Neurosci Lett 1997;225:210–212.
58 Otto M, Wiltfang J, Cepek L, Neumann M, Mollenhauer B, Steinacker P, Ciesielczyk B, Schulz-Schaeffer W, Kretzschmar HA, Poser S: Tau protein and 14-3-3 protein in the differential diagnosis of Creutzfeldt-Jakob disease. Neurology 2002;58:192–197.
59 Kapaki E, Kilidireas K, Paraskevas GP, Michalopoulou M, Patsouris E: Highly increased CSF tau protein and decreased beta-amyloid (1–42) in sporadic CJD: A discrimination from Alzheimer's disease? J Neurol Neurosurg Psychiatry 2001;71:401–403.

60 Kudo T, Mima T, Hashimoto R, Nakao K, Morihara T, Tanimukai H, Tsujio I, Koike Y, Tagami S, Mori H, Nakamura Y, Tanaka T, Mori K, Takeda M: Tau protein is a potential biological marker for normal pressure hydrocephalus. Psychiatry Clin Neurosci 2000;54:199–202.
61 Hesse C, Rosengren L, Andreasen N, Davidsson P, Vanderstichele H, Vanmechelen E, Blennow K: Transient increase in total but not phospho-tau in human cerebrospinal fluid after acute stroke. Neurosci Lett 2001;297:187–190.
62 Zemlan FP, Rosenberg WS, Luebbe PA, Campbell TA, Dean GE, Weiner NE, Cohen JA, Rudick RA, Woo D: Quantification of axonal damage in traumatic brain injury: Affinity purification and characterization of cerebrospinal fluid tau proteins (see comments). J Neurochem 1999;72: 741–750.
63 Mitani K, Furiya Y, Uchihara T, Ishii K, Yamanouchi H, Mizusawa H, Mori H: Increased CSF tau protein in corticobasal degeneration. J Neurol 1998;245:44–46.
64 Urakami K, Mori M, Wada K, Kowa H, Takeshima T, Arai H, Sasaki H, Kanai M, Shoji M, Ikemoto K, Morimatsu M, Hikasa C, Nakashima K: A comparison of tau protein in cerebrospinal fluid between corticobasal degeneration and progressive supranuclear palsy. Neurosci Lett 1999; 259:127–129.
65 Petersen RC, Smith GE, Waring SC, Ivnik RJ, Tangalos EG, Kokmen E: Mild cognitive impairment: Clinical characterization and outcome. Arch Neurol 1999;56:303–308.
66 Maruyama M, Arai H, Sugita M, Tanji H, Higuchi M, Okamura N, Matsui T, Higuchi S, Matsushita S, Yoshida H, Sasaki H: Cerebrospinal fluid amyloid beta(1–42) levels in the mild cognitive impairment stage of Alzheimer's disease. Exp Neurol 2001;172:433–436.
67 Andreasen N, Minthon L, Davidsson P, Vanmechelen E, Vanderstichele H, Winblad B, Blennow K: Evaluation of CSF-tau and CSF-A-beta-42 as diagnostic markers for Alzheimer disease in Clinical practice. Arch Neurol 2001;58:373–379.
68 Hock C, Golombowski S, Naser W, Müller-Spahn F: Increase levels of tau protein in cerebrospinal fluid of patients with Alzheimer's disease – Correlation with degree of cognitive impairment. Ann Neurol 1995;37:414–415.
69 Blomberg M, Jensen M, Basun H, Lannfelt L, Wahlund LO: Increasing cerebrospinal fluid tau levels in a subgroup of Alzheimer patients with apolipoprotein E allele epsilon 4 during 14 months follow-up. Neurosci Lett 1996;214:163–166.
70 Tapiola T, Pirttilä T, Mehta PD, Alafuzoff I, Lehtovirta M, Soininen H: Relationship between apoE genotype and CSF beta-amyloid (1–42) and tau in patients with probable and definite Alzheimer's disease. Neurobiol Aging 2000;21:735–740.
71 Sunderland T, Wolozin B, Galasko D, Levy J, Dukoff R, Bahro M, Lasser R, Motter R, Lehtimaki T, Seubert P: Longitudinal stability of CSF tau levels in Alzheimer patients. Biol Psychiatry 1999; 46:750–755.
72 Galasko D: Biological markers and the treatment of Alzheimer's disease. J Mol Neurosci 2001; 17:119–125.
73 Hyman BT, Trojanowski JQ: Consensus recommendations for the postmortem diagnosis of Alzheimer disease from the National Institute on Aging and the Reagan Institute Working Group on diagnostic criteria for the neuropathological assessment of Alzheimer disease. J Neuropathol Exp Neurol 1997;56:1095–1097.
74 Hardy J, Selkoe JD: The amyloid hypothesis of Alzheimer's disease: progress and problems on the road to therapeutics. Science 2002;297:353–356.
75 Ida N, Hartmann T, Pantel J, Schroder J, Zerfass R, Forstl H, Sandbrink R, Masters CL, Beyreuther K: Analysis of heterogeneous A4 peptides in human cerebrospinal fluid and blood by a newly developed sensitive Western blot assay. J Biol Chem 1996;271:22908–22914.
76 Tamaoka A, Sawamura N, Fukushima T, Shoji S, Matsubara E, Shoji M, Hirai S, Furiya Y, Endoh R, Mori H: Amyloid beta protein 42(43) in cerebrospinal fluid of patients with Alzheimer's disease. J Neurol Sci 1997;148:41–45.
77 Andreasen N, Hesse C, Davidsson P, Minthon L, Wallin A, Winblad B, Vanderstichele H, Vanmechelen E, Blennow K: Cerebrospinal fluid beta-amyloid (1–42) in Alzheimer disease: Differences between early- and late-onset Alzheimer disease and stability during the course of disease. Arch Neurol 1999;56:673–680.

78 Otto M, Esselmann H, Schulz-Shaeffer W, Neumann M, Schroter A, Ratzka P, Cepek L, Zerr I, Steinacker P, Windl O, Kornhuber J, Kretzschmar HA, Poser S, Wiltfang J: Decreased beta-amyloid 1–42 in cerebrospinal fluid of patients with Creutzfeldt-Jakob disease. Neurology 2000; 54:1099–1102.
79 Riemenschneider M, Schmolke M, Lautenschlager N, Guder WG, Vanderstichele H, Vanmechelen E, Kurz A: Cerebrospinal beta-amyloid (1–42) in early Alzheimer's disease: Association with apolipoprotein E genotype and cognitive decline. Neurosci Lett 2000;284:85–88.
80 Hampel H, Fuchsberger T, Teipel SJ, Andreasen N, Engel R, Shen Y, Du Y, Dodel R, Wiltfang J, Otto M, Moeller HJ, Blennow K, Buerger K: CSF β-Amyloid 1–42 and tau-protein as predictors of Alzheimer's Disease in Mild Cognitive Impairment, submitted.
81 Samuels SC, Silverman JM, Marin DB, Peskind ER, Younki SG, Greenberg DA, Schnur E, Santoro J, Davis KL: CSF beta-amyloid, cognition, and APOE genotype in Alzheimer's disease. Neurology 1999;52:547–551.
82 Nitsch RM, Rebeck GW, Deng M, Richardson UI, Tennis M, Schenk DB, Vigo-Pelfrey C, Lieberburg I, Wurtman RJ, Hyman BT: Cerebrospinal fluid levels of amyloid beta-protein in Alzheimer's disease: Inverse correlation with severity of dementia and effect of apolipoprotein E genotype. Ann Neurol 1995;37:512–518.
83 Okamura N, Arai H, Higuchi M, Tashiro M, Matsui T, Itoh M, Iwatsubo T, Tomita T, Sasaki H: Cerebrospinal fluid levels of amyloid beta-peptide l-42, but not tau have positive correlation with brain glucose metabolism in humans. Neurosci Lett 1999;273:203–207.
84 Kohnken R, Buerger K, Zinkowski R, Miller C, Kerkman D, DeBernardis J, Shen J, Möller HJ, Davies P, Hampel H: Detection of tau phosphorylated at threonine 231 in cerebrospinal fluid of Alzheimer's disease patients. Neurosci Lett 2000;287:187–190.
85 Vanmechelen E, Van Kerschaver E, Blennow K, De Deyn PP, Galasko D, Parnetti L, Sindic CJM, Arai H, Riemenschneider M, Hampel H, Pottel H, Valgaeren A, Hulstaert F, Vanderstichele H: CSF-phospho-tau (181P) as a promising marker for discriminating Alzheimer's disease from dementia with Lewy bodies; in Iqbal K, Sisodia S, Winblad B (eds): Alzheimer's disease: advances in etiology, pathogenesis and therapeutics. New York, Wiley, 2001.
86 Vanmechelen E, Vandersticele H, Davidsson P, Van Kerschaver E, Van Der Perre B, Sjogren M, Andreasen N, Blennow K: Quantification of tau phosphorylated at threonine 181 in human cerebrospinal fluid: A sandwich ELISA with a synthetic phosphopeptide for standardization. Neurosci Lett 2000;285:49–52.
87 Hu YY, He SS, Wang X, Duan QH, Grundke-Iqbal I, Iqbal K, Wang J: Levels of nonphosphorylated and phosphorylated tau in cerebrospinal fluid of Alzheimer's disease patients: An ultrasensitive bienzyme-substrate-recycle enzyme-linked immunosorbent assay. Am J Pathol 2002;160: 1269–1278.
88 Vincent I, Zheng JH, Dickson DW, Kress Y, Davies P: Mitotic Phosphoepitopes precede paired helical filaments in Alzheimer's disease. Neurobiol Aging 1998;19:287–296.
89 Augustinack JC, Schneider A, Mandelkow EM, Hyman BT: Specific tau phosphorylation sites correlate with severity of neuronal cytopathology in Alzheimer's disease. Acta Neuropathol (Berl) 2002;103:26–35.
90 Buerger K, Zinkowski R, Teipel SJ, Tapiola T, Arai H, Blennow K, Andreasen N, Hoffmann-Kiefer K, DeBernardis J, Kerkman D, McCulloch C, Kohnken R, Padberg F, Pirttilä T, Scharpiro MB, Rapoport SI, Moeller H-J, Davies P, Hampel H: Differential diagnosis of Alzheimer's disease with CSF tau protein phosphorylated at threonine 231. Arch Neurol 2002;59:1267–1272.
91 Buerger KT, Teipel SJ, Zinkowski R, Blennow K, Arai H, Engel R, Hofmann-Kiefer K, McCulloch C, Ptok U, Heun R, Andreasen N, DeBernardis J, Kerkman D, Moeller HJ, Davies P, Hampel H: CSF tau protein phosphorylated at threonine 231 correlates with cognitive decline in MCI subjects. Neurology 2002;59:627–629.
92 Hampel H, Buerger K, Kohnken R, Teipel SJ, Zinkowski R, Moeller HJ, Rapoport S, Davies P: Tracking of Alzheimer's disease progression with CSF tau protein phosphorylated at threonine 231. Ann Neurol 2001;49:545–546.
93 Reiber H: Dynamics of brain-derived proteins in cerebrospinal fluid. Clinica Chimica Acta 2001;310:173–186.

94 Leon de M, Tarshish CY, DeSanti S, Zionkowkski R, Metha PD, Convit A, Caraos C, Rusinek H, Tsui W, Saint Louis L, DeBernadis J, Kerkman D, Qadri F, Gray A, Lesbre P, Wisniewski T, Porier J, Davies P: Longitudinal tau load increases in mild cognitive impairment. Neurosci Lett, in press.
95 Arai H, Ishiguro K, Ohno H, Moriyama M, Itoh N, Okamura N, Mattsui T, Morikawa Y, Horikawa E, Kohno H, Sasaki H, Imahori K: CSF Phosphorylated tau protein and mild cognitive impairment: A prospective study. Exp Neurol 2000;166:201–203.
96 Schoenknecht P, Hunt A, Volkmann M, Buerger K, Hampel H, Schröder J: Levels of total tau and tau protein phosphorylated at threonin 181 in patients with incipient and manifest Alzheimer's disease. Neurosci Lett, in press.
97 Du Y, Dodel R, Hampel H, Buerger K, Lin S, Eastwood B, Bales K, Gao F, Moeller HJ, Oertel W, Farlow M, Paul, S: Reduced levels of amyloid beta-peptide antibody in Alzheimer disease. Neurology 2001;57:801–805.
98 Gaskin F, Finley J, Fang Q, Xu S, Fu SM: Human antibodies reactive with beta-amyloid protein in Alzheimer's disease. J Exp Med 1993;177:1181–1186.
99 Hyman BT, Trojanowski JQ: Consensus recommendations for the postmortem diagnosis of Alzheimer disease from the National Institute on Aging and the Reagan Institute Working Group on diagnostic criteria for the neuropathological assessment of Alzheimer disease. J Neuropathol Exp Neurol 1997;56:1095–1097.
100 Xu S, Gaskin F: Increased incidence of anti-beta-amyloid autoantibodies secreted by Epstein-Barr virus transformed B cell lines from patients with Alzheimer's disease. Mech Ageing Dev 1997;94: 213–222.
101 Bard F, Cannon C, Barbour R, Burke R, Games D, Grajeda T, Hu K, Huang J, Johnson Wood K, et al: Peripherally administered antibodies against amyloid beta-peptide enter the central nervous system and reduce pathology in a mouse model of Alzheimer's disease. Nat Med 2000;6:371–382.
102 Chen KS, Knox J: A learning deficit related to age and beta-amyloid plaques in a mouse model of Alzheimer's disease. Nature 2000;408:975–979.
103 DeMattos RB, Bales KR, Parsadanian M: Plaque-associated disruption of CSF and plasma amyloid-beta equilibrium in a mouse model of Alzheimer's disease. J Neurochem 2002;81:229–236.
104 Janus CPJ, Mclaurin J: A beta peptide immunization reduces behavioural impairment and plaques in a model of Alzheimer's disease. Nature 2000;408:979–982.
105 Morgan DDD, Gottschall PE: A beta peptide immunization reduces behavioural impairment and plaques in a model of Alzheimer's disease. Nature 2000;408:982–985.
106 Schenk DBR, Dunn W: Immunization with amyloid-beta attenuates Alzheimer-disease-like pathology in the PDAPP mouse. Nature 1999;400:143–150.
107 Weiner HL, Lemere CA, Maron R, Spooner ET, Grenfell TJ, Mori C, Issazadeh S, Hancock WW, Selkoe DJ: Nasal administration of amyloid-beta peptide decreases cerebral amyloid burden in a mouse model of Alzheimer's disease. Ann Neurol 2000;48:567–579.
108 Senior K: Dosing in phase II trial of Alzheimer's vaccine suspended. Lancet Neurol 2002;1:3.
109 Dodel R, Hampel H, Candan D, Lin S, Gao F, Schock S, Jäckel S, Wei X, Buerger K, Höft C, Hemmer B, Möller HJ, Farlow M, Oertel WH, Sommer N, Du Y: Human antibodies against amyloid beta peptide: A potential treatment for Alzheimer's disease. Ann Neurol 2002;52:253–256.
110 DeMattos RB, Bales KR, Cummings DJ, Dodart JC, Paul SM: Peripheral anti-A beta antibody alters CNS and A beta clearance and decreases brain A beta burden in a mouse model of Alzheimer's disease. Proc Natl Acad USA 2001;98:8850–8855.
111 Jellinger KA, Bancher C: AD neuropathology. Neurology 1996;46:1186–1187.
112 Kosunen O, Soininen H, Paljarvi L, Heinonen O, Talasniemi S, Riekkinen PJ Sr: Diagnostic accuracy of Alzheimer's disease: a neuropathological study. Acta Neuropathol (Berl) 1996;91: 185–193.
113 Lim A, Tsuang D, Kukull W, Nochlin D, Leverenz J, McCormick W, Bowen J, Teri L, Thompson J, Peskind ER, Raskind M, Larson EB: Clinico-neuropathological correlation of Alzheimer's disease in a community-based case series. J Am Geriatr Soc 1999;47:564–569.
114 Davies L, Wolska B, Hilbich C, Multhaup G, Martins R, Simms G, Beyreuther K, Masters CL: A4 amyloid protein deposition and the diagnosis of Alzheimer's disease: Prevalence in aged brains

determined by immunocytochemistry compared with conventional neuropathologic techniques. Neurology 1988;38:1688–1693.
115 Morris JC, Price AL: Pathologic correlates of nondemented aging, mild cognitive impairment, and early-stage Alzheimer's disease. J Mol Neurosci 2001;17:101–118.
116 Okamura N, Arai H, Maruyama M, Higuchi M, Matsui T, Tanji H, Seki T, Hirai H, Chiba H, Itoh M, Sasaki H: Combined analysis of CSF tau levels and [^{123}I] iodoamphetamine SPECT in mild cognitive impairment: Implications for a novel predictor of Alzheimer's disease. Am J Psychiatry 2002;159:474–476.
117 Heine G, Zucht HD, Schuhmann MU, Bürger K, Jürgens M, Zumkeller M, Schneekloth CG, Hampel H, Schulz-Knappe P, Selle H: High-Resolution peptide mapping of cerebrospinal fluid – A novel concept for diagnostics and research in CNS diseases. J Chromatogr B 2002;25:782: 353–361.
118 Frank RA, Galasko D, Hampel H, Hardy J, De Leon M, Mehta P, Rogers J, Siemers E, Trojanowski J: Proceedings of a working group; NIA initiative on neuroimaging in Alzheimer's disease. Biological Measures Working group. Neurobiol Aging 2003;24:521–536.
119 Weksler ME, Relkin N, Turkenich R, LaRusse S, Zhou L, Szabo P: Patients with Alzheimer disease have lower levels of serum anti-amyloid peptide antibodies than healthy elderly individuals. Exp Gerontol 2002;37:943–948.
120 Dodel R, Hampel H, Du Y: Immunotherapy for Alzheimer's disease. Lancet Neurology 2003;2:215–220.
121 Du Y, Wei X, Dodel R, Sommer N, Hampel H, Gao F, Ma Z, Zhao L, Farlow M: Human anti-Aβ antibodies block fibril formation and prevent Aβ-induced neurotoxicity. Brain, in press.

Harald Hampel, MD
Dementia Research Section and Memory Clinic
Alzheimer Memorial Center and Geriatric Psychiatry Branch
Department of Psychiatry, Ludwig-Maximilian University
Nussbaumstrasse 7, D–80336 Munich (Germany)
Tel. +49 89 5160 5877, Fax +49895160 5856, E-Mail: hampel@psy.med.uni-muenchen.de

Takeda M, Tanaka T, Cacabelos R (eds): Molecular Neurobiology of Alzheimer Disease and Related Disorders. Basel, Karger, 2004, pp 157–163

Genetic Analysis of Familial Alzheimer's Disease in a Japanese Population

Yosuke Wakutani[a], Yoshiki Adachi[a], Kenji Wada-Isoe[a], Kaoru Yamagata[a], Katsuya Urakami[b], Kenji Nakashima[a]

Departments of [a]Neurology, Institute of Neurological Sciences, and [b]Biological Regulation, School of Health Science, Faculty of Medicine, Tottori University, Yonago, Japan

Alzheimer's disease (AD) is one of the most common neurodegenerative disorders in the elderly population. The pathological hallmarks (amyloid plaque, neurofibrillary tangle and neuronal cell loss) have been well characterized. The causal genes for early-onset familial Alzheimer's disease (FAD) are the presenilin 1 (PS-1) gene on chromosome 14 [1], the presenilin 2 (PS-2) gene on chromosome 1 [2] and the amyloid precursor protein (APP) gene on chromosome 21 [3]. In addition, apolipoprotein E (APOE) allele 4 (ε4), located on chromosome 19, is a well-established genetic risk factor for sporadic AD [4]. Among these genes, mutations in PS-1 seem to be the most common genetic factor underlying the development of early-onset FAD. To date, over 100 missense mutations for PS-1, 8 mutations for PS-2 and 16 mutations for APP are cited in an online database (AD mutation database; http://molgen-www.uia.ac.be/ADMutations/). Several PS-1 mutations and only an APP mutation (V717I) were previously described in the Japanese population (table 1) [2, 5–19]. Eighteen missense mutations in the PS-1 gene were reported in the Japanese familial AD (FAD) pedigrees. No pathogenic mutation of the PS-2 gene has been identified in the Japanese population. In this chapter, we report the results of our most recent studies of these three genes in FAD and sporadic AD patients in a Japanese population.

Subjects and Methods

Patient Samples

Twenty-two Japanese patients were selected from 5 early-onset (<65 years old) FAD patients (mean age of onset: 58.2 years), 7 late-onset (>65 years old) FAD patients

Table 1. APP, PS-1 and PS-2 gene mutations in Japanese FAD and sporadic AD

	Exon	Mutation	Reference
PS-1	5	V96F	5
		E123K	6
	6	H163R[1]	5, 7, 8, 9
	7	E184D	10
		G209R	11
		I213T	5
		G217D	12
		F237I[2]	13
	8	A260V	2, 14
		S266G	15
	8	R269H	9
		E273A	9
		E280A	8
		A285V	2
	9	S290C	16
	11	G384A	9
		N405S	17
	12	A431V	18
APP	17	V717I	19

[1]This mutation was reported in early-onset FAD and early-onset sporadic AD.

[2]This mutation was reported in early-onset FAD with spastic paraparesis.

(mean age of onset: 70.3 years old), and 10 early-onset sporadic AD patients (mean age of onset: 55.4 years). All patients fulfilled the National Institute of Neurological and Communicative Disorders and Stroke-Alzheimer's Disease and Related Disorder Association (NINCDS-ADRDA) criteria for probable and possible AD [20]. These diagnoses were assisted by MRI or CT imaging studies. We defined patients as having FAD if at least 2 members were affected in a family and the difference in age of onset was less than 20 years. Genomic DNA was extracted from peripheral leukocytes using the standard phenol-chloroform method and subjected to PCR amplifications.

Primers and PCR Amplification

Intronic primers were generated to amplify exons 16, 17 and 18 of the APP gene, exons 3–12 of the PS-1 and PS-2 genes. The primer sequences are provided in table 2. In brief, 50–100 ng DNA was amplified using PCR in each 15 μl reaction mixture using 1 mmol of specific primers and 0.8 units of Taq DNA polymerase (TaKaRa, Tokyo, Japan) in supplied 1 × PCR buffer for 35 cycles of 30 s at 94°C for denaturing, 30 s at 58°C for annealing, and 40 s at 72°C for extension.

Table 2. PCR primers (5′→3′)

APP		Size, bp
APP Ex16-F	TAGAAAGAAGTTTTGGGTAGGCTTT	260
APP Ex16-R	AGAGTTAATAGGTCATTTGGCAAGACA	
APP Ex17-F	AATGAAATTCTTCTAATTGCGTTT	258
APP Ex17-R	TTCTCTCATAGTCTTAATTCCCACTT	
APP Ex18-F	TTACTGCTTCTCCATGTTCACC	249
APP Ex18-R	TTCTATAAATGGACACCGATGG	
PS-1		
PS1Ex-3F	TCTTTCCCTTTTCAGAACCTCA	222
PS1-Ex3R	CAGAGGTGAGGGGAGATGATAA	
PS1-Ex4F	TCATAGTGACGGGTCTGTTGTT	423
PS1-Ex4R	CTTCAAGGTGATGATGACATGC	
PS1-Ex5F	TTCTGTGTTGGAGGTGGTAATG	287
PS1-Ex5R	CTGTGACAAGAATACCCAACCA	
PS1-Ex6F	AGTCTGGGCGACAAAGTGAG	237
PS1-Ex6R	TGATAGCTACACAGCACAAAGG	
PS1-Ex7F	GTTTGGGAGCCATCACATTATT	373
PS1-Ex7R	AGATGAGGAAAGAAAACACTCCA	
PS1-Ex8F	CCTTCGTTAATTCCTCCCTACC	248
PS1-Ex8R	CATGTGCTTCAGTTCCGATAAA	
PS1-Ex9F	GGAGAAATGATGGCTTGTTGTT	266
PS1-Ex9R	GGAGTCTATGACCAAAGAAAGACG	
PS1-Ex10F	CAATGACAGCTAGTTACTGTTTCCA	346
PS1-Ex10R	TTCATTTTATTCTCAAAAAGGTTGA	
PS1-Ex11F	ATTCATTGTGGGGTTGAGTAGG	294
PS1-Ex11R	GAACTGCCTTAAAGGGACTGTG	
PS1-Ex12F	CTTGTGATTGAGTTTTGCCTGA	294
PS1-Ex12R	TGTCCTCCCCAGATTTTGTTAT	

PS-2		Size, bp
PS2-Ex3F	GTCTCACAGGAAAGTGGAACAAG	284
PS2-Ex3R	GAATGTCTGGTTTTCATCTGTAAGG	
PS2-Ex4F	TGGAAAGCAACATTCAAACTTC	456
PS2-Ex4R	TTCTTCATCCCTGCTCTTTACC	
PS2-Ex5F	TAGCAGGTCCAGAATCACTCAAG	318
PS2-Ex5R	GTTTTTCTAAAGGCGGCTGTTT	
PS2-Ex6F	GCTTGGGTATCAGTCTCAGGAT	258
PS2-Ex6R	GAGCTCGATGGTCATCTTTCC	
PS2-Ex7F	GAATGGTGGTAAACTGCTAGGC	407
PS2-Ex7R	CTCTGCAGAAAGGGATGCAAG	
PS2-Ex8F	GCTACAGGGCAGGCTCTTCT	206
PS2-Ex8R	ACCCCGAGTCAGGCAGAG	
PS2-Ex9F	GCTTTCTGGGACGCAGACT	311
PS2-Ex9R	TGGTTTTCAACGGACCTTTC	
PS2-Ex10F	ACCCCTTCTTGGAGCTTTGT	282
PS2-Ex10R	GAGCCTCCACCCTCTGTCTC	
PS2-Ex11F	GGGCCAGAGTTTCTCTTCTTTT	281
PS2-Ex11R	CCCTAGGGATCCTGAGACCTG	
PS2-Ex12F	CCCAGGGACTAGACCATGACT	290
PS2-Ex12R	AACTGCATCCAATGAAAATTCC	

Table 3. Identified mutation and polymorphism

	Exon (PCR region)	Polymorphism	Amino acid	NCBI SNP cluster ID
APP	Exon 16	2032 G/A	D678N	–
	Exon 18	IVS17-10 T/C	intron	–
PS-2	Exon 3	69 T/C	A23A	rs11405
		129 C/T	N43N	rs6759
	Exon 4	IVS3-42 G/A	intron	rs1295644[1]
		IVS3-29 T/C	intron	rs1295643
		260 T/C	H87H	rs1046240[1]
	Exon 5	IVS5+30 G/C	intron	rs2236910
	Exon 8	861 C/T	P287P	–
	Exon 9	IVS8-24 G/A	intron	rs2802267
	Exon 11	IVS11+24 G/A	intron	rs2855562

[1]These are linked polymorphisms in our samples.

Single-Strand Conformation Polymorphism Analysis and Sequence Analysis

PCR products of AD samples for screening were subjected to single-strand conformation polymorphism (SSCP) analysis. One microliter of PCR product was denatured in formamide-containing buffer at 95°C for 8 min, quickly chilled on ice, and electrophoresed on a 12% polyacrylamide gel with 10% glycerol at 4°C for 24 h at 200 V. DNA bands were visualized using silver staining. The mobility-shifted band was directly cut from the gel using a freshly prepared razor blade. The eluted band was re-amplified under identical PCR conditions for 45 cycles. The purified PCR product derived from the extra band was subjected to direct sequencing using a Big Dye cycle sequence kit (Amersham Bioscience Japan, Tokyo, Japan) and the ALF automated luminescent sequencer (Applied Biosystems Japan, Tokyo, Japan). APOE genotyping was carried out according to standard procedures [22].

Results

In PCR-SSCP analysis, 2 extra conformers in the APP gene (exon 16, exon 18), none in the PS-1 gene and 9 in the PS-2 gene were observed. Table 3 shows the identified mutation and polymorphisms identified by sequence analysis. No missense mutations in the PS-1 and PS-2 gene were detected in any samples. Except for the IVS17-10 T/C polymorphism of the APP gene and the 861 C/T (P287P) polymorphism of the PS-2 gene, the identified polymorphisms were previously reported in the NCBI SNP database (the APP gene; http://www.ncbi.nlm.nih.gov/SNP/snp_ref.cgi?locusId = 35, the PS-1 gene; http://www.ncbi.nlm.

nih.gov/SNP/snp_ref.cgi?locusId = 5663 and the PS-2 gene; http://www.ncbi.nlm.nih.gov/SNP/snp_ref.cgi?locusId = 5664). We identified an entirely novel APP gene mutation (2032 G/A D678N [APP770 numbering]) in an early-onset FAD pedigree containing 3 affected members with a mean age of onset of 59.7 years and APOE genotype ε3/ε3.

Discussion

In the present study, we systematically conducted mutation screening of the PS-1, PS-2 and APP genes in samples from patients diagnosed with varied forms of AD. No pathogenic mutations of the PS-1 or PS-2 genes were identified. In the APP gene, we identified a novel mutation (D678N) in an early-onset FAD pedigree. This mutation is equivalent to an amino acid substitution of Asp at position 7 of amyloid-β (Aβ) (Asp7-Aβ) with Asn (Asn7-Aβ). We hypothesize that Asn7-Aβ derived from the D678N mutant APP has altered fibrillogenic and/or catabolic properties that increase accumulation of protein and/or the neurotoxic potential of Aβ, eventually leading to AD. In vitro studies will be necessary to characterize the pathogenic impact of the D678 mutation on fibrillogenesis and/or secretase activity.

We identified 2 novel SNPs in the APP gene and 9 SNPs (including 1 novel SNP) in the PS-2 gene. The IVS17-10 T/C polymorphism of the APP gene, identified from an EOSAD patient (age of onset: 59 years), is close to a splicing acceptor site and may raise a possibility to influence splicing efficiency. The genetic case-control study of this polymorphism is currently under way. While 861 C/T (P287P) polymorphism of the PS-2 gene was found in an FAD pedigree of variable age of onset (age 63–75), we have not obtained sufficient segregation data regarding this mutation. Furthermore, since 861 C/T (P287P) is a silent polymorphism, it is unclear whether this polymorphism (or a linked mutation) of the PS-2 gene contributes to the development of this FAD pedigree.

Genetic linkage studies have demonstrated multiple susceptible loci for FAD [22–26]. Additional studies are required to identify as many candidate genes as possible to elucidate the pathomechanisms of AD and improve our strategies for treatment and prevention of AD.

Acknowledgements

The authors thank all the families for their participation and the clinicians for their clinical effort and sample collection. Supported in part by a Grant-in-Aid for Encouragement of Young Scientists from the Japan Society for the Promotion of Science, Japan (Y.W.), a Grant-in-Aid for Scientific Research on Priority Areas (C) – Advanced Brain Science Project – from the Ministry of Education, Culture, Sports, Science and Technology, Japan (K.U.), a Grant-in-Aid for Scientific Research on Priority Areas (C) from the Ministry of Education, Culture, Sports, Science and Technology (K.N.), Japan.

References

1 Sherrington R, Rogaev EI, Liang Y, Rogaeva EA, Levesque G, Ikeda M, Chi H, Lin C, Li G, Holman K, et al: Cloning of a gene bearing missense mutations in early-onset familial Alzheimer's disease. Nature 1995;375:754–760.
2 Rogaev EI, Sherrington R, Rogaeva EA, Levesque G, Ikeda M, Liang Y, Chi H, Lin C, Holman K, Tsuda T, et al: Familial Alzheimer's disease in kindreds with missense mutations in a gene on chromosome 1 related to the Alzheimer's disease type 3 gene. Nature 1995;376:775–778.
3 Goate A, Chartier-Harlin MC, Mullan M, Brown J, Crawford F, Fidani L, Giuffra L, Haynes A, Irving N, James L, et al: Segregation of a missense mutation in the amyloid precursor protein gene with familial Alzheimer's disease. Nature 1991;349:704–706.
4 Saunders AM, Strittmatter WJ, Schmechel D, George-Hyslop PH, Pericak-Vance MA, Joo SH, Rosi BL, Gusella JF, Crapper-MacLachlan DR, Alberts MJ, et al: Association of apolipoprotein E allele epsilon 4 with late-onset familial and sporadic Alzheimer's disease. Neurology 1993;43: 1467–1472.
5 Kamino K, Sato S, Sakaki Y, Yoshiiwa A, Nishiwaki Y, Takeda M, Tanabe H, Nishimura T, Ii K, St George-Hyslop PH, Miki T, Ogihara T: Three different mutations of presenilin 1 gene in early-onset Alzheimer's disease families. Neurosci Lett 1996;208:195–198.
6 Yasuda M, Maeda K, Hashimoto M, Yamashita H, Ikejiri Y, Bird TD, Tanaka C, Schellenberg GD: A pedigree with a novel presenilin 1 mutation at a residue that is not conserved in presenilin 2. Arch Neurol 1999;56:65–69.
7 Tanahashi H, Mitsunaga Y, Takahashi K, Tasaki H, Watanabe S, Tabira T: Missense mutation of S182 gene in Japanese familial Alzheimer's disease. Lancet 1995;346:440.
8 Tanahashi H, Kawakatsu S, Kaneko M, Yamanaka H, Takahashi K, Tabira T: Sequence analysis of presenilin-1 gene mutation in Japanese Alzheimer's disease patients. Neurosci Lett 1996; 218:139–141.
9 Kamimura K, Tanahashi H, Yamanaka H, Takahashi K, Asada T, Tabira T: Familial Alzheimer's disease genes in Japanese. J Neurol Sci 1998;160:76–81.
10 Yasuda M, Maeda S, Kawamata T, Tamaoka A, Yamamoto Y, Kuroda S, Maeda K, Tanaka C: Novel presenilin-1 mutation with widespread cortical amyloid deposition but limited cerebral amyloid angiopathy. J Neurol Neurosurg Psychiatry 2000;68:220–223.
11 Sugiyama N, Suzuki K, Matsumura T, Kawanishi C, Onishi H, Yamada Y, Iseki E, Kosaka K: A novel missense mutation (G209R) in exon 8 of the presenilin 1 gene in a Japanese family with presenile familial Alzheimer's disease. Mutation in brief no. 254. Online. Hum Mutat 1999; 14:90.
12 Takao M, Ghetti B, Hayakawa I, Ikeda E, Fukuuchi Y, Miravalle L, Piccardo P, Murrell JR, Glazier BS, Koto A: A novel mutation (G217D) in the presenilin 1 gene (PSEN1) in a Japanese family: Presenile dementia and parkinsonism are associated with cotton wool plaques in the cortex and striatum. Acta Neuropathol (Berl) 2002;104:155–170.
13 Sodeyama N, Iwata T, Ishikawa K, Mizusawa H, Yamada M, Itoh Y, Otomo E, Matsushita M, Komatsuzaki Y: Very early onset Alzheimer's disease with spastic paraparesis associated with a novel presenilin 1 mutation (Phe237Ile). J Neurol Neurosurg Psychiatry 2001;71:556–557.
14 Poorkaj P, Sharma V, Anderson L, Nemens E, Alonso ME, Orr H, White J, Heston L, Bird TD, Schellenberg GD: Missense mutations in the chromosome 14 familial Alzheimer's disease presenilin 1 gene. Hum Mutat 1998;11:216–221.
15 Matsubara-Tsutsui M, Yasuda M, Yamagata H, Nomura T, Taguchi K, Kohara K, Miyoshi K, Miki T: Molecular evidence of presenilin 1 mutation in familial early onset dementia. Am J Med Genet 2002;114:292–298.

16 Sato S, Kamino K, Miki T, Doi A, Ii K, St George-Hyslop PH, Ogihara T, Sakaki Y: Splicing mutation of presenilin-1 gene for early-onset familial Alzheimer's disease. Hum Mutat 1998(suppl 1); S91–S94.
17 Yasuda M, Maeda S, Kawamata T, Tamaoka A, Yamamoto Y, Kuroda S, Maeda K, Tanaka C: Novel presenilin-1 mutation with widespread cortical amyloid deposition but limited cerebral amyloid angiopathy. J Neurol Neurosurg Psychiatry 2000;68:220–223.
18 Matsushita S, Arai H, Okamura N, Ohmori T, Takasugi K, Matsui T, Maruyama M, Iwatsubo T, Higuchi S: Clinical and biomarker investigation of a patient with a novel presenilin-1 mutation (A431V) in the mild cognitive impairment stage of Alzheimer's disease. Biol Psychiatry 2002; 52:907–910.
19 Matsumura Y, Kitamura E, Miyoshi K, Yamamoto Y, Furuyama J, Sugihara T: Japanese siblings with missense mutation (717Val→Ile) in amyloid precursor protein of early-onset Alzheimer's disease. Neurology 1996;46:1721–1723.
20 McKhan G, Drachman D, Folstein M, Katzman R, Price D, Stadlan E. Clinical diagnosis of Alzheimer's disease: Report of the NINCDS-ARDA Work Group under auspices of Department of Health and Human Services Task Force on Alzheimer's disease. Neurology 1984:34:939–944.
21 Hixson JE, Vernier DT: Restriction isotyping of human apolipoprotein E by gene amplification and cleavage with *Hha*I. J Lipid Res 1990;31:545–548.
22 Myers A, Wavrant-De Vrieze F, Holmans P, Hamshere M, Crook R, Compton D, Marshall H, Meyer D, Shears S, Booth J, Ramic D, Knowles H, Morris JC, Williams N, Norton N, Abraham R, Kehoe P, Williams H, Rudrasingham V, Rice F, Giles P, Tunstall N, Jones L, Lovestone S, Williams J, Owen MJ, Hardy J, Goate A: Full genome screen for Alzheimer disease: Stage II analysis. Am J Med Genet 2002;114:235–244.
23 Blacker D, Bertram L, Saunders AJ, Moscarillo TJ, Albert MS, Wiener H, Perry RT, Collins JS, Harrell LE, Go RC, Mahoney A, Beaty T, Fallin MD, Avramopoulos D, Chase GA, Folstein MF, McInnis MG, Bassett SS, Doheny KJ, Pugh EW, Tanzi RE: Results of a high-resolution genome screen of 437 Alzheimer's disease families. Hum Mol Genet 2003;12:23–32.
24 Kehoe P, Wavrant-De Vrieze F, Crook R, Wu WS, Holmans P, Fenton I, Spurlock G, Norton N, Williams H, Williams N, Lovestone S, Perez-Tur J, Hutton M, Chartier-Harlin MC, Shears S, Roehl K, Booth J, Van Voorst W, Ramic D, Williams J, Goate A, Hardy J, Owen MJ: A full genome scan for late onset Alzheimer's disease. Hum Mol Genet 1999;8:237–245.
25 Pericak-Vance MA, Bass MP, Yamaoka LH, Gaskell PC, Scott WK, Terwedow HA, Menold MM, Conneally PM, Small GW, Vance JM, Saunders AM, Roses AD, Haines JL: Complete genomic screen in late-onset familial Alzheimer disease. Evidence for a new locus on chromosome 12. JAMA 1997;278:1237–1241.
26 Lendon C, Craddock N: Susceptibility gene(s) for Alzheimer's disease on chromosome 10. Trends Neurosci 2001;24:557–559.

Yosuke Wakutani
Department of Neurology, Institute of Neurological Sciences
Faculty of Medicine, Tottori University
Nishimachi 36–1, Yonago, Tottori 683–8504 (Japan)
Tel. +81 859 34 8034, Fax +81 859 34 8083, E-Mail ywaku@grape.med.tottori-u.ac.jp

Takeda M, Tanaka T, Cacabelos R (eds): Molecular Neurobiology of Alzheimer Disease and Related Disorders. Basel, Karger, 2004, pp 164–171

Oxidative Stress in Alzheimer Disease: The Earliest Cytological and Biochemical Feature

Akihiko Nunomura[a], *Shigeru Chiba*[a], *Atsushi Takeda*[b], *Mark A. Smith*[c], *George Perry*[c]

[a] Department of Psychiatry and Neurology, Asahikawa Medical College, Asahikawa,
[b] Department of Neurology, Tohoku University School of Medicine, Sendai, Japan,
[c] Institute of Pathology, Case Western Reserve University, Cleveland, Ohio, USA

Alzheimer Disease and Its Pathological Hallmarks

Alzheimer disease (AD) is defined pathologically by amyloid-β (Aβ) senile plaques and neurofibrillary tangles (NFTs) composed of tau. From the time of their original description, nearly a century ago, a major focus has been to understand the role that these lesions play in the pathogenesis of AD. Although senile plaques and NFTs are pathological hallmarks of AD, it is still questionable whether these pathological alterations cause associated neurodegeneration or behavioral and cognitive deficits that accompany the disease. Senile plaques and NFTs are present in a considerable percentage of brains of cognitively normal elderly subjects. Surprisingly, a study investigating autopsied subjects aged between 69 and 100 who were cognitively normal revealed that 49% of those normal subjects met the Khachaturian criteria for AD based on senile plaque density, 25% met the CERAD criteria based on senile plaque density, and 24% were in stages IV–VI of the Braak and Braak staging of AD based on NFT density [1]. Furthermore, it is well known that there is no correlation or a poor correlation between neuronal loss and senile plaque density as well as between disease severity and senile plaque density in AD [2]. By contrast, neuronal loss and clinical severity correlate with NFT density; however, the amount of neuronal loss largely exceeds the amount of

NFTs [3]. Indeed, neurons with NFTs are estimated to be able to survive for decades [4], which suggests that NFTs themselves are not obligatory for neuronal death in AD.

Therefore, senile plaques and NFTs may not be indispensable for death of vulnerable neurons in AD. We cannot exclude the possibility that the processes of senile plaques and NFTs formation are involved in compensatory changes of the aging brain against the pathogenesis of AD [5].

Aging, Oxidative Stress, and Alzheimer Disease

AD is a disease with a prevalence that increases exponentially throughout aging, with about half of the population afflicted by the age of 95 [6], which strongly supports an association between advancing age and AD. As in other organ systems, cells in the brain encounter a cumulative burden of oxidative and metabolic stress that may be a universal feature of the aging process as well as a major causal factor of senescence. Each of the macromolecules including nucleic acids, proteins, and lipids, is oxidatively modified during aging. Indeed, the brain is especially vulnerable to free radical damage because of its high oxygen consumption rate, abundant lipid content, and relative paucity of antioxidant enzymes compared with other organs [7, 8].

One gene that appears to have an influence on aging in general as well as on the risk of AD is apolipoprotein E (APOE). Individuals with an APOE4 allele have a reduced life span [9] and are at risk of AD [10]. Interestingly, APOE shows allele-specific antioxidant activity, with APOE2 being the most effective and APOE4 being the least effective, which suggests a link between oxidative stress and AD [11]. Actually, in autopsied brains with AD, there are increases in lipid peroxidation, protein oxidation, and DNA oxidation. The representatives of these oxidized products demonstrated in AD are 4-hydroxynonenal, protein carbonyls, and 8-hydroxydeoxyguanosine, respectively [12]. In 1999, we reported increased RNA oxidation in AD by in situ demonstration of an oxidized nucleoside derived from RNA, 8-hydroxyguanosine (8OHG) in vulnerable neurons of AD [13]. A significant increase in 8OHG in vulnerable neurons has also been demonstrated in Parkinson disease and Lewy body dementia, where increases in oxidized products of lipid, protein, and DNA have been shown as well [14, 15]. Certainly, oxidative stress is not a specific feature of AD but a common feature of age-associated neurodegeneration. One would expect that this oxidative damage is a common final product resulting from various pathways of neurodegeneration, but this is not the case, at least in AD.

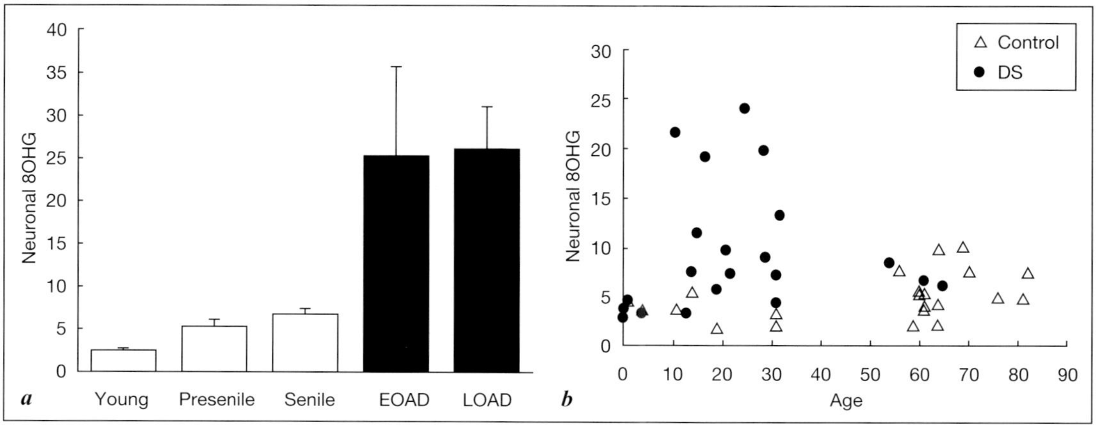

Fig. 1. Relative scale of 8OHG immunoreactivity with 1F7 antibody in AD and DS. Details of the methods for semiquantitative image analysis were described previously [13]. ***a*** The hippocampal subiculum neurons of 27 controls (4 young controls, 10 presenile controls, and 13 senile controls) and 16 cases with AD, i.e. 4 cases of early-onset AD (EOAD) and 12 cases of late-onset AD (LOAD), were examined. Values shown are the means with SE. The difference among all groups is significant by ANOVA ($p < 0.0001$) with post hoc analysis showing significant differences between each control group and EOAD as well as between each control group and LOAD. ***b*** Neurons in layer III of the occipitotemporal gyrus of 23 controls and 22 cases with DS were examined. Individuals with DS show elevation in neuronal 8OHG in their teens and twenties, while controls maintain low levels of neuronal 8OHG between the ages of 4 months and 82 years. In cases with DS, ANOVA followed by post hoc analysis reveals significantly higher 8OHG levels in age groups of the second and third decades compared with the first decade ($p < 0.01$).

Oxidized RNA Nucleoside: A Marker for Steady-State Levels of Oxidative Stress

We selected an in situ approach to identify the oxidized nucleoside 8OHG in AD brains. Immunocytochemically, neurons exhibiting marked immunoreactivity with 8OHG in the cytoplasm were widely distributed in the hippocampal region and cerebral neocortex, whereas neuronal cytoplasm was immunolabeled only faintly in controls. Relative intensity measurements of neuronal 8OHG immunoreactivity showed that there was a significant increase in nucleic acid oxidation in AD compared with controls (fig. 1a). Treatment with nuclease (DNase or RNase) before the immunostaining demonstrated that RNA was a major site of nucleic acid oxidation in AD [13], which was further supported by immunolabeling with ribosomal structures in immunoelectron microscopy for

8OHG [16]. Importantly, we found completely overlapping distributions of neurons showing increased 8OHG and other oxidative stress markers, such as mitochondrial DNA deletion detected with in situ hybridization as well as nitrotyrosine, a protein modification, detected with immunocytochemistry [16]. Because there is no evidence that 8OHG and nitrotyrosine, a non-crosslink protein adduct, are accumulated in cells, the levels of these oxidative markers are expected to reflect steady-state balance rather than a history of oxidative damage [17]. Therefore, an evaluation of the relative levels of these oxidative markers in vulnerable neurons in cases of AD with histopathological alterations of various severity enables us to investigate the relationship between oxidative stress and histological alterations.

Oxidative Stress: The Earliest Event in Alzheimer Disease

To determine whether oxidative stress is an early- or end-stage event in the process of neurodegeneration in AD, we investigated the relationship between the levels of oxidative damage and the extent of Aβ plaques and NFT formation, as well as the duration of dementia. Immunocytochemistry revealed that the levels of neuronal 8OHG and nitrotyrosine, markers of RNA and protein oxidation, were parallel in AD cases [16]. When we measured the area covered by Aβ deposition by image analysis, surprisingly, we found that increases in Aβ deposition were associated with decreased oxidative damage in neurons. Furthermore, a similar pattern of decrease in neuronal 8OHG was noted with increasing disease duration [16]. Moreover, when we investigated the effect of NFTs on the relative amount of 8OHG in neurons, we found that neurons with NFTs showed a 40–60% decrease in relative 8OHG levels compared with neurons free of NFTs [16]. These observations indicate that increased RNA oxidation and protein oxidation are early events in AD.

The early involvement of oxidative stress in the pathological cascade of AD is further supported by an investigation of a series of Down syndrome (DS) brains, in which an AD-like neuropathology is observed as an invariable feature starting in early adulthood. Double immunostaining with 8OHG and Aβ in cases of DS of various ages revealed that strong immunoreaction with 8OHG in neurons was observed in their teens and twenties, whereas the 8OHG immunoreactivity actually decreases with increasing Aβ deposition [18]. Semiquantitative analysis of neuronal 8OHG immunoreactivity and extent of Aβ burden demonstrated that, as a function of age, individuals with DS showed significant elevations in 8OHG in their teens and twenties but showed decreased 8OHG after 30 (fig. 1b), coincident with the accumulation of Aβ. These findings suggest that increased RNA oxidation occurs prior to the onset of Aβ deposition.

In concordance with our data from postmortem brain samples, cerebrospinal fluid from patients with AD showed not only significantly increased levels of 8OHG compared with controls but also greater elevation of 8OHG with shorter disease duration [19]. Furthermore, a recent study using a transgenic mouse model of AD supported the temporal primacy of oxidative stress versus Aβ deposition; in this model, increased lipid peroxidation preceded Aβ plaque formation in a transgenic mouse overexpressing a human AβPP transgene with a double Swedish mutation [20]. Biochemical analyses revealed that levels of urinary and plasma isoprostane, a product of lipid peroxidation were significantly higher in the transgenic than in the wild-type animals as early as 8 months of age. Increased levels of isoprastane were found as well from the age of 8 months in brain homogenates of the total brain cortex or hippocampus but not of the cerebellum. In contrast, pathologically, Aβ plaque formation occurred at 12 months of age in this animal model. The same research group investigated the levels of isoprostane in urine, plasma, and cerebrospinal fluid samples from subjects with AD and mild cognitive impairment (MCI) [21]. In all kinds of samples, levels of isoprostane were significantly higher in MCI than in controls, and significantly higher in AD than in MCI. The rate of progression from MCI to AD is estimated at approximately 12% per year, supporting the concept that MCI, at least in part, represents the prodromal stage of AD. Therefore, the oxidative stress is involved very early or in the preclinical stage of the pathological cascade of AD.

Oxidative Stress: An Attractive Therapeutic Target of Alzheimer Disease

With the notion that increased oxidative stress is one of the earliest changes in the pathogenesis of AD, it is not surprising that agents inhibiting free radical formation reduce both the incidence and the progression of AD. Agents such as vitamin E, selegiline, *Gingko biloba*, estrogen, and nonsteroidal anti-inflammatory drugs have been proven to have an antioxidant activity and to reduce the incidence and/or the rate of progression of AD [22]. According to two recent prospective epidemiological cohort studies [23, 24], higher dietary intake of antioxidants, especially vitamin E, was associated with a lower risk of AD. However, when we consider the complicated system of the fine regulation of cellular redox balance in human body, it is no wonder that extrinsic in vitro antioxidants may show only limited effects on the reduction of oxidative damage in biological systems. In fact, well-known dietary antioxidants such as carotenoids as well as flavonoids can act as prooxidants in certain experimental conditions [25, 26]. Therefore, we should find ways of activating our intrinsic system in order to reduce oxidative damage, which might effectively slow down disease progression, at least in the subclinical and early stage of AD.

It is well known that experimental animals on dietary restriction that lowers steady-state levels of oxidative stress show various signs of retarded aging [27]. Recently, the association between caloric intake and the risk of AD was reported [28]. Compared with individuals in the lowest quartile of total calorie or fat intake, those in the highest quartile had an increased risk of AD. This increased risk was significant only among individuals carrying the APOE4 allele. The hazard ratios of AD for total calorie intake and fat intake are 2.27 and 2.31, respectively. A life-style-related factor other than diet, i.e. lack of exercise is considered to be a risk factor for AD. Friedland et al. [29] evaluated passive, intellectual, and physical activities by using a scale in terms of 'diversity', expressed in total number of activities, and in terms of 'intensity', expressed in hours per month. In patients with AD, the scores were significantly lower in passive diversity, intellectual diversity, and physical diversity as well as in intellectual intensity in early and middle adulthood. Mattson et al. [8] suggest that dietary restriction, intellectual activity, and exercise promote neuronal survival through decreased oxidative stress. They argue that each of them induces mild cellular stress responses, and consequently neurons respond to these stresses by activating on signaling pathways that produce growth factors and protein chaperones. These aspects may be important to prevent AD or at least delay its onset, especially in subjects at high risk of developing AD, such as APOE4 carriers.

Conclusion

The early involvement of oxidative stress in the pathogenesis of AD is supported by cytological and biochemical analyses of human brain samples of AD, DS and MCI as well as a transgenic animal model of AD. Therefore, reducing oxidative stress is an attractive therapeutic target, especially in subjects at high risk of developing AD, such as APOE4 carriers. Not only intake of antioxidants but also calorie restriction as well as maintaining intellectual and physical activities are possible strategies against oxidative stress.

Acknowledgements

Work in the authors' laboratories is supported by funding from the Japan Society for the Promotion of Science (Grant-in-Aid for Scientific Research (c) 14570902).

References

1 Davis DG, Schmitt FA, Wekstein DR, Markesbery WR: Alzheimer neuropathologic alterations in aged cognitively normal subjects. J Neuropathol Exp Neurol 1999;58:376–388.
2 Neve RL, Robakis NK: Alzheimer's disease: A re-examination of the amyloid hypothesis. Trends Neurosci 1998;21:15–19.
3 Gomez-Isla JH, Hollister R, West H, Mui S, Growdon JH, Petersen RC, Parisi JE, Hyman BT: Neuronal loss correlates with but exceeds neurofibrillary tangles in Alzheimer's disease. Ann Neurol 1997;41:17–24.
4 Morsch R, Simon W, Coleman PD: Neurons may live for decades with neurofibrillary tangles. J Neuropathol Exp Neurol 1999;58:188–197.
5 Smith MA, Casadesus G, Joseph JA, Perry G: Amyloid-β and τ serve antioxidant functions in the aging and Alzheimer brain. Free Radic Biol Med 2002;33:1194–1199.
6 United States General Accounting Office's Report to the Secretary of Health and Human Services: Alzheimer's disease: Estimates of prevalence in the United States. 1998;GAO/HEHS-98–16.
7 Coyle JT, Puttfarcken P: Oxidative stress, glutamate, and neurodegenerative disorders. Science 1993;262:689–695.
8 Mattson MP, Chan SL, Duan W: Modification of brain aging and neurodegenerative disorders by genes, diet, and behavior. Physiol Rev 2002;82:637–672.
9 Heijmans BT, Westendorp RG, Slagboom PE: Common gene variants, mortality and extreme longevity in humans. Exp Gerontol 2000;35:865–877.
10 Katsman R: Apolipoprotein E and Alzheimer's disease. Curr Opin Neurobiol 1994;4:703–707.
11 Miyata M, Smith JD: Apolipoprotein E allele-specific antioxidant activity and effects on cytotoxicity by oxidative insults and β-amyloid peptides. Nat Genet 1996;14:55–61.
12 Markesbery WR, Carney JM: Oxidative alterations in Alzheimer's disease. Brain Pathol 1999; 9:133–146.
13 Nunomura A, Perry G, Pappolla MA, Wade R, Hirai K, Chiba S, Smith MA: RNA oxidation is a prominent feature of vulnerable neurons in Alzheimer disease. J Neurosci 1999;19:1959–1964.
14 Zhang J, Perry G, Smith MA, Robertson D, Olson SJ, Graham DG, Montine TJ: Parkinson's disease is associated with oxidative damage to cytoplasmic DNA and RNA in substantia nigra neurons. Am J Pathol 1999;154:1423–1429.
15 Nunomura A, Chiba S, Kosaka K, Takeda A, Castellani RJ, Smith MA, Perry G: Neuronal RNA oxidation is a prominent feature of dementia with Lewy bodies. Neuroreport 2002;13:2035–2039.
16 Nunomura A, Perry G, Aliev G, Hirai K, Takeda A, Balraj EK, Jones PK, Ghanbari H, Wataya T, Shimohama S, Chiba S, Atwood CS, Petersen RB, Smith MA: Oxidative damage is the earliest event in Alzheimer disease. J Neuropathol Exp Neurol 2001;60:759–767.
17 Sayre LM, Perry G, Smith MA: In situ methods for detection and localization of oxidative stress: Application in neurodegenerative disorders; in Wetzel R (ed): Methods of Enzymology. San Diego, Academic Press, 1999, vol 309, pp 133–152.
18 Nunomura A, Perry G, Pappolla MA, Friedland RP, Hirai K, Chiba S, Smith MA: Neuronal oxidative stress precedes amyloid-β deposition in Down syndrome. J Neuropathol Exp Neurol 2000;59:1011–1017.
19 Abe T, Tohgi H, Isobe C, Murata T, Sato C: Remarkable increase in the concentration of 8-hydroxyguanosine in cerebrospinal fluid from patients with Alzheimer's disease. J Neurosci Res 2002;70:447–450.
20 Pratico D, Uryu K, Leight S, Trojanowski JQ, Lee VM-Y: Increased lipid peroxidation precedes amyloid plaque formation in an animal model of Alzheimer amyloidosis. J Neurosci 2001; 21:4183–4187.
21 Pratico D, Clark CM, Liun F, Lee VM-Y, Trojanowski JQ: Increase of brain oxidative stress in mild cognitive impairment. Arch Neurol 2002;59:972–976.
22 Smith MA, Hirai K, Nunomura A, Perry G: Mitochondrial abnormalities: A primary basis for oxidative damage in Alzheimer's disease. Drug Dev Res 1999;46:26–33.
23 Engelhart MJ, Geerlings MI, Ruitenberg A, van Swieten JC, Hofman A, Witteman JCM, Breteler MMB: Dietary intake of antioxidants and risk of Alzheimer disease. JAMA 2002;287:3223–3229.

24 Morris MC, Evans DA, Bienias JL, Tangney CC, Bennett DA, Aggarwal N, Wilson RS, Scherr PA: Dietary intake of antioxidant nutrients and the risk of incident Alzheimer disease in a biracial community study. JAMA 2002;287:3230–3237.
25 Young AJ, Lowe GM: Antioxidant and prooxidant properties of carotenoids. Arch Biochem Biophys 2001;385:20–27.
26 Chan T, Galati G, O'Brien PJ: Oxygen activation during peroxidase catalysed metabolism of flavones or flavanones. Chem Biol Interact 1999;122:15–25.
27 Sohal RS, Weindruch R: Oxidative stress, caloric restriction, and aging. Science 1996;273:59–63.
28 Luchsinger JA, Tang M-X, Shea S, Mayeux R: Caloric intake and risk of Alzheimer disease. Arch Neurol 2002;59:1258–1263.
29 Friedland RP, Fritsch T, Smyth KA, Koss E, Lerner AJ, Chen CH, Petot GJ, Debanne SM: Patients with Alzheimer's disease have reduced activities in midlife compared with healthy control-group members. Proc Natl Acad Sci USA 2001;98:3440–3445.

Akihiko Nunomura, MD, PhD
Department of Psychiatry and Neurology, Asahikawa Medical College
Higashi 2–1–1–1, Midorigaoka, Asahikawa 078–8510 (Japan)
Tel. +81 166 68 2473, Fax +81 166 68 2479, E-Mail nuno@asahikawa-med.ac.jp

Takeda M, Tanaka T, Cacabelos R (eds): Molecular Neurobiology of Alzheimer Disease and Related Disorders. Basel, Karger, 2004, pp 172–182

Neurogenesis: A Promising Therapeutic Target for Alzheimer Disease and Related Disorders

Inge Grundke-Iqbal[a], *Yoshitaka Tatebayashi*[b], *Moon H. Lee, Liang Li*[c], *Khalid Iqbal*[a]

[a] New York State Institute for Basic Research in Developmental Disabilities, Staten Island, New York, N.Y., USA,
[b] Laboratory for Alzheimer's Disease, Brain Science Institute, Riken, Wako-shi, Saitama, Japan,
[c] Department of Pathology, Capital University of Medical Sciences, Beijing, PR China

The clinical characteristic of Alzheimer disease (AD), the major cause of dementia in middle-aged to old-individuals, is a progressive decline in cognitive abilities accompanied by behavioral abnormalities. Although considerable progress has been made, especially during the last three decades, in understanding the various molecular mechanisms of neurodegeneration of the Alzheimer type, to date, no therapeutic drugs that directly inhibit and or reverse the disease process are available. While one approach has rightfully devoted most efforts to developing drugs that can inhibit neurodegeneration, another approach is to enhance regeneration of the brain, which can possibly be achieved either through direct brain progenitor cell transplantation or through promotion of neurogenesis. This chapter discusses the rationale and the promise of neurogenesis as a therapeutic approach for disorders of learning and memory such as AD.

Dentate Gyrus Neurogenesis in AD

The two histopathological hallmarks of AD are intraneuronal tangles of paired helical filaments (PHF) and neuritic amyloid-β plaques which contain

PHF in dystrophic and degenerating neurites. The major protein subunit of PHF is abnormally hyperphosphorylated tau [1, 2]. In early stages of the disease, plaques and neurofibrillary tangles (NFTs) are found predominantly in the hippocampus, especially the entorhinal cortex. The dentate gyrus is the projecting target of the perforant pathway which is the major cortical input from layer II of the entorhinal cortex to the hippocampus. Both the animal and the human dentate gyrus have the unique property of persistent adult neurogenesis [3].

Presently, very little is known about events that lead to increased neurogenesis. One of these events is cell death induced by injury where new neuronal cells are probably generated through the release of neurotrophins from the surviving cells even in an area like the neocortex that does not normally undergo neurogenesis [4, 5]. The other known condition that leads to neurogenesis is environmental enrichment. Rats held under a combination of both complex inanimate and social stimulation show enhanced dentate gyrus neurogenesis as well as dendritic arborization [6–9]. Enhancement of neurogenesis by an enriched environment appears to be mediated mainly by the inhibition of spontaneous apoptosis or prolonged survival of the progenitors [7]. Environmental enrichment not only enhances spatial learning but also protects the brain from various insults like kainate-induced seizures and excitotoxic injury in rodents [9]. In humans, clinical evidence shows that higher education reduces the risk of AD [10], suggesting that environmental stimulation might be protective even against AD-type cerebral insults.

Fibroblast growth factor-2 (FGF-2), the levels of which are increased in AD brains especially in the limbic area [11, 12], induces upregulation of the expression of tau and glycogen synthase kinase-3 (GSK-3) and GSK-3-mediated phosphorylation of tau in cultured neural progenitor cells derived from adult rat hippocampus (adult hippocampal progenitor cells; AHPs [13]). FGF-2, at the concentrations used in the above study, facilitates mainly proliferation and inhibits neuronal differentiation of AHPs [14], giving rise to the possibility that high levels of FGF-2 in AD might inhibit neurogenesis.

The expression of Musashi-1, a marker of undifferentiated progenitors [15], shows that the distribution of Musashi-1-positive immature progenitor cells in the dentate gyrus of AD and control cases is similar. However, the immunohistochemical staining for MAP2, a marker of neuronal differentiation, is dramatically decreased in the granular layer in AD brains [unpublished data]. The decrease in MAP2 levels is associated with a marked increase in the expression of tau and FGF-2, suggesting a dendritic to axonal polarity shift in association with increased expression of FGF-2 in AD.

Reversal of the FGF-2-Mediated Axonal Polarity Shift and Inhibition of Spontaneous Apoptosis by Cerebrolysin, an Anti-Dementia Peptidergic Preparation

FGF-2 induces increased expression of phosphorylated tau in cultured AHPs and increases the expression and the phosphorylation of tau at the Tau-1 site in AHPs, probably through increased expression of GSK-3β [13, 16]. In contrast, FGF-2 negatively regulates the expression of the dendritic marker MAP2. The negative effect of FGF-2 on MAP2 expression is alleviated pharmacologically [16] by the neurotrophic anti-dementia drug cerebrolysin (CL; Ebewe Pharmaceuticals, Austria) [17]. CL consists of amino acids and small peptides derived from porcine brain by proteolytic digestion. At 5 or 10 μl/ml, CL increases MAP2 and synapsin expression in AHPs. At 20 μl/ml, CL loses its effect, indicating that CL probably has a bell–shaped dose action. CL most probably contains a factor(s) that can counteract FGF-2 and increases the expression of neuronal markers in early passages of AHPs. In multiply passaged cells (>P6), CL fails to induce MAP2 expression, indicating that long exposure to FGF-2 in culture alone may change cellular commitment. CL also counteracts the effect of FGF-2 on the expression of tau. Addition of CL decreases tau levels significantly even in multiply passaged cells. Consistent with these findings, CL induces more mature neuron–like phenotypes in early passage cultures with increased MAP2 and decreased tau levels in association with their proper compartmentalization. In the CL-treated cultures some clearly differentiated neurons with long tau-dominated axon-like processes can be seen. These studies [16] indicate that FGF-2, which is mitogenic, might drive the polarity of AHPs to an increase in the tau:MAP2 ratio and that CL can counteract this effect.

In vitro BrdU-labeling of dividing AHPs for 24 h shows that the percentage of BrdU-positive cells in the FGF-2-deprived condition (4 days) is 9.9 ± 1.2% (mean ± SE). This percentage increases to 78.1 ± 1.9% following the addition of FGF-2 but not of CL for the last 24 h, suggesting that, unlike FGF-2, CL has no direct mitotic effect on AHPs and increases the cell numbers probably by inhibiting spontaneous apoptosis in AHPs. This is confirmed by TUNEL staining. In the absence of CL, FGF-2 decreases the proportion of TUNEL-positive cells dose-dependently. When CL is added, these proportions are further decreased.

Enhancement of Dentate Gyrus Neurogenesis and Associated Spatial Learning and Memory

A striking difference is found between CL-treated (CL rats) and control rats (only BrdU-treated) in the distribution and the number of BrdU-positive

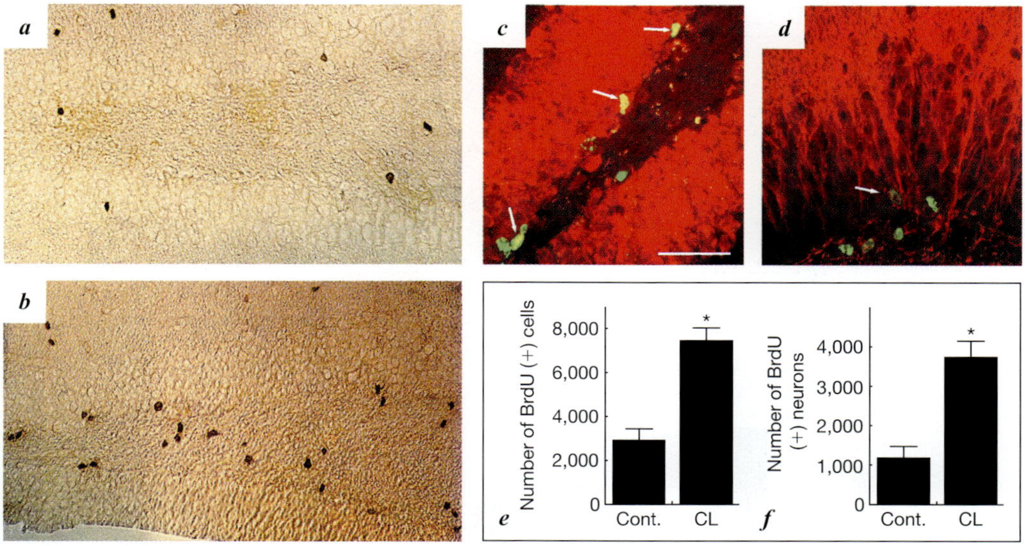

Fig. 1. Enhancement of neurogenesis in dentate gyrus by CL in normal adult rats. BrdU immunohistochemistry and phenotyping in the dentate gyrus of the hippocampus. ***a, b*** Comparison of BrdU-labeling in controls (***a***) and CL rats (***b***) 1 day after the last injection. ***c, d*** Confocal microscopic images of CL rats: arrows indicate double–labeling of BrdU-positive cells (green and/or yellow) with NeuN (red) (***c***) and MAP2 (red) (***d***). ***c*** Some BrdU-positive cells are NeuN-negative (green only). Scale bar, 50 μm. ***e, f*** Numbers of BrdU-positive (***e***) and NeuN-BrdU-positive cells (***f***) in total GSL per brain. The reference volumes used were $1.39 \pm 0.074\,mm^3$ for CL and $1.30 \pm 0.029\,mm^3$ for control hippocampi ($p < 0.05$; Mann-Whitney U test); *$p < 0.01$ (t test). Reproduced with permission from Tatebayashi et al. [17].

cells, with a high frequency of pairs of BrdU-positive cells, most probably two daughter cells from one parent cell in CL rats (fig. 1). Quantitative analysis in the dentate gyrus granular/subgranular layer (GSL) shows that CL rats have about 2.5 times more BrdU-positive cells than control animals. The number of newborn neurons (BrdU- and NeuN-positive) in the CL rats is almost three times higher than in the controls (fig. 1). Furthermore, small numbers of weakly MAP2-positive newborn cells are found mostly in CL rats.

Both CL and control rats show a progressively reduced escape latency to reach the hidden platform in the Morris water maze (fig. 2). However, analysis of probe trials given immediately after 4 days of training shows that CL rats perform better than the controls: the search by CL rats is more concentrated in the quadrant where the platform had been located previously ($p < 0.05$). In addition, CL rats cross the precise location of platform position more often than the controls ($p < 0.05$).

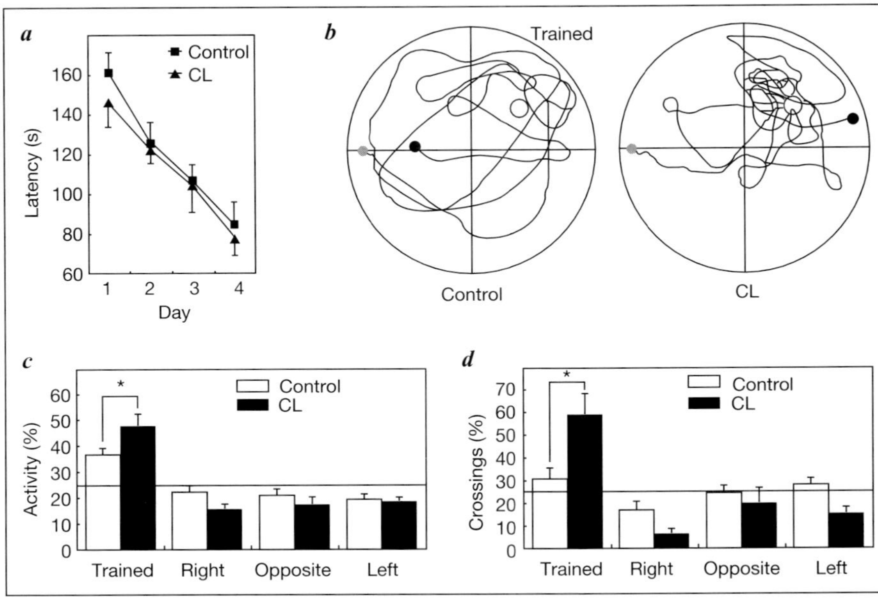

Fig. 2. Enhanced spatial memory of CL rats. ***a*** Escape latency across 4 days of training. ANOVA shows the expected training day effect ($F_{1, 14} = 17.38$, $p < 0.001$). ***b*** Representative swimming paths of the best learners of each group in 60-second trials, the empty circle in the upper right quadrant represents the position of the platform during the last training session. ***c*** Percent activity in each quadrant during the trials. ANOVA with treatment as between-subject factors and quadrant activity as within-subject factors revealed a significant main effect of quadrant ($F_{3, 42} = 30.56$, $p < 0.001$) and an interaction between treatment and quadrant ($F_{3, 42} = 3.25$, $p < 0.05$). Planned comparisons showed that whereas both groups spent more time in trained quadrant (Q1) than any other quadrants ($p < 0.05$), CL rats showed a significantly higher activity than controls in Q1 (*$p < 0.05$). ***d*** Percent crossings during the trials. CL rats showed significantly higher crossings than controls over the trained platform position than other putative platform areas (*$p < 0.05$). Reproduced with permission from Tatebayashi et al. [17].

Therapeutic Promise of the Dentate Gyrus Neurogenesis in AD

Elucidation of the functions, regulatory mechanisms and the development of clinically effective pharmacological methods for the enhancement of adult neurogenesis are one of the most promising directions of the neural stem cell technologies for the treatment of neurodegenerative and neuropsychiatric disorders and developmental disabilities. Neurogenesis in the brain is strongly affected by the composition of hormonal and neurotrophic factors and neurotransmitters in the microenvironment of the neural progenitors [6]. FGF-2, which is twice as high in AD brains as in normal brains [12], dose-dependently increases the expression of tau and its phosphorylation in progenitor cells isolated from adult rat hippocampus [13]. We observed the presence of Musashi-1-positive early neural progenitor cells in the dentate gyrus of both aged controls and AD patients. These cells resemble the Musashi-1-stained cells previously described in the granular layer of rat hippocampus [15], but which are morphologically distinct from the BrdU-labeled progenitors in the dentate gyrus of aged cancer patients after BrdU injection [3]. In the latter case, these cells were already neuronally committed whereas the Musashi-1-positive dentate gyrus progenitors as well as the isolated AHPs probably represent a more immature stage of precursor with the capability to differentiate either into neurons or astroglia [13, 15]. The presence of these precursors in all layers of the human dentate gyrus and even in the pyramidal layer of the hippocampus (unpublished results) indicate a much wider potential for regeneration than was previously assumed.

Previous in vivo studies in rats and monkeys have shown the development from precursors to morphologically, immunologically and electrophysiologically mature dentate gyrus granular cells within 4 weeks [18]. We observed an almost complete lack of the dendritic marker MAP2 in the granular neurons of AD hippocampi, which seem to have at least as many neural progenitor cells as control hippocampi. In light of our findings [16] on AHPs, in which we found that FGF-2 dose-dependently depressed the expression of MAP2 and increased that of tau, it is possible that due to increased FGF-2 levels the cells of the dentate gyrus might have undergone similar changes in polarity, leading to dysregulation of neurogenesis in AD. These cells because of their relatively young age might possess a higher plasticity than the other cells of the hippocampus and therefore be more susceptible to changes in their microenvironment. The reversal of mature oligodendrocytes to multipotent progenitors by changes in neurotrophic factors has been shown previously [19]. Thus, the FGF-2-mediated disturbance of neurogenesis might be a polarity shift and or inhibition of maturation. Studies in AHPs and 8- to 12-month old rats show that the neurotrophic Alzheimer drug CL, a proteolytic digest of brain proteins, can counteract the effect of FGF-2 on cell polarity in AHPs and enhance both neurogenesis and normal memory in rats [16]. Thus, in the AD brain, the modulation of its altered microenvironment through systemically applied trophic factors or their peptides might lead to improved cognition through enhanced neurogenesis including neuronal maturation.

CL appears to enhance neurogenesis probably by the direct inhibition of the spontaneous apoptosis of progenitors. CL decreases the number of TUNEL-positive AHPs, but unlike FGF-2, it has no mitotic effect on the cultured cells. Consistent with these findings, BrdU-positive newborn cells are found more frequently as pairs in CL-treated rats than in controls (fig. 1), indicating that CL probably rescues one of the daughter cells of a parent progenitor which would probably die by apoptosis in controls. The contrasting effects of FGF-2 and CL on tau and MAP2 might be ascribed to their effect(s) on polarity formation, an essential event during developmental patterning [20], asymmetrical cell division [21], and neuronal maturation [22]. Our findings that CL increases neuron-like differentiated AHPs, through upregulation of MAP2 and downregulation of tau, indicate the involvement of polarity formation in adult neurogenesis and raise the possibility that dysregulation of polarity such as that induced by high levels of FGF-2 might perturb neurogenesis. FGF-2 is a mitogen for adult progenitors [23], but its role in neurogenesis has not been fully elucidated. The upregulation of tau and downregulation of MAP2 by FGF-2 observed by us suggest that normally FGF-2 might also be an axogenic stimulant for neural progenitors. The normal adult brain contains relatively high levels of FGF-2 [12] (10–11 ng/mg in Brodmann areas 10, 11, 21 and 22), but most of it appears to be inactive, binding with extracellular matrix and basement membranes [24–26]. In the peripheral tissue, its action is stringently regulated and becomes activated mostly upon tissue injury and inflammation [27–29]. In the spinal cord, injury facilitates the proliferation of adult progenitors but only with glial differentiation [30]. Brain injury also shows a similar tendency [31], indicating that the increased levels of active FGF-2 after injury might reach an inhibitory level for adult neurogenesis. Furthermore, we have also found that long-term treatment (multiple passages) of AHPs with FGF-2 dose dependently induces upregulation of GSK-3β [14], a signaling component downstream of Notch that inhibits the neuronal commitment of progenitors during development [32, 33]. These data suggest that long-term exposure of progenitors to high levels of FGF-2 might inhibit adult neurogenesis by keeping progenitors 'uncommitted' (or giving them more chance to alter their lineage potential to 'gliogenic') [14, 15] or, even if neuronally committed, keeping them in cell cycle and/or preventing them from being fully functional probably through the polarity shift. Similar events may also play a role in the AD dentate gyrus, where the decrease in MAP2 as well as the increase in tau might be a consequence of the effect of the elevated levels of FGF-2 on neurogenesis and neuronal differentiation.

The clinical action of CL in AD as well as our observation of its action on the enhancement of both dentate gyrus neurogenesis and spatial memory in rats support the concept put forward previously [34, 35] that the dentate gyrus neurogenesis may play a role in cognitive performance and mood in humans. Furthermore, CL has been known to have an additive effect when used in conjunction with antidepressants which have recently been shown to increase dentate gyrus neurogenesis in rodents presumably via their common target serotonin. Therefore, impaired dentate gyrus neurogenesis in AD might cause cognitive impairment and depression. A prospective clinical study has recently revealed that motivation-related depressive symptoms (i.e., lack of interest, psychomotor change, loss of energy, concentration difficulties) occur frequently in the preclinical phase of AD [36] with incipient mild cognitive impairment.

The pathogenic potential of polarity shift in AD might be profound since several findings indicate its involvement in AD neuropathology. For example, both tau and amyloid precursor protein, which are extensively involved in AD histopathology, are axonal proteins [37, 38]. The preferential occurrence of tau pathology in the apical dendrites of affected neurons [2, 37] might also be a case in point. We have also found a dramatic shift of polarity formation in the AD dentate gyrus [unpubl. data]. The levels of FGF-2, an axogenic stimulant, are twice as high in AD brain as in normal brains [13]. Moreover, strongly FGF-2-immunoreactive structures have been observed in the limbic areas of AD brains, including neurofibrillary tangles of the entorhinal cortex and cells around the dentate gyrus [12]. Furthermore, decreased levels of apolipoprotein E (ApoE), a stimulant for dendritic arborization, have been reported as a risk factor for AD [39]. For example, the APOE4 allele has been reported to be associated with decreased ApoE protein levels in AD brain [40]. Consistent with this concept, a loss of ApoE signaling in ApoE-knockout or ApoE receptor and 2-VLDL receptor double-knockout mice has been reported to induce hyperphosphorylation of tau [41, 42]. Extracellular amyloid β might possibly promote this shift both by sequestering apoE in its fibrillar deposits [43] and by stimulating the expression of FGF-2 in astrocytes [44]. In conclusion, the polarity shift might play a role in AD neuropathology, leading to increased levels and abnormal compartmentalization of hyperphosphorylated tau in the neurons and functional abnormalities associated with suppressed dentate gyrus neurogenesis. Therefore, agents that neutralize this shift by increasing neurogenesis and dendritic arborization, such as CL and probably an enriched environment, might be protective against AD. Moreover, elevated levels of FGF-2 also might be a promising target of AD therapy.

Acknowledgements

We thank Drs. T. Beach and L.I. Sue of Sun Health Research Institute, Sun City, AZ for the sections of human hippocampi, Dr. H. Okano, Keio University School of Medicine, Tokyo for the gift of the Musashi 1 antibody and Ms. Yunn-Chyn Tung for excellent technical assistance. We gratefully acknowledge Ms. J. Biegelson and Ms. S. Warren for typing the manuscript. This work was supported in part by the New York State Office of Mental Retardation and Developmental Disabilities, T.L.L. Temple Foundation Discovery

Award for Alzheimer's Disease Research, Alzheimer's Association, USA, and NIH grant AG19158.

References

1 Grundke-Iqbal I, Iqbal K, Quinlan M, Wisniewski HM, Binder L: Microtubule-associated protein tau: A component of Alzheimer paired helical filaments. J Biol Chem 1986;261: 6084–6089.
2 Grundke-Iqbal I, Iqbal K, Tung YC, Quinlan M, Wisniewski HM, Binder LI: Abnormal phosphorylation of the microtubule-associated protein (tau) in Alzheimer cytoskeletal pathology. Proc Natl Acad Sci USA 1986;93:4913–4917.
3 Eriksson PS, Perfilieva E, Björk-Eriksson T, Alborn AM, Noedborg C, Peterson DA, Gage FH: Neurogenesis in the adult human hippocampus. Nat Med 1998;4:1313–1317.
4 Wang Y, Sheen VL, Macklis JD: Cortical interneurons upregulate neurotrophins in vivo in response to targeted apoptotic degeneration of neighboring pyramidal neurons. Exp Neurol 1998;154: 389–402.
5 Magavi SS, Leavitt BR, Macklis JD: Induction of neurogenesis in the neocortex of adult mice. Nature 2000;405:951–955.
6 Gould E, Beylin A, Tanapat P, Reeves A, Shors TJ: Learning enhances adult neurogenesis in the hippocampal formation. Nat Neurosci 1999;2:260–265.
7 Kempermann G, Kuhn HG, Gage FH: More hippocampal neurons in adult mice living in an enriched environment. Nature 1997;386:493–495.
8 Nilsson M, Perfilieva E, Johansson U, Orwar O, Erikson PS: Enriched environment increases neurogenesis in the adult rat dentate gyrus and improves spatial memory. J Neurobiol 1999;39:569–578.
9 Young D, Lawlor PA, Leone P, Dragunow M, During MJ: Environmental enrichment inhibits spontaneous apoptosis, prevents seizures and is neuroprotective. Nat Med 1999;5:448–453.
10 Evans DA, Hebert LE, Beckett LA, Scherr PA, Albert MS, Chown MJ, Pilgrim DM, Taylor JO: Education and other measures of socioeconomic status and risk of incident Alzheimer disease in a defined population of older persons. Arch Neurol 1997;54:1399–1405.
11 Cummings BJ, Su JH, Cotman CW: Neuritic involvement within bFGF immunopositive plaques of Alzheimer's disease. Exp Neurol 1993;124:315–325.
12 Stopa EG, Gonzalez AM, Chorsky R, Corona RJ, Alzarez J, Bird ED, Baird A: Basic fibroblast growth factor in Alzheimer's disease. Biochem Biophys Res Commun 1990;171:690–696.
13 Tatebayashi Y, Iqbal K, Grundke-Iqbal I: Dynamic regulation of expression and phosphorylation of tau by fibroblast growth factor-2 in neural progenitor cells from adult rat hippocampus. J Neurosci 1999;19:5245–5254.
14 Johe KK, Hazel TG, Muller T, Dugich-Djordjevic MM, McKay RDG: Single factors direct the differentiation of stem cells from fetal and adult central nervous system. Genes Dev 1996;10: 3129–3140.
15 Kaneko Y, Sakakibara S, Imai T, Suzuki A, Nakamura Y, Sawamoto K, Osawa Y, Toyama T, Miyata T, Okano H: Musashi 1: An evolutionary conserved marker for CNS progenitor cells including neural stem cells. Dev Neurosci 1999;22:139–153.
16 Tatebayashi Y, Lee MH, Li L, Iqbal K, Grundke-Iqbal I: The dentate gyrus neurogenesis: A therapeutic target for Alzheimer's disease. Acta Neuropathol (Berl) 2003;105:225–232.
17 Rüther E, Ritter R, Apecechea M, Freytag S, Gmeinbauer R, Windisch M: Sustained improvements in patients with dementia of Alzheimer's type (DAT) 6 months after termination of cerebrolysin therapy. J Neural Transm 2000;107:815–829.
18 van Praag H, Schinder AF, Christie BR, Toni N, Palmer TD, Gage FH: Functional neurogenesis in the adult hippocampus. Nature 2002;415:1030–1034.
19 Kondo T, Raff M: Oligodendrocyte precursor cells reprogrammed to become multipotential CNS stem cells. Science 2000;289:1754–1757.

20 Anderson KV: Pinning down positional information: Dorsal-ventral polarity in the Drosophila embryo. Cell 1998;95:439–442.
21 Jan YN, Jan LY: Polarity in cell division: What frames thy fearful asymmetry? Cell 2000;100: 599–602.
22 Craig AM, Banker G: Neuronal polarity. Annu Rev Neurosci 1994;17:267–310.
23 Gage FH, Coates PW, Palmer TD, Kuhn HG, Fisher LJ, Suhonen JO, Peterson DA, Suhr ST, Ray J: Survival and differentiation of adult neuronal progenitor cells transplanted to the adult brain. Proc Natl Acad Sci USA 1995;92:11879–11883.
24 Bashkin P, Doctrow S, Klagsbrun M, Svahn CM, Folkman J, Vlodavsky I: Basic fibroblast growth factor binds to subendothelial extracellular matrix and is released by heparitinase and heparin-like molecules. Biochemistry 1989;28:1737–1743.
25 Folkman J, Klagsbrun M, Sasse J, Wadzinski M, Ingber D, Vlodavsky I: A heparin-binding angiogenic protein – basic fibroblast growth factor – is stored within basement membrane. Am J Pathol 1988;130:393–400.
26 Gonzalez AM, Buscaglia M, Ong M, Baird A: Distribution of basic fibroblast growth factor in the 18-day rat fetus: Localization in the basement membranes of diverse tissues. J Cell Biol 1990;110: 753–765.
27 Flaumenhaft R, Rifkin DB: The extracellular regulation of growth factor action. Mol Biol Cell 1992;3:1057–1065.
28 Frautschy SA, Walicke PA, Baird A: Localization of basic fibroblast growth factor and its mRNA after CNS injury. Brain Res 1991;553:291–299.
29 Kato M, Wang H, Kainulainen V, Fitzgerald ML, Ledbetter S, Ornitz DM, Bernfield M: Physiological degradation converts the soluble syndecan-1 ectodomain from an inhibitor to a potent activator of FGF-2. Nat Med 1998;4:691–697.
30 Johansson CB, Momma S, Clarke DL, Risling M, Lendahl U, Frisen J: Identification of a neural stem cell in the adult mammalian central nervous system. Cell 1999;96:25–34.
31 Logan A, Frautschy SA, Gonzalez AM, Baird A: A time course for the focal elevation of synthesis of fibroblast growth factor and one of its high-affinity receptors (flg) following a localized cortical brain injury. J Neurosci 1992;12:3828–3837.
32 Marcus EA, Kintner C, Harris W: The role of GSK3β in regulating neuronal differentiation in xenopus laevis. Mol Cell Neurosci 1998;12:269–280.
33 Ruel L, Bourouis M, Heitzler P, Pantesco V, Simpson P: Drosophila shaggy kinase and rat glycogen synthase kinase-3 have conserved activities and act downstream of Notch. Nature 1993;362: 557–560.
34 Brown ES, Rush AJ, McEwen BS: Hippocampal remodeling and damage by corticosteroids: Implications for mood disorder. Neuropsychopharmacology 1999;21:474–484.
35 Jacobs BL, Praag H, Gage FH: Adult brain neurogenesis and psychiatry: A novel theory of depression. Mol Psychiatry 2000;5:262–269.
36 Berger AK, Fratiglioni L, Forsell Y, Winblad B, Backman L: The occurrence of depressive symptoms in the preclinical phase of AD: A population-based study. Neurology 1999;53: 1998–2002.
37 Braak E, Braak H, Mandelkow EM: A sequence of cytoskeleton changes related to the formation of neurofibrillary tangles and neuropil threads. Acta Neuropathol (Berl) 1994;87: 554–567.
38 Koo EH, Sisodia SS, Archer DR, Martin LJ, Weidemann A, Beyreuther K, Fischer P, Masters CL, Price DL: Precursor of amyloid protein in Alzheimer disease undergoes fast anterograde axonal transport. Proc Natl Acad Sci USA 1990;87:1561–1565.
39 Masliah E, Mallory M, Ge N, Alford M, Beinbergs I, Roses AD: Neurodegeneration in the central nervous system of apoE-deficient mice. Exp Neurol. 1995;136:107–122.
40 Bertrand P, Poirier J, Oda T, Finch CE, Pasinetti GM: Association of apolipoprotein E genotype with brain levels of apolipoprotein E and apolipoprotein J (clusterin) in Alzheimer disease. Brain Res Mol Brain Res 1995;33:174–178.
41 Genis I, Fisher A, Michaelson DM: Site-specific dephosphorylation of tau of apolipoprotein E-deficient and control mice by M1 muscarinic agonist treatment. J Neurochem 1999;72: 206–213.

42 Hiesberger T, Trommsdorff M, Howell BW, Goffinet A, Mumby MC, Cooper JA, Herz J: Direct binding of Reelin to VLDL receptor and ApoE receptor 2 induces tyrosine phosphorylation of disabled-1 and modulates tau phosphorylation. Neuron 1999;24:481–489.
43 Bales KR, Verina T, Dodel RC, Du Y, Altstiel L, Bender M, Hyslop P, Johnstone EM, Little SP, Cummins DJ, Piccardo P, Ghetti B, Paul SM: Lack of apolipoprotein E dramatically reduces amyloid beta-peptide deposition. Nat Genet 1997;17:263–264.
44 Araujo DM, Cotman CW: Beta-amyloid stimulates glial cells in vitro to produce growth factors that accumulate in senile plaques in Alzheimer's disease. Brain Res 1992;569:141–145.

Inge Grundke-Iqbal, PhD
1050 Forest Hill Road
Staten Island, New York, NY 10314 (USA)
Tel. +1 718 494 5263, Fax +1 718 494 1080, E-Mail i_g_iqbal@yahoo.com

Takeda M, Tanaka T, Cacabelos R (eds): Molecular Neurobiology of Alzheimer Disease and Related Disorders. Basel, Karger, 2004, pp 183–194

Learning Deficits in N279K Tau Transgenic Mice and an Assembly Model of Tau Protein

Taizo Taniguchi[a], *Shogo Matsuyama*[b], *Katsuhiko Minoura*[e], *Hiroyuki Iso*[c], *Masahiro Sasaki*[a], *Koji Tomoo*[e], *Toshimasa Ishida*[e], *Hiroshi Mori*[d], *Chikako Tanaka*[a]

[a] Hyogo Institute for Aging Brain and Cognitive Disorders, Himeji,
[b] Department of Genome Sciences, Kobe University Graduate School of Medicine, Kobe,
[c] Department of Behavior Science, Hyogo College of Medicine, Nishinomiya,
[d] Department of Neuroscience, Osaka City University Medical School, Osaka,
[e] Osaka University of Pharmaceutical Sciences, Takatsuki, Japan

Neurofibrillary lesions composed of tau protein are neuropathological hallmarks of an expanding family of neurodegenerative diseases, such as Alzheimer's disease, Pick's disease, progressive supranuclear palsy and corticobasal degeneration, now collectively known as tauopathies. The discovery of a missense mutation in the tau gene in frontotemporal dementia with parkinsonism linked to chromosome 17 (FTDP-17) has demonstrated that tau dysfunction can cause neurodegeneration and dementia and provided unique opportunities to elucidate the mechanisms underlying the tauopathies. In the present work, we generated transgenic mice that overexpressed human tau with one of the FTDP-17 mutations (N279K) and investigated the synaptic plasticity and learning ability in these mice. We also proposed an assembly model by analyzing the conformation of 3MBD, i.e., dimer formation by the hydrophilic interactions and then the molecular aggregation of these dimer structures by hydrophobic interactions.

Experimental Procedures

Transgenic Constructs and Animals

The human pathogenic mutation N279K was introduced into the cDNA encoding human tau (383 residues) [1] using the Quick Change Site-Directed Mutagenesis Kit

(Stratagene). Transgenic constructs were inserted into the MoPrP vector [2]. Transgenic mice were produced by microinjection of DNA into pronuclei of C57BL/6 × SJL F2 hybrid mouse eggs. Founder animals were intercrossed with C57BL/6 mice to establish lines. Male homozygotes of transgenic mice were used for behavioral and electrophysiological analysis at 5–7 months of age. Transgenic animal production and care were carried out in cooperation with the Behavioral and Medical Sciences Research Consortium (BMRC) and animal handling was done in accordance with the *Guidelines for Animal Experimentation at BMRC*. The numbering of amino acid residues was always based on the longest isoform of human tau (441 residues).

Immunoanalysis

A polyclonal antibody specific for human tau (htaurg2) was raised in a rabbit against a synthetic peptide, TYGLGDRKDQGG. The brains were dissected into 7 regions; the olfactory bulb, cerebral cortex, striatum, hippocampus, diencephalon, brain stem (including mesencephalon, pons and medulla oblongata) and cerebellum. Tissue homogenates were separated on 10% SDS-PAGE and subjected to Western blot analysis as described [3]. The brains were fixed with 4% paraformaldehyde, and frozen sections were subjected to immunohistochemistry as described [4].

Electrophysiology

In vivo long-term potentiation (LTP) recording in the dentate gyrus was done in anesthetized mice (SJLB9, n = 7; UBJAP18, n = 7 and non-Tg, n = 7) as described [5]. A saline-filled glass recording electrode was placed in the dentate granule cell layer. Initial responses were obtained using a cathodal stimulation (6.0–8.0 V, 0.1 Hz, 0.1 ms duration) of the perforant path. After population responses were obtained, the preparation was allowed to stabilize for 60 min prior to baseline recording. Voltage was reduced so that the baseline spike amplitude was one-third of the maximum asymptotic value. The LTP-inducing voltage used was the lowest voltage level that could evoke maximum asymptotic spike amplitude. The parameters for tetanus to induce LTP consisted of eight 0.4-ms 400-Hz pulses. We plotted only the population spike without the EPSP slope during LTP to minimize animal suffering, because the potentiated change of population spikes is similar to that of the EPSP slope during LTP in this procedure [5]. The amplitude of the population spike was measured from initial positivity to peak negativity. The population spikes induced by 5 successive stimulations at 5-min intervals were averaged. Data are expressed as mean ± SEM from 7 mice in each group. Multigroup means were compared using the Scheffé test following one-factor or two-factor factorial analysis of variance (ANOVA).

Behavioral Analysis

Behavioral tests (1 and 2) were done to avoid order effects as described [SJLB9, n = 6; UBJAP18, n = 11 (n = 10 for the second test because of a failure of the data recording) and non-Tg, n = 11] [6]. For data analysis, a mixed-type ANOVA was done and the group difference was tested by a post-hoc test.

Open-Field Test. Animals were allowed to search freely in a square acrylic box (30 × 30 cm) for 20 min. On each X and Y bank of the open field, two infrared ray beams were attached 2 cm above the floor at 10-cm intervals making a flip-flop circuit between two beams. The total number of beam crossings was counted as the traveling behavior of the animal (locomotion). On the X bank, the other 12 infrared ray beams were set 5 cm above the floor at 3-cm intervals, and the total number of beam crossings was counted as the rearing behavior (rearing). After the session, the number of feces was counted.

Morris Water Maze Learning. A swimming pool of 96 cm in diameter was used. Nontoxic India ink was added to make the water opaque. The water was maintained at 20–24°C. A platform of 10 cm in diameter was set in a constant place of a swimming pool. It was submerged 5 mm from the surface of the water in case of maze training, and appeared upon the surface of the water in case of visible training. In the maze and visible learning trial, a mouse was put at one of three quadrants of the pool, where the platform was not positioned. The experimenter counted the swimming time until reaching the platform (latency) with a stopwatch. When the animal reached to the platform, it was allowed to stay there for 10 s, then was returned to a waiting box. If the mouse could not reach to the platform within 60 s, it was put on the platform by the experimenter and the latency was counted as 60 s. The intertrial interval was 30 s.

Structural Analysis

CD Measurements. The 31-residue peptide of 3MBD (corresponding to 306–336 amino acids) was synthesized by American Peptide Company (Calif., USA). It was characterized by mass spectrometry and was pure to >95%. A sample solution of 3MBD was prepared using water or trifluoroethanol (TFE): 2.0×10^{-2} m*M* in water and 4.0×10^{-2} m*M* in TFE. All measurements at 25°C were made with a JASCO J-820 spectrometer in a cuvette with 2 mm of path length. For each experiment under the flow of N_2 gas, the measurement from 190 to 260 nm was repeated eight times and summed up, and the molar ellipticity was determined after normalizing for the protein concentrations. The same experiments were performed at least three times using the newly prepared samples, and their averaged values are given in this chapter. Data were expressed in terms of [θ], the molar ellipticity, in degrees cm^2 $dmol^{-1}$.

NMR Measurements. The peptide was used without further purification by dissolving in TFE-d_2 to prepare the sample solution. ^{1}H-NMR spectra were recorded on a Varian Unity INOVA500 spectrometer with a variable temperature control unit. ^{1}H chemical shifts were referenced to 0 ppm for TSP. From the comparison of the respective NMR spectra under different pH and temperature set-ups, the condition for the usual NMR measurements was finally determined as follows: concentration = 2 m*M*, temperature = 298 K except for special cases, and pH = 3.9 (a solution of pH > 5.0 leads to decreased solubility of the peptide). The pH value was adjusted by adding HCl or NaOH. For the concentration dependence experiment, the chemical shift of each NH proton was measured under three different concentrations (0.5, 1.0 and 2.0 m*M*), because of the solubility problem. For the temperature dependence experiment, the chemical shift of each NH proton was measured in the range of 20–60°C (10°C intervals), respectively.

The two-dimensional spectra of DQF-COSY, TOCSY and NOESY were acquired in the phase-sensitive mode by using standard pulse programs available in the Varian software library. In order to follow direct single- and multiple-relayed through-bond connectivity, TOCSY spectra were successively recorded with mixing times of 40 and 100 ms. The NOESY spectra were also measured with mixing times of 100, 200 and 300 ms. Assuming the same correlation time for all protons, the offset dependence of the NOESY cross-peaks was used for the estimation of proton-proton distances. The NOE intensities were classified into three groups (strong, medium, and weak).

On the other hand, vicinal coupling constants by DQF-COSY measurements were used to estimate possible torsion angles: $^3J_{HNC\alpha H} = 1.9 - 1.4\cos\theta + 6.4\cos^2\theta$, where $\Phi = |\theta - 60|°$ for Φ torsion angle around the C'_{i-1}–N_i–$C\alpha_i$–C'_i bond sequence [7].

Results and Discussion

FTDP-17 with a missense mutation N279K is characterized as familial dementia and parkinsonism with tau pathology in the frontal and temporal cortex, hippocampus, amygdala, caudate nucleus, subthalamic nucleus, globus pallidus, substatia nigra, striatum and pontine tegmentum [8]. To investigate the effects of the N279K mutation, we expressed either human N279K or wild-type tau cDNA containing exon 10 and lacking exons 2 and 3 [1] which encodes 4-repeat tau without amino-terminal inserts (383 amino acids). We generated two homozygous N279K mutation lines (SJLB9 and SJLB2) and two homozygous wild lines (UBJAP4 and UBJAP18). SJLB9 (N279K mice) and UBJAP18 (WILD mice) mice used in this study expressed transgenic tau at a roughly equal level (fig. 1a). Western blots using an antibody specific for human tau proteins revealed that transgenic tau protein was highly expressed in the hippocampus and striatum, moderately in the cerebral cortex, olfactory bulb, diencephalon and cerebellum, and slightly in the brain stem of N279K and WILD mice (fig. 1a). Hyperphosphorylated tau loses the tubulin assembly-promoting activity [9], and abnormal phosphorylation of tau protein may be a crucial step preceding neurofibrillary formation [10]. We investigated the phosphorylation states of tau protein in the brains of these mice. In N279K mice, strong immunoreactivities of hyperphosphorylated tau were detected in the striatum, hippocampus and diencephalon on Western blotting with AT8, and in the cerebral cortex, striatum and hippocampus with AT270. Weak signals were also detected in the hippocampus and striatum with AT180. We also observed phosphorylated tau in these brain regions, especially in the hippocampus of WILD mice, but the signal intensities were much weaker than those in the N279K mice. Phosphorylated tau was virtually not detected by AT8, AT180 and AT270 in nontransgenic mice (non-Tg mice) (fig. 1b). Although light-microscopic histochemical analysis demonstrated diffuse tau immunoreactivities throughout the brain, no apparent neuronal loss and no neurofibrillary tangles were observed in these brain regions of N279K and WILD mice at up to 8 months of age (data not shown).

N279K and WILD mice were viable and had normal gross motor-posture patterns, circadian activity, feeding and body weight. The open-field test revealed no significant differences in locomotion and rearing between the transgenic and non-Tg mice at 5–7 months of age (fig. 2). In mice expressing human tau and its mutant, it has not been demonstrated if transgenic mice undergo changes in synaptic plasticity or learning before the formation of neurofibrillary tangles in the brain.

Spatial learning and memory require integrative control functions of the hippocampus. A process known as LTP which can be induced by a short burst

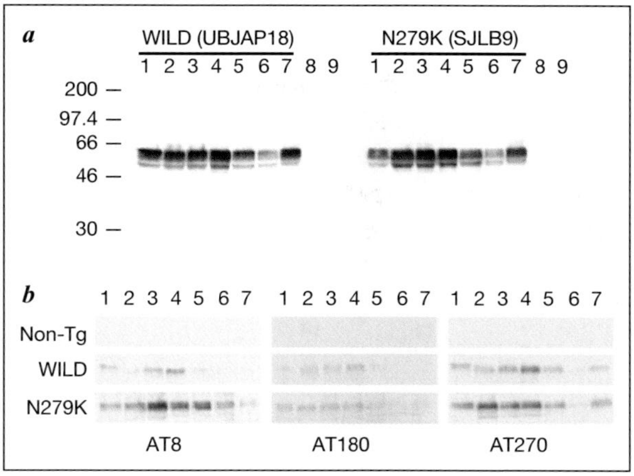

Fig. 1. ***a*** Human tau expression in transgenic mice. Protein samples from olfactory bulb (lane 1), cerebral cortex (lane 2), striatum (lane 3), hippocampus (lane 4), diencephalon (lane 5), brain stem (lane 6), cerebellum (lane 7), non-Tg whole brain (lane 9) and blank (lane 8) were separated by 10% SDS-PAGE and subjected to Western blot analysis with human tau specific antibody (htaurg2). ***b*** Tau phosphorylation in transgenic mice. Protein samples same as ***a*** were separated on 10% SDS-PAGE and tau phosphorylation was analyzed by Western blot analysis using anti-phosphotau antibodies AT8, AT180 and AT270.

of high-frequency stimulation is assumed to underlie plastic changes associated with learning and memory [11]. We determined if LTP in the dentate gyrus of the anesthetized mouse is affected by expressing a 4-repeat isoform of human tau or its N279K mutant. As shown in figure 3, LTP of the population spike in the dentate gyrus was induced by tetanic stimulation of the perforant path in non-Tg mice. In contrast, LTP was significantly suppressed in transgenic mice (fig. 3a, b). Cumulative histograms of the population spike amplitude for 120 min after tetanic stimulation reveal that LTP is reduced to 78% of non-Tg mice in WILD mice and to 53% in N279K mice (fig. 3c). The results provide evidence that expression of human tau and its mutant in these animals induces an impairment of synaptic plasticity in the hippocampus.

We next characterized their behavior in the Morris water maze in our transgenic mice [12]. Non-Tg mice learned to escape to the hidden platform within approximately 20 s after 5 daily 5-trial sessions. In contrast, N279K mice required significantly longer to escape than WILD mice and non-Tg mice (fig. 4). Escape latency was also longer in WILD mice than in non-Tg mice, but the difference was not statistically significant. The performance deficit was due to increased path lengths rather than to decreased swimming velocity in N279K mice (data not shown). Learning was established in the visible-platform task,

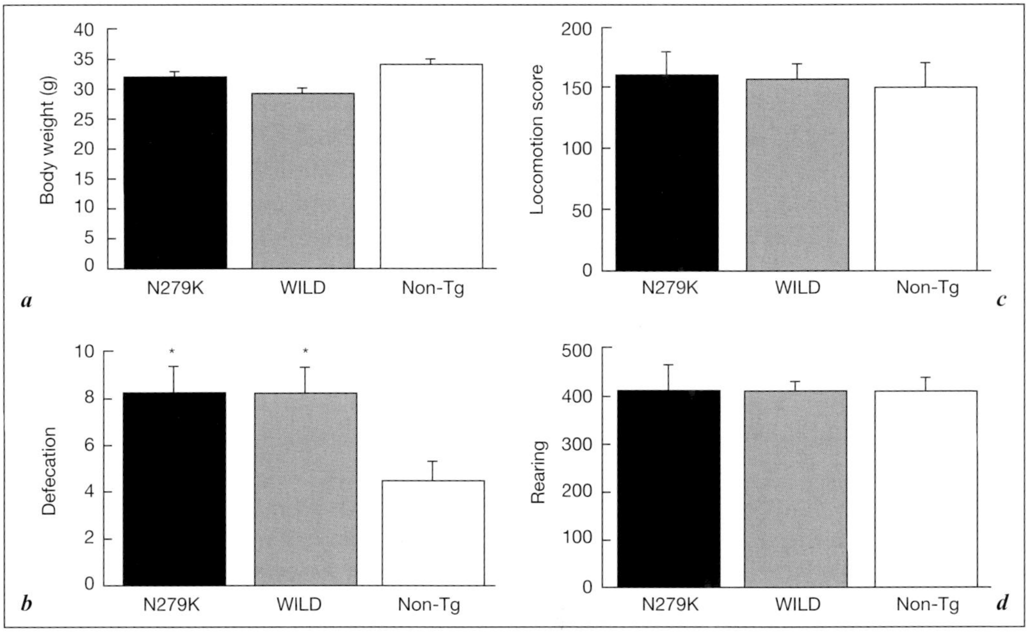

Fig. 2. The results of the open-field test. ***a*** Body weight. ***b*** Defecation. ***c*** Locomotive behavior. ***d*** Rearing during a 20-min test (*significant difference compared with non-Tg mice by Student's t test, $p < 0.05$). Although the transgenic mice showed lighter body-weight and greater defecation than non-Tg mice, there were no significant differences in locomotion and rearing among them.

indicating that they were able to find the platform. In rodents, disruption of NMDA receptor function appears to impair LTP induction as well as performance on the hidden platform Morris water maze task [13]. Our results indicate that expression of the transgene may impair the hippocampal NMDA receptor function.

The P301L mutations also lie in the extra repeat of tau, and induced a reduced ability to promote microtubule assembly. Mice expressing mutant P301L tau show motor disturbance, amyotrophy and neurofibrillary tangles in the brain and spinal cord [14]. However, minimal or no neurofibrillary tangle pathology was observed in wild-type tau transgenic mice [15] and N279K mice. The N279K mutation increased 4-repeat tau [16], but did not reduce the potential to promote microtubule assembly [17]. N279K mice and WILD mice expressed a roughly equal level of 4-repeat tau protein, nevertheless N279K mice showed more severe deficits in spatial learning and in LTP production. These data suggest a specific role for N279K mutation, besides overproduction of 4-repeat tau.

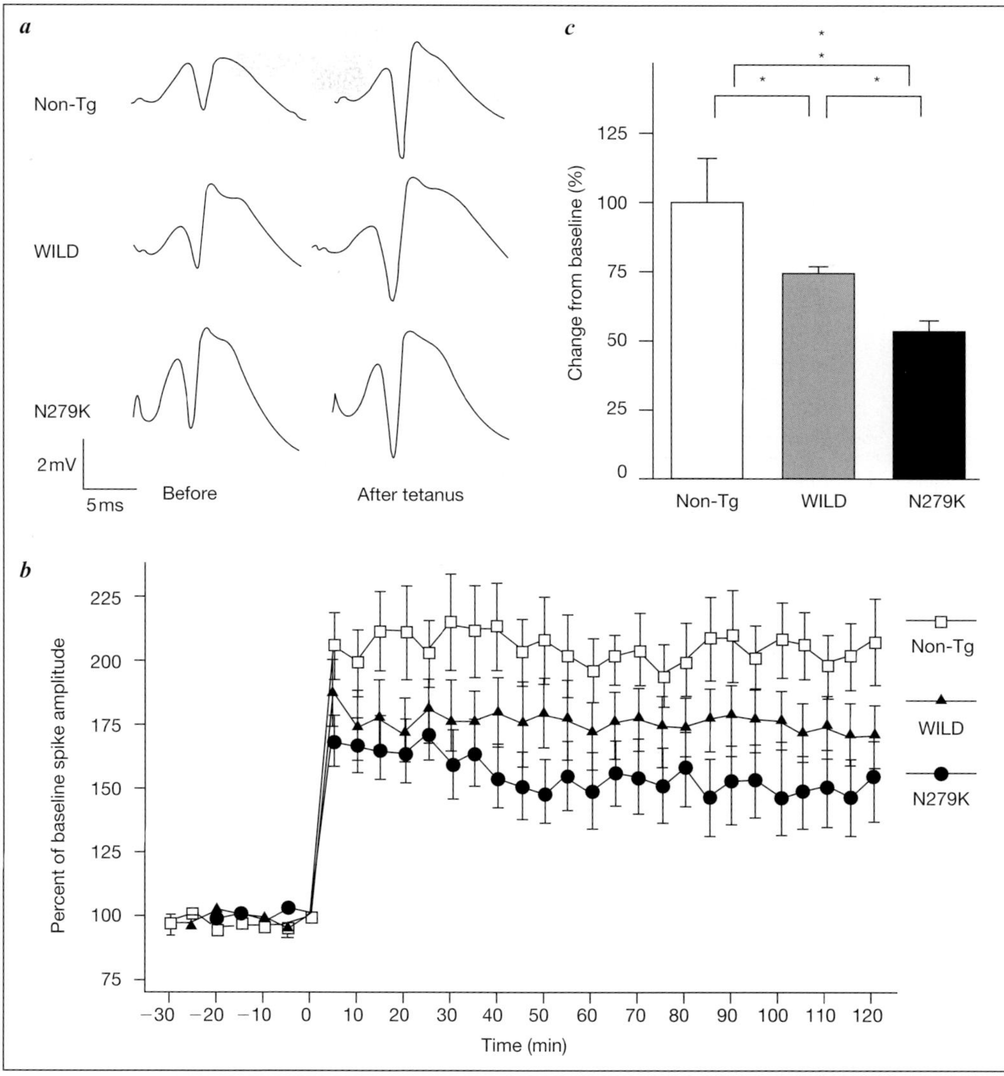

Fig. 3. In vivo LTP in the dentate gyrus. ***a*** Representative traces of the N279K mice, WILD mice and non-Tg mice are recorded from the granule cell layer before and after tetanus. ***b*** Time course of LTP in N279K mice (closed circles), WILD mice (closed triangles) and non-Tg (open squares) mice are shown for 120 min after tetanus at 0 min. Each point represents the mean ± SEM percentage of basal population spike amplitude at 0 min. ***c*** Mean ± SEM of population spikes over 120 min after tetanus in the indicated mice (significant difference by Scheffé's test following two-factor factorial analysis of variance, $*p < 0.05$; $**p < 0.01$).

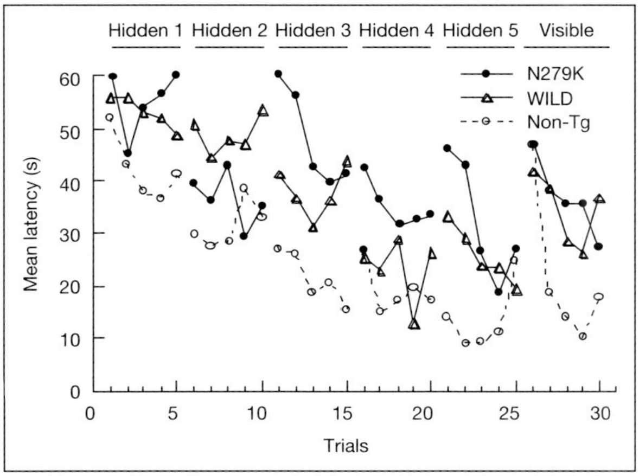

Fig. 4. Water maze learning. Five 5-trial maze training sessions were conducted daily, and followed by a 5-trial visible platform test on the next day. The main effects of group (G), day (D) and trial (T) were significant [$F(2, 25) = 6.38$, $p = 0.0058$, $F(4, 100) = 23.925$, $p < 0.0001$, and $F(4, 100) = 6.16$, $p = 0.0002$, respectively]. The interaction of D × G was slightly significant [$F(8, 100) = 1.98$, $p = 0.0562$], but the other interactions were not significant. The post-hoc test showed that N279K mice learned the maze significantly slower than the non-Tg mice. In the visible platform test, the effect of trial (T) was significant [$F(4, 100) = 7.55$, $p < 0.0001$], but the effect of group (G) and the interaction T × G were not significant [$F(2, 25) = 2.60$, $p = 0.0939$, and $F(8, 100) = 1.597$, respectively].

To study the overall conformation of tau protein, we measured the CD spectra of 3MBD (corresponding to 306–336 amino acids) in water (pH 7.0 and 4.2) and TFE (pH 3.9 and 7.2) (fig. 5). The spectra of 3MBD in TFE are indicative of an α-helical structure characterized by two negative peaks around 209 and 222 nm, the conformation of which is little affected by pH variation, whereas those in water showed a random conformation characterized by a negative peak around 197 nm. This result indicates the solvent-sensitive behavior of 3MBD conformation [18].

We further analyzed the solution conformation of 3MBD in TFE solvent by ^{1}H-NMR spectroscopy. Superposition of the backbone structures of the most stable 20 conformers indicates that the N-terminal Val1-Lys6 and Leu10-Leu20 fragments form extended and α-helical structures, respectively, whereas the C-terminal moiety is flexible and does not show any defined three-dimensional structure (fig. 6). The VQIVYK sequence in 3MBD has been reported to be necessary for the assembly of tau protein into Alzheimer paired helical filaments [19], the extended structure of this sequence may be important as a trigger for self-aggregation. On the other hand, the sequence

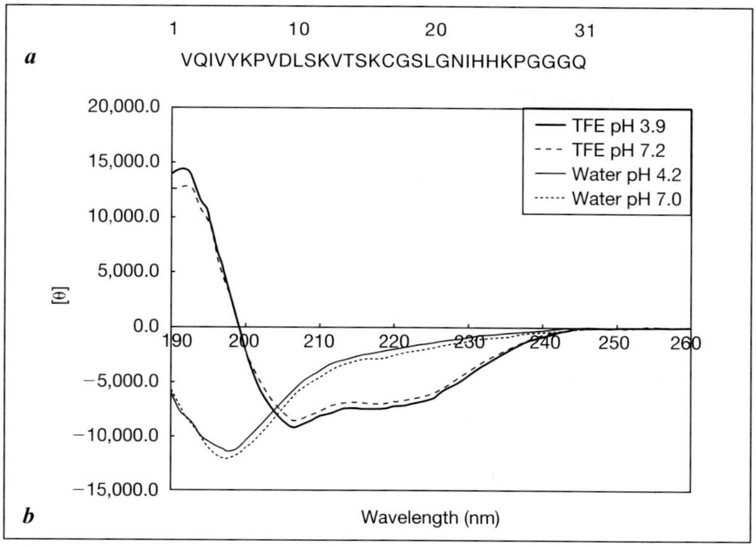

Fig. 5. ***a*** Amino acid sequence of 3MBD. ***b*** CD Spectra of 3MBD in water and TFE solutions.

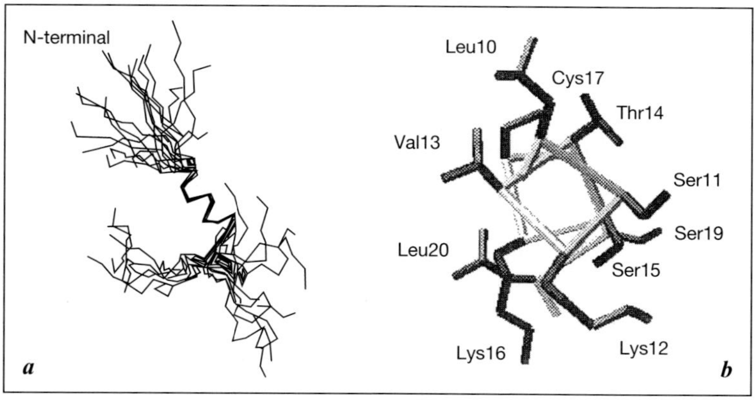

Fig. 6. ***a*** Superposition of the most stable 20 conformers of 3MBD. Each conformer is projected so as to be superimposable on the Leu10-Leu20 sequence. ***b*** Helical wheel drawing of the Leu10-Leu20 sequence of the most stable conformer of 3MBD, viewed down from the N-terminal side.

of Leu10-Leu20 forms a well-defined α-helical structure (the RMS deviation of backbone is 0.55 Å). Interestingly, the helix wheel drawing of this sequence showed the amphipathic distribution of the respective amino acid residues (fig. 6). The hydrophobic residues of Leu10, Val13 and the

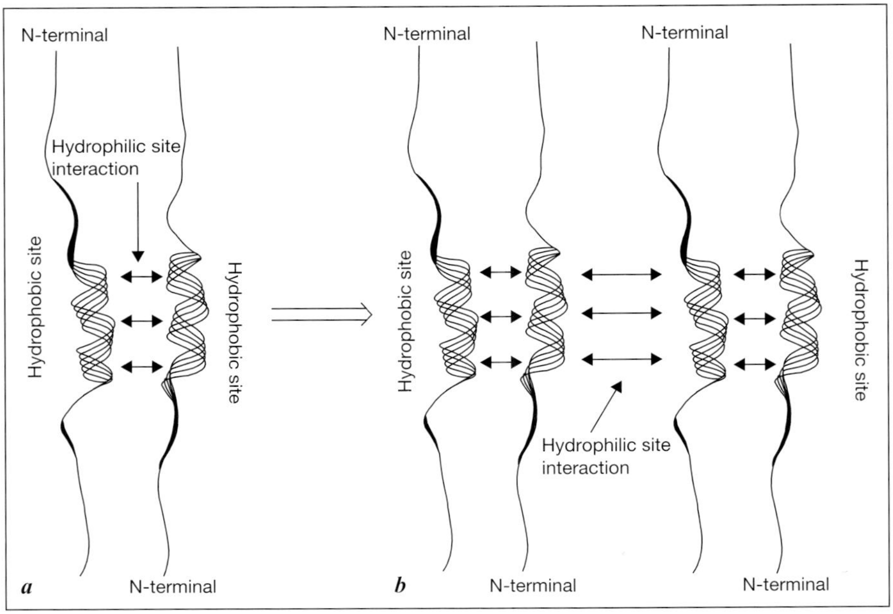

Fig. 7. A possible assembly model of 3MBD. Self-assembly is performed via anti-parallel dimer formation due to the hydrophilic interactions between polar amino acid residues (***a***) and molecular assembly of dimer structures due to the hydrophobic interactions between nonpolar amino acid residues (***b***).

hydrophilic residues of Ser11, Thr14, Ser15, and Ser19 are arranged on the two sides of the helix axis, respectively, and the polar residues of Cys17, Lys12, and Lys16 are located at the interfacing region between the two sides. Although the biological significance of such an orientation is unknown at present, we would propose an assembly model via the helical structure of 3MBD, i.e., the dimer formation by the hydrophilic interactions and then the molecular aggregation of these dimer structures by the hydrophobic interactions (fig. 7) [18].

In conclusion, the present work provides an in vivo transgenic model and an in vitro assembly model for elucidating mechanisms underlying cognitive deficits not only in FTDP-17 but also diverse other tauopathies.

Acknowledgements

This work was supported in part by the Uehara Memorial Foundation. We thank Prof. D. Borchelt (Department of Pathology, Johns Hopkins School of Medicine) for providing an MoPrP vector and Dr. T. Sakaue (Senju Pharmaceutical Co., Ltd.) for technical advice on behavioral analysis and M. Sumida, M. Moriwaki, D. Akagi and H. Kitamura (BMRC) for animal maintenance and antibody development.

References

1 Mori H, Hamada Y, Kawaguchi M, Honda T, Kondo J, Ihara Y: A distinct form of tau is selectively incorporated into Alzheimer's paired helical filaments. Biochem Biophys Res Commun 1989; 159:1221–1226.

2 Borchelt DR, Davis J, Fischer M, Lee MK, Slunt HH, Ratovitsky T, Regard J, Copeland NG, Jenkins NA, Sisodia SS, Price DL: A vector for expressing foreign genes in the brains and hearts of transgenic mice. Genet Anal 1996;13:159–163.

3 Taniguchi T, Kawamata T, Mukai H, Hasegawa H, Isagawa T, Yasuda M, Hashimoto T, Terashima A, Nakai M, Mori H, Ono Y, Tanaka C: Phosphorylation of tau is regulated by PKN. J Biol Chem 2001;276:10025–10031.

4 Kawamata T, Taniguchi T, Mukai H, Kitagawa M, Hashimoto T, Maeda K, Ono Y, Tanaka C: A protein kinase, PKN, accumulates in Alzheimer neurofibrillary tangles and associated endoplasmic reticulum-derived vesicles and phosphorylates tau protein. J Neurosci 1998;18:7402–7410.

5 Namgung U, Valcourt E, Routtenberg A: Long-term potentiation in vivo in the intact mouse hippocampus. Brain Res 1995;689:85–92.

6 McIlwain KL, Merriweather MY, Yuva-Paylor LA, Paylor R: The use of behavioral test batteries: Effects of training history. Physiol Behav 2001;73:705–717.

7 Bystrov VF: Spin-spin coupling and the conformational states of peptide systems. Prog Nucl Magn Reson Spectrosc 1976;10:41–81.

8 Yasuda M, Kawamata T, Komure O, Kuno S, D'Souza I, Poorkaj P, Kawai J, Tanimukai S, Yamamoto Y, Hasegawa H, Sasahara M, Hazama F, Schellenberg GD, Tanaka C: A mutation in the microtubule-associated protein tau in pallido-nigro-luysian degeneration. Neurology 1999;53: 864–868.

9 Alonso AC, Zaidi T, Grundke-Iqbal I, Iqbal K: Role of abnormally phosphorylated tau in the breakdown of microtubules in Alzheimer disease. Proc Natl Acad Sci USA 1994;91:5562–5566.

10 Morishima-Kawashima M, Hasegawa M, Takio K, Suzuki M, Yoshida H, Titani K, Ihara Y: Proline-directed and non-proline-directed phosphorylation of PHF-tau. J Biol Chem 1995;270: 823–829.

11 Bliss TV, Collingridge GL: A synaptic model of memory: Long-term potentiation in the hippocampus. Nature 1993;361:31–39.

12 Morris RG, Garrud P, Rawlins JN, O'Keefe J: Place navigation impaired in rats with hippocampal lesions. Nature 1982;297:681–683.

13 Morris RG, Anderson E, Lynch GS, Baudry M: Selective impairment of learning and blockade of long-term potentiation by an N-methyl-*D*-aspartate receptor antagonist, AP5. Nature 1986; 319:774–776.

14 Lewis J, McGowan E, Rockwood J, Melrose H, Nacharaju P, Van Slegtenhorst M, Gwinn-Hardy K, Paul Murphy M, Baker M, Yu X, Duff K, Hardy J, Corral A, Lin WL, Yen SH, Dickson DW, Davies P, Hutton M: Neurofibrillary tangles, amyotrophy and progressive motor disturbance in mice expressing mutant (P301L) tau protein. Nat Genet 2000;25:402–405.

15 Goedert M, Hasegawa M: The tauopathies: Toward an experimental animal model. Am J Pathol 1999;154:1–6.

16 Clark LN, Poorkaj P, Wszolek Z, Geschwind DH, Nasreddine ZS, Miller B, Li D, Payami H, Awert F, Markopoulou K, Andreadis A, D'Souza I, Lee VM, Reed L, Trojanowski JQ, Zhukareva V, Bird T, Schellenberg G, Wilhelmsen KC: Pathogenic implications of mutations in the tau gene in pallido-ponto-nigral degeneration and related neurodegenerative disorders linked to chromosome 17. Proc Natl Acad Sci USA 1998;95:13103–13107.

17 Hong M, Zhukareva V, Vogelsberg-Ragaglia V, Wszolek Z, Reed L, Miller BI, Geschwind DH, Bird TD, McKeel D, Goate A, Morris JC, Wilhelmsen KC, Schellenberg GD, Trojanowski JQ,

Lee VM: Mutation-specific functional impairments in distinct tau isoforms of hereditary FTDP-17. Science 1998;282:1914–1917.

18 Minoura K, Tomoo K, Ishida T, Hasegawa H, Sasaki M, Taniguchi T: Amphipathic helical behavior of the third repeat fragment in the tau microtubule-binding domain, studied by (1)H NMR spectroscopy. Biochem Biophys Res Commun 2002;294:210–214.

19 von Bergen M, Friedhoff P, Biernat J, Heberle J, Mandelkow EM, Mandelkow E: Assembly of tau protein into Alzheimer paired helical filaments depends on a local sequence motif ((306)VQIVYK(311)) forming beta structure. Proc Natl Acad Sci USA 2000;97:5129–5134.

Dr. T. Taniguchi
Laboratory of Molecular Pharmacology, Biosignal Research Center
Kobe University
1–1 Rokkodai-cho, Nada-ku, Kobe 657–8501 (Japan)
Tel. +81 78 803 5962, Fax +81 78 803 5971, E-Mail tanigu@kobe-u.ac.jp

Takeda M, Tanaka T, Cacabelos R (eds): Molecular Neurobiology of Alzheimer Disease and Related Disorders. Basel, Karger, 2004, pp 195–204

Animal Models of Tauopathies

Takeshi Ishihara, Hanae Nakashima

Department of Neuropsychiatry, Okayama University Graduate School of Medicine and Dentistry, Okayama, Japan

Tau is an abundant microtubule (MT)-associated protein in the central nervous system (CNS) that is implicated in the pathogenesis of frontotemporal dementia with parkinsonism linked to chromosome 17 (FTDP-17), progressive supranuclear palsy (PSP), amyotrophic lateral sclerosis/parkinsonism-dementia complex (ALS/PDC) of Guam, Alzheimer's disease (AD) and a number of other neurodegenerative tauopathies [1–6]. Tauopathies are characterized neuropathologically by abundant inclusions formed by aggregated paired helical filaments (PHFs) and/or straight filaments composed of aberrantly phosphorylated tau proteins (PHF-tau) in selectively vulnerable neurons and glial cells throughout widespread regions of the CNS [1–6].

Six alternatively spliced tau isoforms are expressed primarily by neurons in the adult human CNS, and are localized predominantly in axons, although glial cells also contain small amounts of tau [7–9]. Tau proteins bind to MTs, promote the assembly of MTs and stabilize MTs in the polymerized state [10, 11], but the formation of PHF-tau results in a loss of these important functions [12, 13]. Moreover, unlike normal tau, PHF-tau is insoluble, accumulates in the somatodendritic domain of neurons and assembles into filaments [14] that aggregate as NFTs in tauopathies [1–6]. In addition to PHF-tau, other cytoskeletal proteins, i.e. neurofilament (NF) subunits, are also found in many NFTs as is ubiquitin [1, 15, 16].

The massive degeneration of neurons and extensive gliosis associated with the progressive accumulation of PHF-tau lesions provided circumstantial evidence implicating filamentous tau pathology in the onset/progression of neurodegenerative disease. However, the discovery of multiple pathogenic mutations in the *tau* gene of many distinct FTDP-17 families demonstrated directly and unequivocally that tau abnormalities cause neurodegenerative disease. These pathogenic FTDP-17 mutations are located at topographically distinct sites in exons and introns of the *tau* gene and they include exonic missense substitutions, in-frame deletions and intronic substitutions [3, 17–23]. Significantly, emerging evidence suggests that topographically separate mutations cause FTDP-17 by differential mechanisms that specifically alter the functions or levels of tau isoforms in the CNS [18, 22, 24].

Recently, we and other researchers tested the hypothesis that neurodegenerative disease can result from altered expression levels of normal tau isoforms by generating transgenic (Tg) mice that overexpressed specific isoforms of human tau in CNS neurons. We reported that these Tg mice develop progressive age-dependent accumulations of intraneuronal filamentous inclusions accompanied by neurodegeneration, gliosis and tau protein abnormalities [25, 26]. Since overexpression of normal tau in these Tg mice causes a neurodegenerative disease that partially recapitulates human tauopathies, these mice will be useful in studies to elucidate mechanisms of brain degeneration in tauopathies and those studies will result in novel treatments for testing in animal models that can then be translated into new treatments for human neurodegenerative tauopathies.

Progressive Tauopathy in Tau Tg Mice

Generation of Tg Mice That Overexpress the Shortest Human Tau Isoform

To generate Tg mice expressing human tau, a cDNA corresponding to the shortest human brain tau isoform ('fetal tau') was cloned into an expression plasmid MoPrP·Xho. This vector was used here because it enables relatively high levels of transgene expression in CNS neurons [27]. After screening potential Tg mouse lines by Southern blots, we identified 3 stable Tg lines that were shown by Western blot analysis to variably overexpress human tau. Using a polyclonal antibody (17026) that recognizes human and mouse tau in quantitative Western blot studies, we showed that the heterozygous Tg mouse lines 7, 43 and 27 overexpressed tau proteins at approximately 5-, 10- and 15-fold higher levels, respectively, than endogenous mouse tau. Since Tg line 27 mice were not viable beyond 3 months, and homozygous Tg mice generated from any of these 3 lines of Tg mice died either in utero or within 3 months postnatally, the observations summarized below come from studies conducted on 1- to 24-month-old heterozygous tau Tg mice from lines 7 and 43.

Tau Tg Mice Acquire CNS Tau Inclusions with Advancing Age Similar to Human Tauopathies

Tg mice and their wild-type (WT) littermates from lines 7 and 43 between 1 and 12 months of age were subjected to histological studies that revealed widespread expression of human tau in neurons and their processes throughout the CNS of the Tg, but not the WT mice. In Tg mouse spinal cords, T14, a human tau-specific monoclonal antibody (MAb) stained spheroidal intraneuronal inclusions that were observed initially at 1 month of age. The size and number of these inclusions increased up to 6–9 months, but then subsequently decreased in abundance by 12 months. Notably, many vacuolar lesions that were the same size or larger than the tau inclusions also were observed in the older Tg mice, which may reflect degeneration of affected axons. The inclusions were about the size of medium to large spinal cord neurons, and some appeared to arise within proximal axons of spinal cord neurons. Although they occurred in gray and white matter at all spinal cord levels, these tau-positive inclusions were most frequent at the gray-white junction.

Spinal cord sections were probed with a panel of antibodies to tau and other neuronal cytoskeletal proteins, and the inclusions were immunostained by a MAb, Alz50, an antibody that detects a conformation change in the tau protein found in PHFs, as well as by other antibodies specific for phosporylated PHF-tau epitopes, including PHF1 (phosphoserine 396 and 404; numbering according to the largest human brain tau), PHF6 (phosphothreonine 231), T3P (phosphoserine 396), AT8 (phosphoserine 202 and 205), AT270 (phosphothreonine 181), and 12E8 (phosphoserine 262) [25]. Therefore, these lesions contain hyperphosphorylated tau similar to PHF-tau. Significantly, these inclusions were also stained strongly with anti-NF protein antibodies specific to the low (NFL)-, middle (NFM)-, and high (NFH)-molecular-weight NF proteins. Both phosphorylated and nonphosphorylated NFM and NFH were observed in these lesions. Indirect immunofluorescence double labeling confirmed the colocalization of tau and NFs in these inclusions. In addition, anti-tubulin antibodies also immunostained these inclusions.

In the brainstem and cortex of the Tg mice, tau-positive intraneuronal aggregates were also detected, but they were smaller and appeared later than the spinal cord inclusions. They were first seen in the pontine neurons of 1-month-old animals, and emerged in the cerebral cortex at about 6 months of age. The immunohistochemical profile of these brain aggregates was similar to that of the spinal cord lesions. However, the morphological features of these inclusions indicate that some are similar to the spinal cord axonal lesions; however, others occur in the somatodentritic compartment of cortical neurons and resemble NFTs and dystrophic neuritis observed in tauopathies. Notably, the brain and the spinal cord inclusions stained similarly to human NFTs by the Bodian silver method. However they were thioflavin S negative, and they were not stained by antibodies to α-internexin, peripherin, ubiquitin and synucleins. Tg mouse line 43 expressed higher levels of human tau than line 7, and similar tau-rich inclusions were also observed to accumulate in the spinal cord and brain of Tg line 43 in an age-dependent manner, but they were larger and more abundant than in line 7. These results demonstrate a transgene dose-dependent, as well as age-dependent, accumulation of tau-rich inclusions in the Tg mice.

We also examined 12- to 24-month-old Tg mice. These Tg mice develop congophilic and NFT-like tau inclusions in several brain regions including the hippocampus, amygdala, and entorhinal cortex. These inclusions were first detected in Tg mice at 18–20 months of age and they were detected by histochemical dyes that bind specifically to crossed β-pleated sheet structures (e.g., Congo red, Thioflavin S). Moreover, ultrastructurally these lesions contained straight tau filaments comprised of both mouse and human tau proteins but not other cytoskeletal proteins (e.g., NFs and MTs). Isolated tau filaments were also recovered from detergent-insoluble tau fractions and insoluble tau proteins accumulating in the brain in an age-dependent manner (this biochemical aspect is discussed in the next section).

Insoluble Tau Protein Progressively Accumulates in the CNS of Tau Tg Mice

To determine whether there was an age-dependent accumulation of insoluble tau similar to that described in human tauopathies, we extracted tau from brain and spinal cord samples using buffers with increasing extraction strengths. In the WT mice, about 90% of endogenous mouse tau was largely re-assembly buffer (RAB)-soluble and no tau immunoreactivity was detected in the 70% formic acid (FA)-soluble fraction. Although the RAB-soluble tau from the Tg mice remained relatively constant at around 75–80% with increasing age, RAB-insoluble tau represented by the RIPA and FA fractions progressively accumulated in both the brain and the spinal cord of the Tg mice. The accumulated RAB-insoluble tau was mainly fetal human tau expressed from the transgene, and the time course of accumulation correlated with the emergence of the tau inclusions in the Tg mice. In addition, RAB-insoluble tau, especially that in the FA fraction, was more pronounced in the spinal cord than in the brain, consistent with more abundant tau aggregates in the spinal cord.

The Phosphorylation of Tau in the Tau Tg Mice Recapitulates That in Human Tauopathies

Western blot studies of soluble and insoluble tau extracted from the cerebral cortex of Tg mice were performed using MAb T1 (specific for a nonphosphorylated tau epitope that is located within amino acids 189–207), and this antibody did not recognize PHF-tau, but was immunoreactive with human adult normal tau, fetal tau, and both soluble and insoluble Tg tau. This indicates that tau from the Tg mice is partially dephosphorylated at the T1 epitope. However, several phosphorylation-dependent antibodies, which reacted with PHF-tau and fetal tau, but not with normal adult tau, also recognized both soluble and insoluble tau from the Tg mice. These antibodies include PHF1 (phosphoserine 396 and 404), T3P (phosphoserine 396), PHF6 (phosphothreonine 231), AT8 (phosphoserine 202 and 205), AT270 (phosphothreonine 181), and 12E8 (phosphoserine 262). Therefore the phosphorylation state of Tg tau recapitulates that of PHF-tau found in human tauopathies including ALS/PDC, PSP, FTDP-17 and AD.

Tau Tg Mice Develop Gliosis, Axon Degeneration and Reduced Fast Axonal Transport Linked to Progressive Motor Weakness

To detect astrocytosis, we immunostained Tg and WT mouse brain and spinal cord with an MAb to glial fibrillary acidic protein. Numerous reactive astrocytes were detected in the brain and spinal cord of Tg but not WT mice, indicating the presence of profound gliosis in regions with neuronal damage. Furthermore, the astrocytosis was almost undetectable at 1 month of age but progressed thereafter with age. Since inclusions in the proximal axons of affected neurons could cause disease by damaging axons, we examined the morphology of spinal cord ventral root axons. In semithin sections, the normal L5 ventral root of WT mice contained many large and small myelinated axons, but the ventral root of a 6-month-old Tg mouse primarily contained irregularly shaped axons, and at 12 months of age, the endoneurial space appeared to increase. This is consistent with the removal of degenerated axons in these axons. There was also evidence of axonal degeneration in the ventral roots of Tg mice. We compared the axon numbers in L5 ventral roots of Tg and WT mice based on the analysis of photomicrographs of semithin sections. A 20% decrease in the number of axons was seen in 12-month-old Tg mice relative to their WT counterparts. We also showed that despite a significant reduction in MT density in the 12-month-old Tg mice, the NF density remained unchanged when compared with age-matched WT mice. This finding correlated with the biochemical analysis of β-tubulin and NF subunits in the proximal sciatic nerve, which showed a progressive decrease in β-tubulin levels in the Tg mice and relatively constant levels of NF subunits.

To assess whether or not the neuropathology in ventral roots of 12-month-old tau Tg mice compromised axonal transport, we measured radiolabeled proteins transported in the fast component of axonal transport following microinjection of [^{35}S]methionine into the L5 ventral horn of tau Tg and age-matched WT mice. Significantly, these studies showed that the fast axonal transport of radiolabeled proteins was retarded in the tau Tg mice. Finally, the spinal cord pathology in the tau Tg mice was associated with the development of a progressive motor weakness as assessed by their impaired ability to stand on a slanted surface and by retraction of their hind limbs when lifted by their tails. These impairments correlate with the fact that the Tg mice weighed about 30–40% less than age-matched WT littermates.

Discussion

As summarized here, our previous studies [25, 26] provide compelling evidence that the overexpression of normal human fetal tau protein in Tg mice causes a CNS neurodegenerative tauopathy that recapitulates key aspects of human tauopathies. For example, we observed a progressive, age-dependent accumulation of argyrophilic, tau-immunoreactive inclusions in neurons of spinal cord, brainstem and neocortex similar to human tauopathies. Since the inclusions in the young Tg mice were most abundant in spinal cord neurons, the tauopathy in these mice most closely resembles ALS/PDC wherein tangles are abundant in the spinal cord [1, 4, 28, 29]. Significantly, ALS/PDC patients who present with motor weaknesses do so about a decade earlier than those who present with parkinsonism and dementia. Thus, the fact that tau aggregates accumulate later in the brains of our Tg mice than in the spinal cord may mirror disease progression in ALS/PDC patients who present with motor weakness. Moreover, as shown here and reported earlier, these tau tangles are also immunostained by antibodies to NF proteins and tubulin [30], as are the inclusions in our Tg mice. Finally, ALS/PDC is associated with a progressive motor weakness similar to that observed in the Tg mice.

By contrast, age-related NFTs are found in small numbers and primarily in the hippocampus and related limbic neurons as normal individuals reach an advanced age. Thus, we propose that the mouse tangles that form in Tg mice as they advance to the terminal phase of the murine life span recapitulate key features of age-related NFTs in humans, since the location of these Tg mouse tangles as well as their frequency of occurrence in aged Tg mice parallel those found in humans. Age-related NFTs have never been detected in rodents, perhaps due to their relative short life span. However, the overexpression of the human tau protein may change the dynamic equilibrium of tau in aging hippocampal neurons of the mouse such that the excess human tau protein can assemble with the endogenous mouse tau into structures very similar to authentic human tau tangles.

Although the major tau pathology is detected in the spinal cord of Tg mice and most closely resembles that found in ALS/PDC, PSP and some FTDP-17 syndromes, filamentous tau aggregates in Tg mice share many characteristics with authentic NFTs in AD and other tauopathies. First, like highly insoluble PHF-tau in AD NFTs [31], a substantial fraction of tau proteins from the Tg mice is extracted only with RIPA and FA despite the fact that normal tau is an extremely soluble protein. Second, the amount of insoluble tau protein increases with age and disease progression in Tg mice similar to AD and other tauopathies. Third, PHF-tau proteins in human NFTs are hyperphosphorylated as are soluble and insoluble tau proteins recovered from Tg mice [14, 32].

Finally, although AD NFTs contain mostly PHFs, straight filaments similar to those found in the inclusions of the Tg mice are also present [1–6].

Since the accumulation of filamentous tau inclusions in spinal cord neurons was associated with the degeneration of ventral root axons in the tau Tg mice, we hypothesize that this reflects a gain of toxic function by the overexpressed tau and several lines of evidence support this hypothesis. First, previously described tau Tg mice that expressed lower levels (<2 fold) of tau protein did not develop filamentous tau inclusions or neurodegeneration [33, 34]. Second, we observed a dose-dependent increase in the size and number of tau aggregates in our 2 lines of tau Tg mice. Thus, one plausible explanation to account for the axonal degeneration in these Tg mice is a gain of toxic function by the excess tau proteins that cannot bind MTs, and consequently aggregate in the neuronal cytoplasm, block axonal transport, and lead to the degeneration of affected axons.

The reduced numbers of MTs and the reduced levels of tubulin but not NFs or NF proteins, in the remaining axons of the degenerating ventral roots also imply a loss of the MT-stabilizing function of tau. Since overexpressed human tau could aggregate with endogenous mouse tau leading to progressive insolubility and hyperphosphorylation of both human and mouse tau in the Tg mice, this could impair the ability of endogenous mouse tau to perform an MT-stabilizing function. Indeed, the observed reduction in fast axonal transport in 12-month-old Tg mice is consistent with a loss of MT function although the loss of axons in the Tg mice may contribute to this. Based on indirect evidence from studies of human tauopathies, we and others have proposed that both gains of toxic functions and losses of normal tau functions could be involved mechanistically in causing neurodegenerative disease [18, 22, 24], and the data presented here support both of these hypotheses.

Although a dose effect of the transgene is observed in 2 different Tg mouse lines, the distribution of the tau-rich lesions within the CNS is not completely dependent upon the expression levels in the different regions. For example, the expression in the spinal cord is less than in the brain, but there are more abundant and larger inclusions that develop at a younger age in the spinal cord neurons than in other brain regions. This could be explained by the metabolic differences among diverse types of neurons, and excess tau may aggregate at a lower concentration in spinal cord neurons under the influence of local factors or pathological chaperones, such as high concentrations of NF proteins [35]. Similarly, the selective distribution of tau pathology in different human tauopathies is likely due to other as yet unidentified local vulnerability factors. Indeed, the findings described here parallel the well-known, but enigmatic 'selective vulnerability' that is a constant feature of most human neurodegenerative diseases and Tg mouse models thereof [36].

Since the Tg tau mice described here exhibit the key neuropathological features of human tauopathies, they will accelerate efforts to discover drugs to prevent the development of tau pathology or disrupt and eliminate existing tau pathology in patients with tauopathies. Indeed, experiments using various agents that influence tau pathology are now under way. Those agents include inhibitors of GSK-3ββ one of the major tau protein kinases (e.g., lithium) and antioxidants (e.g., vitamin E, curcumin). These studies will hopefully be translated into new treatments for human neurodegenerative tauopathies.

References

1 Ginsberg SD, Schmidt ML, Crino PB, Eberwine JH, Lee VMY, Trojanowski JQ: Molecular pathology of Alzheimer's disease and related disorders; in Peters A, Morrison JH (eds): Cerebral Cortex: Neurodegenerative and Age-Related Changes in Structure and Function of Cerebral Cortex. New York, Kluwer Academic/Plenum Publishers, 1999, vol 14, pp 603–654.

2 Hof PR, Bouras C, Morrison JH: Cortical neuropathology in aging and dementing disorders: Neuronal typology, connectivity, and selective vulnerability; in Peters A, Morrison JH (eds): Cerebral Cortex: Neurodegenerative And Age-Related Changes in Structure and Function of Cerebral Cortex. New York, Kluwer Academic/Plenum Publishers, 1999, vol 14, pp 175–312.

3 Hong M, Trojanowski JQ, Lee VMY: Tau-based neurofibrillary lesions; in Clark CM, Trojanowski JQ (eds): Neurodegenerative Dementias: Clinical Features and Pathological Mechanisms. New York, McGraw-Hill, 2000, pp 161–175.

4 Nakano I, Hirano A: Ultrastructural changes in dementing illnesses; in Peters A, Morrison JH (eds): Cerebral Cortex: Neurodegenerative and Age-Related Changes in Structure and Function of Cerebral Cortex. New York, Kluwer Academic/Plenum Publishers, 1999, vol 14, pp 399–432.

5 Forman MS, Lee VMY, Trojanowski JQ: New insights into genetic and molecular mechanisms of brain degeneration in tauopathies. J Chem Neuroanat 2000;20:225–244.

6 Lee VMY, Geodert M, Trojanowski JQ: Neurodegenerative Tauopathies. Annu Rev Neurosci 2001;24:1121–1159.

7 Goedert M, Spillantini MG, Jakes R, Rutherford D, Crowther RA: Multiple isoforms of human microtubule-associated protein tau: Sequences and localization in neurofibrillary tangles of Alzheimer's disease. Neuron 1989;3:519–526.

8 Andreadis A, Brown WM, Kosik KS: Structure and novel exons of the human tau gene. Biochem 1992;31:10626–10633.

9 Binder LI, Frankfurter A, Rebhun LI: The distribution of tau in the mammalian central nervous system. J Cell Biol 1985;101:1371–1378.

10 Weingarten MD, Lockwood AH, Hwo SY, Kirschner MW: A protein factor essential for microtubule assembly. Proc Natl Acad Sci USA 1975;72:1858–1862.

11 Drechsel DN, Hyman AA, Cobb MH, Kirschner MW: Modulation of the dynamic instability of tubulin assembly by the microtubule-associated protein tau. Mol Biol Cell 1992;3:1141–1154.

12 Bramblett GT, Goedert M, Jakes R, Merrick SE, Trojanowski JQ, Lee VMY: Abnormal tau phosphorylation at Ser396 in Alzheimer's disease recapitulates development and contributes to reduced microtubule binding. Neuron 1993;10:1089–1099.

13 Yoshida H, Ihara Y: Tau in paired helical filaments is functionally distinct from fetal tau: Assembly incompetence of paired helical filament-tau. J Neurochem 1993;61:1183–1186.

14 Lee VMY, Balin BJ, Otvos L Jr, Trojanowski JQ: A68: A major subunit of paired helical filaments and derivatized forms of normal tau. Science 1991;251:675–678.

15 Perry G, Rizzuto N, Autilio-Gambetti L, Gambetti P: Paired helical filaments from Alzhemier's disease patients contain cytoskeletal components. Proc Natl Acad Sci USA 1985;82:3916–3920.

16 Schmidt ML, Lee VMY, Trojanowski JQ: Relative abundance of tau and neurofilament epitopes in hippocampal neurofibrillary tangles. Am J Pathol 1990;136:1069–1075.
17 Clark LN, Poorkaj P, Wszolek ZK, Geschwind DH, Nasreddine ZS, Miller B, Payami H, Awert F, Markopoulou K, D'Souza I, Lee VMY, Reed L, Trojanowski JQ, Zhukareva V, Bird T, Schellenberg G, Wilhelmsen KC: Pathogenic implications of mutations in the tau gene in pallido-ponto-nigral degeneration and related chromosome 17-linked neurodegenerative disorders. Proc Natl Acad Sci USA 1998;95:13103–13107.
18 Hutton M, Lendon CL, Rizzu P, Baker M, Froelich S, Houlden H, Pickering-Brown S, Chakraverty S, Isaacs A, Grover A, Hackett J, Adamson J, Lincoln S, Dickson D, Davies P, Petersen RC, Stevens M, Graaff E, Wauters E, van Baren J, Hillebrand M, Joosse M, Kwon JM, Nowotny P, Che LK, Norton J, Morris JC, Reed LA, Trojanowski JQ, Basun H, Lannfelt L, Neystat M, Fahn S, Dark F, Tannenberg T, Dodd PR, Hayward N, Kwok JBJ, Schofield PR, Andreadis A, Snowden J, Craufurd D, Neary D, Owen F, Oostra BA, Hardy J, Goate A, van Swieten J, Mann D, Lynch T, Heutink P: Association of missense and 5′-splice-site-mutations in tau with the inherited dementia FTDP-17. Nature 1998;393:702–705.
19 Iijima M, Tabira T, Poorkaj P, Schellenberg GD, Trojanowski JQ, Lee VMY, Schmidt ML, Takahashi K, Nabika T, Matsumoto T, Yamashita Y, Yoshioka S, Ishino H: A distinct familial presenile dementia with a novel missense mutation in the tau gene. Neuroreport 1999;10:497–501.
20 Poorkaj P, Bird TD, Wijsman E, Nemens E, Garruto RM, Anderson L, Andreadis A, Wiederholt WC, Raskind M, Schellenberg GD: Tau is a candidate gene for chromosome 17 frontotemporal dementia. Ann Neurol 1998;43:815–825.
21 Spillantini MG, Murrell JR, Goedert M, Farlow MR, Klug A, Ghetti B: Mutation in the tau gene in familial multiple system tauopathy with presenile dementia. Proc Natl Acad Sci USA 1998; 95:7737–7741.
22 D'Souza I, Poorkaj P, Hong M, Nochlin D, Lee VMY, Bird TD, Schellenberg GD: Missense and silent mutations in tau cause FTDP-17 by altering alternative splicing. Proc Natl Acad Sci USA 1999;96:5598–5603.
23 Rizzu P, Van Swieten JC, Joosse M, Hasegawa M, Stevens M, Tibben A, Niermeijer MF, Hillebrand M, Ravid R, Oostra BA, Goedert M, van Duijn CM, Heutink P: High prevalence of mutations in the microtubule-associated protein tau in a population study of frontotemporal dementia in the Netherlands. Am J Hum Genet 1999;64:414–421.
24 Hong M, Zhukareva V, Vogelsberg-Ragaglia V, Wszolek Z, Reed L, Miller BI, Geschwind DH, Bird TD, McKeel D, Goate A, Morris JC, Wilhelmsen KC, Schellenberg GD, Trojanowski JQ, Lee VMY: Mutation-specific functional impairments in distinct tau isoforms of hereditary FTDP-17. Science 1998;282:1914–1917.
25 Ishihara T, Hong M, Zhang B, Nakagawa Y, Lee MK, Trojanowski JQ, Lee VMY: Age-dependent emergence and progression of a tauopathy in transgenic mice overexpressing the shortest human tau isoform. Neuron 1999;24:751–762.
26 Ishihara T, Zhang B, Higuchi M, Yoshiyama Y, Trojanowski JQ, Lee VMY: Age-dependent induction of congophilic neurofibrillary tau inclusions in tau transgenic mice. Am J Pathol 2001; 158:555–562.
27 Borchelt DR, Davis J, Fischer M, Lee MK, Slunt HH, Ratovitsky T, Regard J, Copeland NG, Jenkins NA, Sisodia SS, Price DL: A vector for expressing foreign genes in the brains and hearts of transgenic mice. Genetic Anal 1996;13:159–163.
28 Hirano A, Malamud N, Kurland LT: Parkinsonism dementia complex: An endemic disease on the island of Guam – pathological features. Brain 1961;84:642–661.
29 Matsumoto S, Hirano A, Goto S: Spinal cord neurofibrillary tangles of Guamanian amyotrophic lateral sclerosis and parkinsonism-dementia complex: An immunohistochemical study. Neurology 1990;40:975–979.
30 Shankar SK, Yanagihara R, Garruto RM, Grundke-Iqbal I, Kosik KS, Gajdusek DC: Immunocytochemical characterization of neurofibrillary tangles in amyotrophic lateral sclerosis and parkinsonism-dementia of Guam. Ann Neurol 1989;25:146–151.
31 Bramblett GT, Trojanowski JQ, Lee VMY: Regions with abundant neurofibrillary pathology in human brain exhibit a selective reduction in levels of binding-competent tau and accumulation of abnormal tau-isoforms (A68 proteins). Lab Invest 1992;66:212–222.

32 Matsuo ES, Shin RW, Billingsley ML, Van deVoorde A, O'Connor M, Trojanowski JQ, Lee VMY: Biopsy-derived adult human brain tau is phosphorylated at many of the same sites as Alzheimer's disease paired helical filament tau. Neuron 1994;13:989–1002.
33 Brion JP, Tremp G, Octave JN: Transgenic expression of the shortest human tau affects its compartmentalization and its phosphorylation as in the pretangle stage of Alzheimer's disease. Am J Pathol 1999;154:255–270.
34 Goetz J, Probst A, Spillantini MG, Schaefer T, Jakes R, Buerki K, Goedert M: Somatodendritic localization and hyperphosphorylation of tau protein in transgenic mice expressing the longest human brain tau isoform. EMBO J 1995;14:1304–1313.
35 Ishihara T, Higuchi M, Zhang B, Yoshimasa Y, Hong M, Trojanowski JQ, Lee VMY: Attenuated neurodegenerative disease phenotype in tau transgenic mouse lacking neurofilaments. J Neurosci 2001;21:6026–6035.
36 Tu PH, Gurney ME, Julien JP, Lee VMY, Trojanowski JQ: Oxidative stress, mutant SOD1, and neurofilament pathology in transgenic mouse models of human motor neuron disease. Lab Invest 1997;76:441–456.

Dr. Takeshi Ishihara
Department of Neuropsychiatry
Okayama University Graduate School of Medicine and Dentistry
2–5–1 Shikata-cho, Okayama 700–8558, Okayama (Japan)
Tel. +81 86 235 7242, Fax +81 86 235 7246, E-Mail takeshiishihara@yahoo.com

Takeda M, Tanaka T, Cacabelos R (eds): Molecular Neurobiology of Alzheimer Disease and Related Disorders. Basel, Karger, 2004, pp 205–214

Aberrant Splicing of Tau Transcripts in Frontotemporal Dementia with Parkinsonism Linked to Chromosome 17

Noriaki Yamamoto, Shinichi Kondo*, Shuta Yoshino, Masayo Okumura, Kazunori Imaizumi*

Division of Structural Cellular Biology, Nara Institute of Science and Technology, Takayama, Ikoma, Japan

Alternative splicing represents mechanisms that typically underlie gene expression regulation in eukaryotic cells [1, 2]. Exon selection results in the production of different protein isoforms from the same gene, isoforms that may share functions with the original form. Alternatively, variant protein isoforms may either lack function or confer novel characteristics on their cellular environment. In fact, two splicing defects lacking exon 4 [3, 4] and exon 9 [5] of the presenilin 1 (PS1) transcript have been identified in familial Alzheimer's disease, and tau splicing mutations that increase or decrease 4-repeat isoforms containing exon 10 of tau were found in frontotemporal dementia with parkinsonism linked to chromosome 17 (FTDP-17) [6–9]. In addition, aberrant transcripts of the excitatory amino acid transporter-2 gene are commonly present in patients with sporadic amyotrophic lateral sclerosis [10].

FTDP-17 is an autosomal-dominant disease with variable clinical and neuropathological features [11]. Personality changes, sometimes with psychosis, hyperorality, reduced speech output, and loss of executive function are observed as symptoms [12–18]. Neuropathologic changes include frontotemporal atrophy, sometimes with atrophy of the basal ganglion, substantia nigra, and amygdala. FTDP-17 is caused by mutations in the gene for tau, a microtubule-associated protein that normally functions to promote microtubule assembly and stability. In FTDP-17, tau aggregates in the brain to form abnormal filamentous structures including neurofibrillary tangles (NFTs) [19, 20], neuropil threads, glial tangles,

*These authors contributed equally to this work.

and dense intracellular deposits [15]. The type and location of tau pathology vary among FTDP-17 families.

Tau mutations cause FTDP-17 by at least two different mechanisms. First, tau mutations cause impairment of tau protein function. Tau with G272V, P301L, V337M, or R406W mutations exhibits reduced affinity and capacity for microtubule binding and a reduced ability to facilitate microtubule polymerization when compared with normal tau [21]. Second, tau mutations cause aberrant splicing of tau exon 10. Some mutations in tau exon 10, or intron 10 immediately adjacent to the 3′ end of alternative spliced exon 10, increase the inclusion of tau exon 10 [6–9]. Tau exon 10 encodes 1 of 4 microtubule-binding motifs found in the longer isoforms of tau. When exon 10 is included, isoforms with 4 microtubule-binding domains (4-repeat tau) are produced, and when exon 10 is excluded, tau isoforms with 3 microtubule repeats (3-repeat tau) are produced. However, the regulatory mechanisms of splicing of tau exon 10, which contains the FTDP-17 related mutations, have not been elucidated.

In this report, we examined the splicing patterns of tau exon 10 by exon trap systems and attempted to identify the regulators of its splicing by an RNA-protein interaction assay.

Materials and Methods

Cell Cultures

COS-7, HEK293T and SK-N-SH cells were used for in vivo splicing assays. COS-7 and HEK293T cells were grown in 10% fetal calf serum (FCS)/Dulbecco's modified Eagle's medium (DMEM), and SK-N-SH cells were cultured in Alpha Minimum Essential Medium (α-MEM) with 10% FCS. Prior to transfection, cells were plated at a density of 60–80% confluency on 3.5-cm dishes.

Exon Trapping Systems

Tau mini-genes (both wild and mutants) containing intron 9, exon 10, and intron 10 were inserted into the Exon Trapping Vector pSPL3 (Life Technologies) [22, 23] or a modified exon-trapping vector driven by a CMV promoter. All constructs were sequenced before use in experiments. The constructs were transfected into COS-7 cells, and total cellular RNA was isolated 24 h after transfection using the RNeasy Mini Kit (Qiagen). Aliquots of 3.0 μg RNA were reverse-transcribed using the SA2 primer (5′-ATC TCA GTG GTA TTT GTG AGC-3′) and MMLV reverse transcriptase (Life Technologies). Splicing products were detected by PCR using pSPL3 vector-specific primer sets, SD6 (5′-TCT GAG TCA CCT GGA CAA CC-3′) and SA2. PCR was performed in a total volume of 40 μl. Amplification was conducted as follows: 1 min at 94°C, 1 min at 60°C, 1 min at 72°C for 30 cycles followed by 72°C for 5 min. The PCR products were electrophoresed in a 5% acrylamide gel.

RNA-Protein Binding Assay

For the collection of nuclear extracts, SK-N-SH cells were homogenized in 50 vol of 10 m*M* HEPES-NaOH (pH 7.9) containing 10 m*M* KCl, 1 m*M* EDTA, 1 m*M* EGTA, 5 m*M* dithiothreitol (DTT) and 1 m*M* (*p*-amidinophenyl) methanesulfonyl fluoride (PMSF) at 4°C.

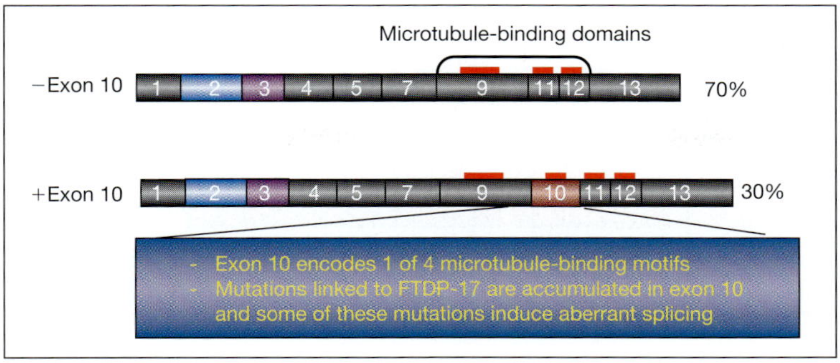

Fig. 1. The structure of tau mRNA. Tau is a microtubule-associated protein that normally functions to promote microtubule assembly and stability. Tau mRNA is composed of 13 exons, and exon 10 encodes 1 of 4 microtubule-binding motifs. In the normal adult brain, the ratio of the expression of 3-repeat tau, which excludes exon 10, and 4-repeat tau, which includes tau exon 10, is approximately 70:30.

Buffers and all other solutions used in this study were sterilized before each use by filtration through a Steritop (Millipore Corporation, USA) with a pore size of 220 nm. Following the addition of 10% Nonidet P-40 to a final concentration of 0.6%, homogenates were centrifuged at 15,000 rpm for 5 min. Pellets were resuspended in 10 vol of 20 m*M* Tris-HCl (pH 7.5) containing 400 m*M* KCl, 1 m*M* EDTA, 1 m*M* EGTA, 1 m*M* DTT and 1 m*M* PMSF, followed by centrifugation at 15,000 rpm for 5 min. The supernatants thus obtained were stored at −80°C as nuclear extracts for pre-mRNA binding assays.

Pre-mRNA binding assays were performed as described [23, 24]. Briefly, sense strand RNAs were transcribed and uniformly labeled with [α-^{32}P] UTPs in vitro by T7 RNA polymerase (TaKaRa) from templates of 177-bp sequences containing exon 10 of wild or mutant tau. The RNAs were incubated on a heat block at 25°C with SK-N-SH cell nuclear extracts. The RNA-protein complexes were UV-irradiated (300,000 μJ/cm^2) at room temperature for 5 min, digested with 10 μg of RNase A (Boehringer Mannheim, Germany) at 37°C for 30 min, and resolved by 10% SDS-PAGE.

Results and Discussion

In vivo Splicing

Tau transcript is composed of 13 exons (fig. 1). Tau exon 10 encodes 1 of 4 microtubule-binding motifs. In adult brain, the ratio of 3- and 4-repeat tau is 70:30. Mutations linked to FTDP-17 are accumulated in exon 10, and some of these mutations affect the splicing of tau exon 10. The resultant aberrant splicing causes a change in the ratio of 3- and 4-repeat tau; a change that is believed to associate with disease onset.

Mutations in the tau gene that were used in this experiment are shown in figure 2. These exon 10 fragments containing flanking intron sequences were

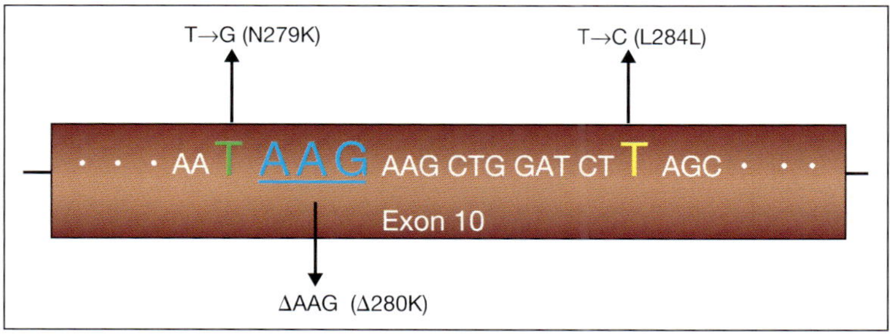

Fig. 2. Mutations in the tau gene that were used in the present study. Tau pre-mRNA containing the missense mutation, N279K, silent mutation, L284L, and deletion mutation, Δ280K, was cloned into the exon trap vectors. Further, tau exon 10 fragments containing these mutations and flanking intron sequences were used for RNA-protein binding assay.

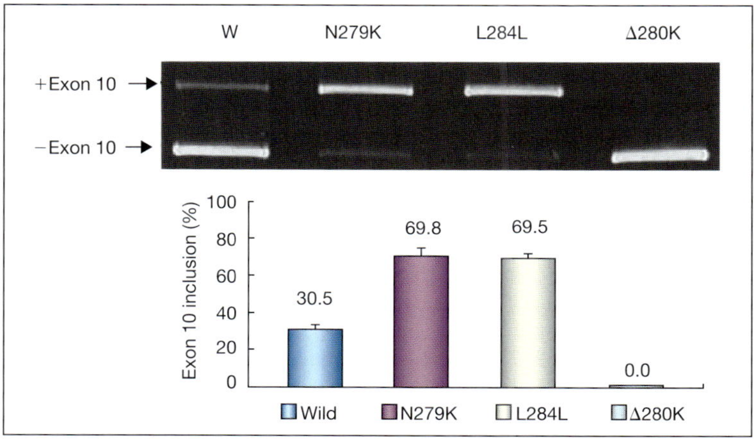

Fig. 3. Alternative splicing of tau exon 10 in an exon trap system.The upper panel shows RT-PCR products from an exon trap system. The lower panel shows the quantitative analysis of the splicing patterns of each construct. Each band intensity was determined using the Densitography Program and the percentages of exon 10 inclusion relative to the total transcripts are represented (means ± SD of 4 analyses). Transfection with the constructs with N279K and L284L mutations increase the inclusion of exon 10 compared with the wild construct. In contrast, Δ280K mutation results in no detectable inclusion of exon 10.

inserted between HIV tat exons, the constructs transfected into neuroblastoma SK-N-SH cells, and then transcribed in the cells by a CMV promoter. Twenty-four hours after transfection, total cellular RNA was isolated. Splicing products were detected by RT-PCR using the SD6 and SA2 promoter set as described in Materials and Methods. When wild-type exon 10 was tested, transcripts with both inclusion and exclusion of exon 10 were produced (fig. 3). The ratio of inclusion type was about 30% of the total, and it was almost equal to that of the endogenous 4-repeat tau per total tau. Transfection of tau exon 10 containing FTDP-17-linked mutations, N279K and L284L, resulted in an increase of the inclusion of exon 10 to approximately 70% of the total. In contrast, transfection with the Δ280K mutation led to no detectable inclusion of tau exon 10. These results suggest that mutations linked to FTDP-17 affect the splicing of tau exon 10. We also examined the splicing patterns of tau exon 10 in nonneuronal cells such as HEK293T or COS-7 cells. The results were basically the same as those in the neuronal cell lines, indicating that the splicing of tau exon 10 is regulated by the same mechanisms in both neuronal and nonneuronal cells.

Splicing Machinery and Its Binding Sites on the Pre-mRNA

One of the most remarkable features of the mammalian pre-mRNA splicing machinery is its ability to select precisely correct pairs of splice sites, but inappropriate sites. The removal of intron sequences by splicing occurs by two sequential transesterification reaction steps that are catalyzed by components of a large RNA-protein complex, termed the spliceosome. The formation of the major spliceosome involves the stepwise assembly of 4 small nuclear ribonucleoproteins (snRNPs, U1, U2, U4/U6 and U5) and many non-snRNP splicing factors on a pre-mRNA. The components of the splicing machineries and their binding sites on the pre-mRNA are shown in figure 4 [25, 26]. U1 snRNP binds to the 5′ splice site and U2 snRNP binds to the branching point of the pre-mRNA. The binding of U2 snRNP to the branching point is assisted by U2AF35 and 65 through its binding of the polypyrimidine tract. These complex formations on the exon-intron junction sites of the pre-mRNA promote intron processing. Exonic splicing enhancer (ESE) or silencer (ESS) is known to be sequenced within exons that promote or suppress both constitutive and regulated splicing, respectively. SR proteins, rich in serine and arginine residues, directly bind to the ESE sequence. The binding of SR-related proteins to the ESE promotes the recruitment of U1 and U2 snRNP on the pre-mRNA, and intron processing is accelerated.

RNA-Protein Binding Assay

We examined whether nuclear proteins bind to pre-mRNA sequences containing tau exon 10 to detect its splicing regulators. For this, we performed a pre-mRNA binding assay. Sense strand RNAs were transcribed and labeled with ^{35}S-UTP by T7 RNA polymerase from templates of 177-bp tau cDNA containing the exon 10 as shown in figure 5a. The labeled RNAs were incubated with nuclear extracts of SK-N-SH cells. The RNA-protein complexes were UV-crosslinked and digested with RNase A, and then subjected to SDS-PAGE and autoradiography.

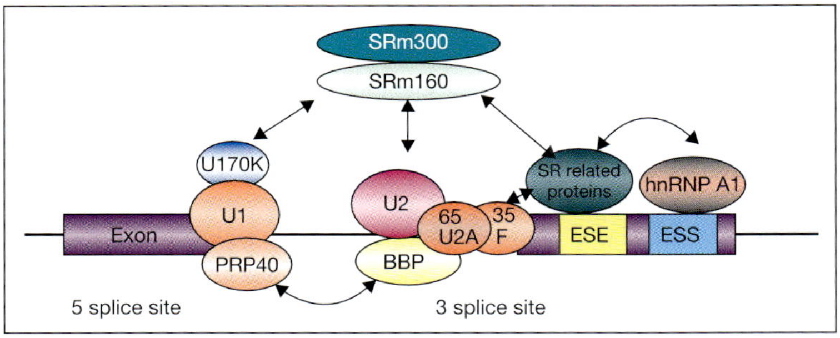

Fig. 4. A splicing model that includes splicing machineries and their binding sites. An SR protein (serine- and arginine-residue-rich protein) binds to an exonic splicing enhancer (ESE) through its RNA-recognition motifs (RRM) and contacts the splicing factor U2AF35 and/or U1–70k at the adjacent splice sites through its RS domain. The U2AF splicing factor consists of two subunits (U2AF65 and U2AF35), and the large subunit binds to the polypyrimidine tract. U2AF65 also promotes binding of U2 snRNP to the branching point. U2AF35 recognize the 3′ splice site AG dinucleotide. The U1 snRNP binds to the upstream and downstream 5′ splice sites through base pairing of the U1 snRNA. The three sets of splicing-factor-pre-mRNA interactions (U2AF-3′ splice site, U1 snRNP-5′ splice site, and SR protein-ESE) are strengthened by the protein-protein interactions that are mediated by the RS domain. For some ESE-dependent pre-mRNA, indirect interactions are bridged by the splicing coactivator Srm 160, which stimulates splicing of some ESE-dependent pre-mRNAs and also interacts with the U2 snRNP.

Several kinds of proteins were detected to bind to each probe in the same pattern, but between 30- and 35-kD ranges; the patterns of the proteins binding to the wild tau probe were different from those of mutant probes, that is, the wild-type probe bound to three proteins in this range (fig. 5b). In mutant probes, N279K and L284L, the motilities of these bands are slightly different from those of wild-type probes. In the Δ280K mutation, only two bands were detected in this range. These results suggest that mutations linked to FTDP-17 cause the changes of the binding patterns of the splicing factors to the tau pre-mRNA containing exon 10.

To determine the binding sites of 30- to 35-kD proteins to the tau exon 10, we carried out a binding assay using various deletion mutants of tau pre-mRNA (fig. 6a). Among these probes, probes 5–8 which were deleted 5′ sites of exon 10 sequences, could not bind to the 30- to 35-kD proteins (fig. 6b), suggesting that 30- to 35-kD proteins bind to the 5′ site of tau exon 10. It has been reported that the region may contain the ESE sequence, and furthermore, the mutations that are linked to FTDP-17 are accumulated in the site.

The present study shows that mutations linked to FTDP-17 affect the splicing patterns of tau exon 10, either an increase or a decrease in splicing of tau exon 10. Mutation of N279K increases exon 10 inclusion, presumably by

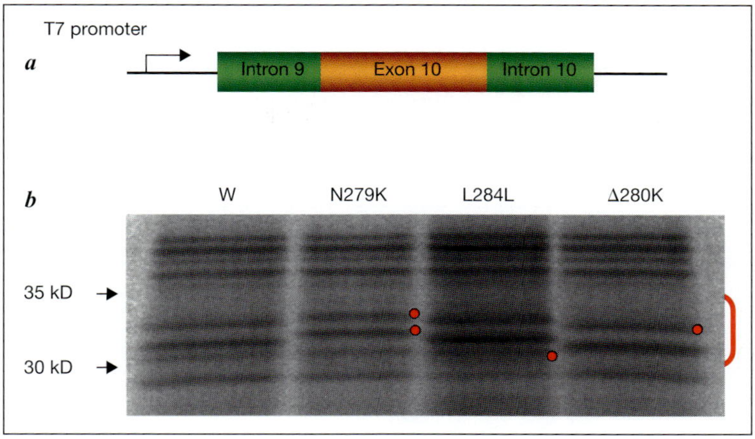

Fig. 5. RNA protein binding assay using nuclear extracts and tau pre-mRNA including exon 10. ***a*** Construction of the probes. Sense strand RNAs were transcribed and uniformly labeled with [α-^{32}P]UTPs in vitro by T7 RNA polymerase from templates of 177-bp sequences containing exon 10. RNA probes were also synthesized from the templates containing FTDP-17-linked mutations. ***b*** Nuclear extracts from SK-N-SH cells were incubated with RNA probes, UV-irradiated, and digested with RNase A as described under Materials and Methods. Mutations linked to FTDP-17 result in changes of binding patterns of proteins to the tau pre-mRNA probes at the ranges from 30 to 35 kD, although several proteins are bound to the probes in the same patterns except this range. The bands (red circles) are specific in each probe.

enhancing ESE. The N279K mutation may change the purine content of this ESE sequence or create a GAR repeat (R, purine), and results in enhancement of the splicing of tau exon 10. In contrast, deletion mutation Δ280K abolished exon 10 splicing. The reason is probably that the Δ280 mutation may destroy the same ESE activated by the N279K mutation. L284L mutation increases the inclusion of tau exon 10, presumably by affecting the ESS in the tau exon 10. Thus, the tight regulation of the proper ratio of 3 and 4-repeat tau is very important for normal functioning in neurons, and even subtle changes in the efficiencies of exon 10 inclusion may lead to disease development.

We detected some of RNA-binding proteins that directly associated with tau pre-mRNA, and these proteins showed different binding patterns between wild and mutant tau exon 10 pre-mRNA. Furthermore, the binding site was the 5′ site of tau exon 10, which was reported to contain ESE sequences and where the mutations linked to FTDP-17 are accumulated [27, 28]. Taken together, the 30- to 35-kD RNA-binding proteins that we detected in the present study could be splicing factors that regulate tau exon 10, such as SR-related proteins. We are currently attempting to identify these binding proteins. The identification of

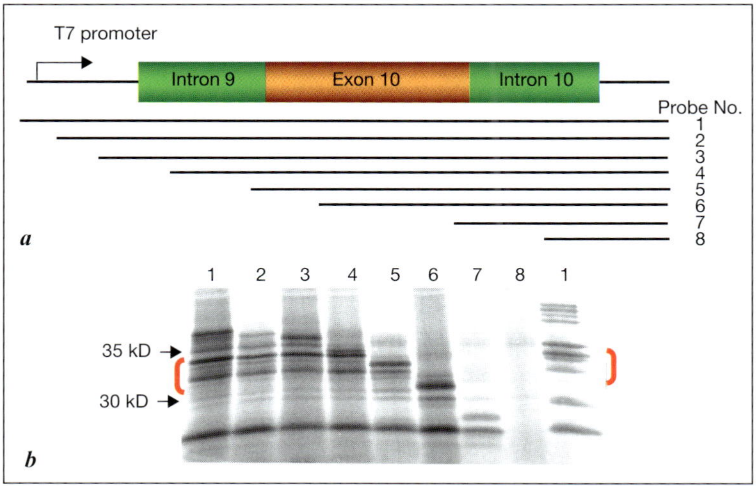

Fig. 6. Binding sites of 30- to 35-kD proteins on the tau exon 10 pre-mRNA. ***a*** Various probes that are deleted from the 5′ site of the tau pre-mRNA. ***b*** Binding patterns of each probe. The 30- to 35-kD proteins did not bind to probes 5–8, indicating that the proteins bind to the 5′ site of the tau exon 10.

splicing factors may be one of key factors for understanding deregulation of the splicing of tau exon 10 containing FTDP-17-linked mutations.

Acknowledgements

We thank Mrs. K. Otori for her technical support in this study. This work was partly supported by the Toray Sciences Foundation, and grants from the 'Spatiotemporal Network of RNA Information Flow' of the Ministry of Education, Culture, Sports, Science and Technology (# 15030233), and Grant-in-Aid for Scientific Research (A) (# 14208093).

References

1 Smith CW, Patton JG, Nadal-Ginard B: Alternative splicing in the control of gene expression. Annu Rev Genet 1989;23:973–977.
2 Maniatis T: Mechanisms of alternative pre-mRNA splicing. Science 1991;251:33–34.
3 Tysoe C, Whittaker J, Xuereb J, Cairns NJ, Cruts M, Van Broeckhoven C, Wilcock G, Rubinsztein DC: A presenilin-1 truncating mutation is present in two cases with autopsy-confirmed early-onset Alzheimer disease. Am J Hum Genet 1998;62:70–76.
4 De Jonghe C, Cruts M, Rogaeva EA, Tysoe C, Singleton A, Vanderstichele H, Meschino W, Dermaut B, Vanderhoeven I, Backhovens H, Vanmechelen E, Morris CM, Hardy J, Rubinsztein DC, St George-Hyslop PH, Van Broeckhoven C: Aberrant splicing in the presenilin-1 intron 4 mutation causes presenile Alzheimer's disease by increased Abeta42 secretion. Hum Mol Genet 1999;8:1529–1540.

5 Perez-Tur J, Froelich S, Prihar G, Crook R, Baker M, Duff K, Wragg M, Busfield F, Lendon C, Clark RF, et al: A mutation in Alzheimer's disease destroying a splice acceptor site in the presenilin-1 gene. Neuroreport 1995;7:297–301.
6 Clark LN, Poorkaj P, Wszolek Z, Geschwind DH, Nasreddine ZS, Miller B, Li D, Payami H, Awert F, Markopoulou K, Andreadis A, D'Souza I, Lee VM, Reed L, Trojanowski JQ, Zhukareva V, Bird T, Schellenberg G, Wilhelmsen KC: Pathogenic implications of mutations in the tau gene in pallido-ponto-nigral degeneration and related neurodegenerative disorders linked to chromosome 17. Proc Natl Acad Sci USA 1998;95:13103–13107.
7 Hutton M, Lendon CL, Rizzu P, Baker M, Froelich S, Houlden H, Pickering-Brown S, Chakraverty S, Isaacs A, Grover A, Hackett J, Adamson J, Lincoln S, Dickson D, Davies P, Petersen RC, Stevens M, de Graaff E, Wauters E, van Baren J, Hillebrand M, Joosse M, Kwon JM, Nowotny P, Heutink P, et al: Association of missense and 5′-splice-site mutations in tau with the inherited dementia FTDP-17. Nature 1998;393:702–705.
8 Spillantini MG, Murrell JR, Goedert M, Farlow MR, Klug A, Ghetti B: Mutation in the tau gene in familial multiple system tauopathy with presenile dementia. Proc Natl Acad Sci USA 1998; 95:7737–7741.
9 Hasegawa M, Smith MJ, Goedert M: Tau proteins with FTDP-17 mutations have a reduced ability to promote microtubule assembly. FEBS Lett 1998;437:207–210.
10 Lin CL, Bristol LA, Jin L, Dykes-Hoberg M, Crawford T, Clawson L, Rothstein JD: Aberrant RNA processing in a neurodegenerative disease: The cause for absent EAAT2, a glutamate transporter, in amyotrophic lateral sclerosis. Neuron 1998;20:589–602.
11 Foster NL, Wilhelmsen K, Sima AA, Jones MZ, D'Amato CJ, Gilman S: Frontotemporal dementia and parkinsonism linked to chromosome 17: A consensus conference. Conference Participants. Ann Neurol 1997;41:706–715.
12 Lynch T, Sano M, Marder KS, Bell KL, Foster NL, Defendini RF, Sima AA, Keohane C, Nygaard TG, Fahn S, et al: Clinical characteristics of a family with chromosome 17-linked disinhibition-dementia-parkinsonism-amyotrophy complex. Neurology 1994; 44:1878–1884.
13 Wilhelmsen KC, Lynch T, Pavlou E, Higgins M, Nygaard TG: Localization of disinhibition-dementia-parkinsonism-amyotrophy complex to 17q21–22. Am J Hum Genet 1994;55: 1159–1165.
14 Wijker M, Wszolek ZK, Wolters EC, Rooimans MA, Pals G, Pfeiffer RF, Lynch T, Rodnitzky RL, Wilhelmsen KC, Arwert F: Localization of the gene for rapidly progressive autosomal dominant parkinsonism and dementia with pallido-ponto-nigral degeneration to chromosome 17q21. Hum Mol Genet 1996;5:151–154.
15 Spillantini MG, Goedert M, Crowther RA, Murrell JR, Farlow MR, Ghetti B: Familial multiple system tauopathy with presenile dementia: A disease with abundant neuronal and glial tau filaments. Proc Natl Acad Sci USA 1997;94:4113–4118.
16 Bird TD, Wijsman EM, Nochlin D, Leehey M, Sumi SM, Payami H, Poorkaj P, Nemens E, Rafkind M, Schellenberg GD: Chromosome 17 and hereditary dementia: Linkage studies in three non-Alzheimer families and kindreds with late-onset FAD. Neurology 1997;48:949–954.
17 Heutink P, Stevens M, Rizzu P, Bakker E, Kros JM, Tibben A, Niermeijer MF, van Duijn CM, Oostra BA, van Swieten JC: Hereditary frontotemporal dementia is linked to chromosome 17q21-q22: A genetic and clinicopathological study of three Dutch families. Ann Neurol 1997;41: 150–159.
18 Yamaoka LH, Welsh-Bohmer KA, Hulette CM, Gaskell PC Jr, Murray M, Rimmler JL, Helms BR, Guerra M, Roses AD, Schmechel DE, Pericak-Vance MA: Linkage of frontotemporal dementia to chromosome 17: clinical and neuropathological characterization of phenotype. Am J Hum Genet 1996;59:1306–1312.
19 Spillantini MG, Crowther RA, Goedert M: Comparison of the neurofibrillary pathology in Alzheimer's disease and familial presenile dementia with tangles. Acta Neuropathol (Berl) 1996; 92:42–48.
20 Reed LA, Grabowski TJ, Schmidt ML, Morris JC, Goate A, Solodkin A, Van Hoesen GW, Schelper RL, Talbot CJ, Wragg MA, Trojanowski JQ: Autosomal dominant dementia with widespread neurofibrillary tangles. Ann Neurol 1997;42:564–572.

21 Hong M, Zhukareva V, Vogelsberg-Ragaglia V, Wszolek Z, Reed L, Miller BI, Geschwind DH, Bird TD, McKeel D, Goate A, Morris JC, Wilhelmsen KC, Schellenberg GD, Trojanowski JQ, Lee VM: Mutation-specific functional impairments in distinct tau isoforms of hereditary FTDP-17. Science 1998;282:1914–1917.
22 Church DM, Stotler CJ, Rutter JL, Murrell JR, Trofatter JA, Buckler AJ: Isolation of genes from complex sources of mammalian genomic DNA using exon amplification. Nat Genet 1994;6: 98–105.
23 Miyajima H, Miyaso H, Okumura M, Kurisu J, Imaizumi K: Identification of a cis-acting element for the regulation of SMN exon 7 splicing. J Biol Chem 2002;277:23271–23277.
24 Zhu J, Mayeda A, Krainer AR: Exon identity established through differential antagonism between exonic splicing silencer-bound hnRNP A1 and enhancer-bound SR proteins. Mol Cell 2001;8:1351–1361.
25 Blencowe BJ: Exonic splicing enhancers: mechanism of action, diversity and role in human genetic diseases. Trends Biochem Sci 2000;25:106–110.
26 Cartegni L, Chew SL, Krainer AR: Listening to silence and understanding nonsense: Exonic mutations that affect splicing. Nat Rev Genet 2002;3:285–298.
27 D'Souza I, Poorkaj P, Hong M, Nochlin D, Lee VM, Bird TD, Schellenberg GD: Missense and silent tau gene mutations cause frontotemporal dementia with parkinsonism-chromosome 17 type, by affecting multiple alternative RNA splicing regulatory elements. Proc Natl Acad Sci USA 1999;96:5598–5603.
28 D'Souza I, Schellenberg GD: Determinants of 4-repeat tau expression. Coordination between enhancing and inhibitory splicing sequences for exon 10 inclusion. J Biol Chem 2000;275: 17700–17709.

Kazunori Imaizumi
Division of Structural Cellular Biology
Nara Institute of Science and Technology (NAIST)
8916–5 Takayama, Ikoma, Nara 630–0101 (Japan)
Tel. +81 743 72 5411, Fax +81 743 72 5419, E-Mail imaizumi@bs.aist-nara.ac.jp

Takeda M, Tanaka T, Cacabelos R (eds): Molecular Neurobiology of Alzheimer Disease and Related Disorders. Basel, Karger, 2004, pp 215–224

Tau Filament Formation and Associative Memory Deficit in Aged Mice Expressing Mutant (R406W) Human Tau

Tomohiro Miyasaka[a], *Yoshitaka Tatebayashi*[a], *De-Hua Chui*[a], *Takumi Akagi*[b], *Ken-ichi Mishima*[c], *Katsunori Iwasaki*[c], *Michihiro Fujiwara*[c], *Kentaro Tanemura*[a], *Miyuki Murayama*[a], *Koichi Ishiguro*[d], *Emmanuel Planel*[a], *Shinji Sato*[a], *Tsutomu Hashikawa*[b], *Akihiko Takashima*[a]

[a] Laboratory for Alzheimer's Disease and
[b] Neural Architecture, Brain Science Institute, The Institute of Physical and Chemical Research (RIKEN), Wako-shi,
[c] Department of Physiology and Pharmacology, Faculty of Pharmaceutical Sciences, Fukuoka University, Jonan-Ku, and
[d] The Mitsubishi Kasei Institute of Life Sciences, Machida, Japan

Neurofibrillary tangles (NFTs) composed of abnormally hyperphosphorylated microtubule-associated protein tau are prominent in certain types of neurodegenerative diseases including: Alzheimer's disease (AD), progressive supranuclear palsy, corticobasal degeneration, Pick's disease, and frontotemporal dementia and parkinsonism linked to chromosome 17 (FTDP-17) [for reviews, see ref. 1, 2]. The discovery of multiple tau gene mutations in FTDP-17 provides evidence that tau abnormalities alone can cause neurodegenerative diseases [1, 2].

More than 20 intronic and exonic mutations have been identified in the tau genes of FTDP-17 families, but the pathogenic mechanisms of such mutants are still not understood [1, 2]. Although the clinical phenotypes of these mutants were variable, the R406W mutant has characteristic features. First, the R406W mutations localizing in constitutively expressed exons affect all 6 tau isoforms and result in NFTs similar to those present in secondary tauopathies, such as AD [3, 4]. This mutation causes AD-like clinical symptoms (e.g., progressive memory loss and personality change) mostly without amyloid-β deposition [3]. Furthermore, immunocytochemical and biochemical analysis of brains from R406W patients using antibodies specific for R406W tau revealed that both mutant and wild-type tau were deposited in NFTs and recovered from insoluble fractions [5]. Thus, some of the adverse properties associated with the R406W mutation might explain some early clinicopathological events in AD; therefore the development of model mice would help us decipher the detailed mechanisms of neurodegeneration in the brains of primary and secondary tauopathies.

To analyze early clinicopathological events of the AD-like neurodegeneration with NFT formation, we generated transgenic (Tg) mice expressing the longest human tau with the R406W mutation. The expression of the human tau isoform with this mutation was driven under the control of the α-calcium-calmodulin-dependent kinase II (CaMK-II) promoter, which is expressed in neurons of the forebrain [6]. Tg mice develop filamentous tau inclusions as early as 18 months of age with impairment of associative memory.

Results

Generation of Tg Mice Expressing R406W Human Tau

Tg mouse lines expressing R406W human tau were generated as described previously [7, 8], except that the CaMK-II promoter [6] was used. A cDNA construct of R406W human tau containing myc (EQKLISEEDL) and FLAG (DYKDDDDK) tags at the N- and C-terminal ends, respectively, was inserted into the CaMK-II chain expression vector at the *Xho*I and *Not*I sites. A 4.3-kb *Bgl*II and *Nae*I fragment containing the CaMK-II promoter, R406W human tau cDNA, and a 3′ untranslated sequence were used as the transgene to create Tg mice on a B6SJL background. An 8.5-kb portion of the CaMK-II promoter containing upstream control regions and the transcriptional initiation site are known to drive the expression postnatally in forebrain neurons [6]. Western blot analysis revealed that mutant tau expression was greatest in the hippocampus (17.8 ± 0.35% of corresponding endogenous tau levels), with progressively decreasing amounts in the neocortex (17.0 ± 0.92%), olfactory bulb (14.3 ± 1.07%), striatum (14.1 ± 0.60%) and thalamus (7.59 ± 0.41%). Mutant tau was barely detectable in the mid brain, cerebellum, or spinal cord (fig. 1a, 5 months). Immunohistochemical analysis with anti-human tau (H-150, fig. 1b) antibody further confirmed this regional specificity; it selectively stained mostly forebrain neurons of Tg brains (5 months).

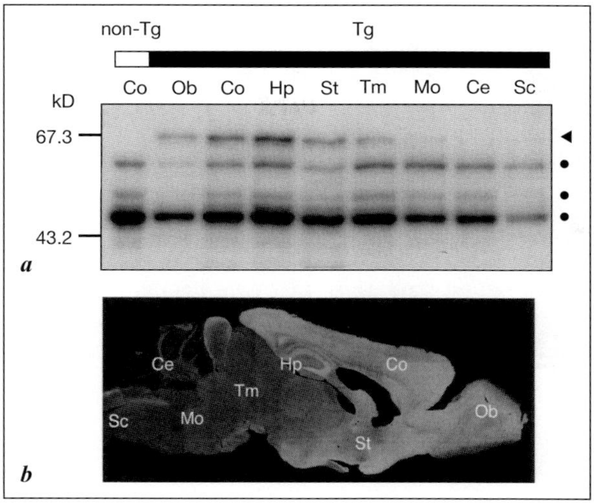

Fig. 1. Expression profile of mutant tau in Tg mice. ***a*** Western blot analysis of tau expression in Tg mice (5 months). Equal amount of proteins of TBS-soluble fractions from cortex (Co), olfactory bulb (Ob), hippocampus (Hp), striatum (St), thalamus and mid brain (Tm), medulla oblongata (Mo), cerebellum (Ce), and spinal cord (Sc) were dephosphorylated as described previously [5], and subjected to SDS-PAGE. Total amounts of tau were labeled by Tau-1. An arrowhead and dots indicate the position of mutant and endogenous mouse tau, respectively. ***b*** Fluorescence photomicrograph showing the expression pattern of transgenic mutant tau. A 13-μm sagittal-cryo-section from a Tg brain was immunostained with anti-human tau antibody (H-150). Note that mutant tau is expressed robustly in forebrain regions, but weakly in the cerebellum and brainstem.

Histopathological Studies

We next examined aged Tg mice (>18 months) for neuropathological changes. Formalin-fixed and paraffin- embedded sections (4–6 μm) were used for histopathological analyses performed as described previously [7, 8]. Immunostaining with anti-myc antibodies confirmed the selective neuronal expression and accumulation of mutant tau in some aged Tg brains. We observed occasional intense myc immunoreactivity in neurons from the hippocampus (fig. 2a), striatum, amygdala, neocortex, and olfactory bulb. (We did not observe any myc-positive cells in the brains of non-Tg littermates (fig. 2b).) Myc-positive cells were also stained with a series of tau antibodies that recognize phosphorylated (phospho-Ser199, 202, 396, 404, 413, 422, Thr205, fig. 2c, d, data not shown) tau epitopes [9]. In a subset of these neurons, we sometimes identified congophilic inclusions by thioflavin-S, Congo red and Gallyas silver staining (fig. 2e–l).

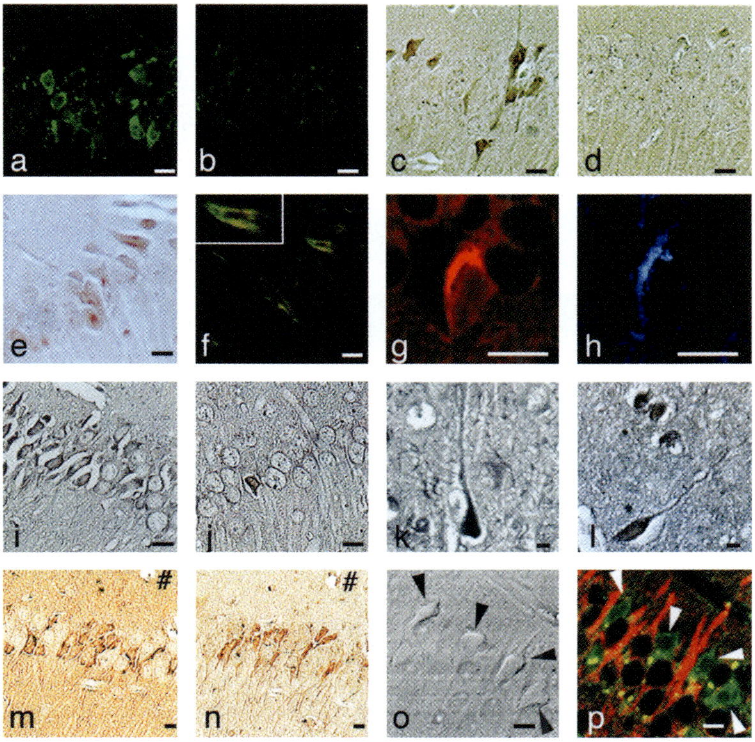

Fig. 2. Characterization of irregularly shaped neurons in the brains of aged Tg mice (***a–d***) Immunostaining of the hippocampal region (CA1–CA2) of a 19-month-old Tg mouse (***a***, ***c***) and a non-Tg littermate (***b***, ***d***) with monoclonal anti-myc antibody (***a***, ***b***) or PS199 (***c***, ***d***). ***e***, ***f*** Photomicrographs of sections stained with Congo Red. The images were photographed without (***e***) or with (***f***) a polarizing filter set at different angles. Inset indicates high magnification view of Congo Red birefringent neuron. ***g***, ***h*** A neuron of the Tg hippocampus stained with polyclonal anti-tau antibody (JM); ***g*** was further stained with thioflavin-S (***h***). ***i–l*** Photomicrographs showing sections of the hippocampus stained with Gallyas silver (***i***), neocortex (***k***), and amygdala (***l***) from a Tg mouse and the hippocampus from a non-Tg littermate (***j***). ***m***, ***n*** Photomicrographs of two serial sections stained with AR406 (***m***) and AW406 (***n***). Note that both antibodies stained the same irregularly shaped neurons in the CA1 region of the aged Tg hippocampus. The same blood for each part of adjacent sections is indicated (#). ***o***, ***p*** Photomicrographs showing double immunostaining with AR406 and anti-α-tubulin antibodies. Note that irregularly shaped neurons (arrowheads in ***o***; phase contrast) in the aged Tg hippocampus are stained with AR406 (green) but lack tubulin immunoreactivity (red) (marge in ***p***). (scale bars; 10 μm).

A striking feature of R406W mutants is that both mutant and wild-type tau are deposited in NFTs and recovered from insoluble fractions [5]. To characterize how endogenous mouse tau is involved in these tau inclusions, we employed polyclonal antibodies AR406 and AW406 raised against wild-type and mutant human tau, respectively [5]. Immunostaining of serial sections with these antibodies revealed that most of the AW406-positive irregularly shaped neurons were also immunoreactive for AR406 (fig. 2m, n).

Because a disruption of microtubule structures is inevitably found in NFT-containing neurons [10], we studied the microtubule structures in aged Tg hippocampi. Double immunostaining with anti-α-tubulin and AR406 antibodies revealed that the only irregularly shaped neurons that were positive for AR406 lacked cytosolic tubulin immunoreactivity (fig. 2o, p). Ultrastructural analysis further revealed that these abnormal neurons were electron-dense, filled with ribosomes, mitochondria and lipofuscin but devoid of microtubules (data not shown). These data suggest that microtubule disruption occurred in the vulnerable neurons in association with tau inclusions and, probably, abnormal organelle transport.

Biochemical Studies

We next assessed insolubility of tau in aged Tg brains. Tris-buffered saline (TBS)-soluble fractions were collected from Tg and non-Tg brain tissues, and the resultant insoluble pellets were refractionated by solubility for Sarkosyl according to the procedure developed by Greenberg and Davies [11]. In Sarkosyl-insoluble fractions, we often found tau recovered only from aged Tg brains (fig. 3a–c). The proportion of mutant to endogenous tau in Sarkosyl-insoluble fractions (12.2 $\pm$ 1.29%) was almost the same as that in TBS-soluble fractions (11.6 $\pm$ 0.49%), suggesting that the mutant and the wild-type tau were proportionally incorporated into the insoluble fractions in Tg brains. When Sarkosyl-insoluble materials were subjected to ultrastructural analysis, we observed bundles of tau filaments that were mostly straight ($\sim$10 nm in diameter) and immunostained with anti-tau antibody JM (fig. 3d).

Behavioral Studies

To gain more direct insight into the effects of these changes on pathophysiology, we subjected aged (16–23 months) Tg mice and their non-Tg littermates to a comprehensive battery of behavioral tests. We found impairment in associative memory retrieval without detectable sensorimotor deficits (data not shown). R406W Tg mice showed significant impairments in the contextual and cued fear-conditioning test (fig. 4). Fear conditioning is a simple associative form of learning in which both a novel environment and a tone are paired with a foot shock on the training day. Memory is assessed by the duration of

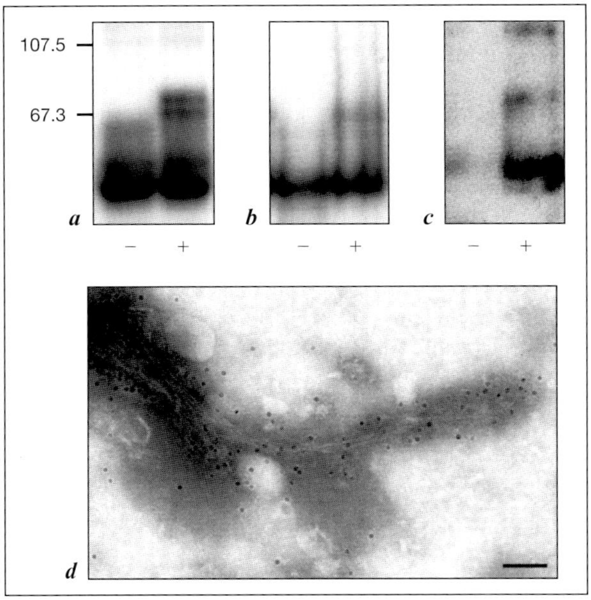

Fig. 3. Tau filament formation in aged Tg mice. ***a–c*** Western blot analysis of Sarkosyl-insoluble tau. TBS-soluble (***a***), Sarkosyl-soluble (***b***), and Sarkosyl-insoluble (***c***) fractions from 22-month-old control (–) and Tg (+) brains were subjected to SDS-PAGE and stained with pan-tau antibody JM. ***d*** Immunoelectron photomicrograph of Sarkosyl-insoluble tau. Abnormally thin filaments, intensely labeled by JM, were recovered from Sarkosyl-insoluble fractions of the Tg brain but not from those of control brain. Scale bar: 50 μm.

freezing (i.e., the fear response) elicited by either the same environment (contextual test, 24 h or 15 days later) or the same tone (cued test, 48 h later) in the different environment. In the contextual version of the task, the mouse is selectively impaired by lesions in the hippocampus, and in the combined cued and contextual versions of the task, it is impaired by lesions in the amygdala [12].

During the conditioning period, Tg mice showed significantly lower levels of freezing after foot shocks (Conditioning; data not shown) (genotype effect, $F(1, 15) = 6.834$, $p = 0.0195$). The distance traveled just before (2 s), during (2 s) and immediately after (2 s) the foot shocks was almost same for Tg and non-Tg mice (data not shown). This excludes the possibility that reduced sensitivity to foot shock is responsible for reduced freezing times in Tg mice. When the conditioned stimulus (tone) was presented in an altered context 48 h after conditioning (i.e., cued testing), Tg mice showed reduced freezing responses (fig. 4a, genotype effect, $F(1, 15) = 5.387$; $p = 0.0359$). As no significant difference was found in the acoustic startle response or open-field tests

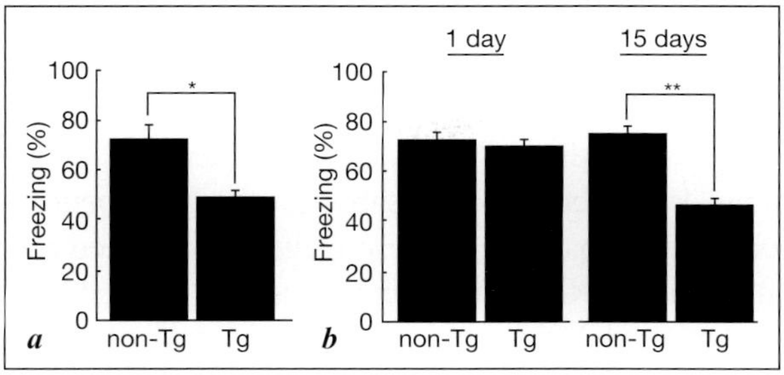

Fig. 4. Impaired associative memory in aged Tg mice. ***a*** Tg mice (19 months of age) showed reduced freezing durations during cued testing (*p = 0.0359) conducted 48 h after conditioning. ***b*** Tg mice showed significantly reduced levels of freezing during contextual testing conducted 15 days after conditioning (**p = 0.0072). There was no significant difference in freezing between genotypes during contextual testing conducted 1 day after conditioning (p = 0.6291). Values are expressed as means ± SEM.

(data not shown), it is unlikely that this reduction was due to the reduced acoustic sensitivity or increased locomotor activity of the aged Tg mice.

During contextual testing conducted 24 h after conditioning, Tg mice showed the same levels of freezing compared to their non-Tg littermate controls (fig. 4b, 24 h, genotype effect, $F(1, 15) = 0.244$; $p = 0.6291$), suggesting that contextual fear memory developed both in Tg and non-Tg mice. In the contextual testing conducted 15 days after conditioning, however, freezing was significantly reduced in Tg mice (fig. 4b, 15 days, genotype effect, $F(1, 15) = 10.447$; $p = 0.0072$). This suggests that memory loss may be more pronounced after a longer period of time in aged Tg mice.

Discussion

In the present study, we report for the first time Tg mouse lines expressing modest levels of R406W human tau confined predominantly to forebrain neurons. Importantly, these neurons develop congophilic tau inclusions, and the Tg mice develop associative memory deficits. This finding suggests that modest levels of mutant (R406W) tau expression in neurons can cause neurodegenerative disease in mice. In contrast to previously reported Tg mice expressing wild-type or mutant human tau in both brain and spinal cord, our mice expressed mutant tau mostly in forebrain neurons under the control of the CaMK-II promoter. This regional specificity enabled us to exclude, with a high degree of confidence, possible confounding motor dysfunctions associated with neurogenic muscle atrophy (data not shown). As neurogenic motor symptoms can impair memory performance in behavioral tests that are unrelated to learning and memory function per se, our mice provide an animal model of CNS tauopathy for evaluating the relationship between mutant tau expression, NFT formation and neuronal disturbance leading to memory impairment.

Congophilic tau inclusions developed only in a subset of forebrain neurons from aged Tg mice as early as 18 months of age. These inclusions fulfilled several criteria for the detection of human NFTs, including the ability to be stained by histochemical dyes that bind crossed β-pleated sheet structures, such as Congo Red and thioflavin S, the ability to be impregnated by the Gallyas silver method, and the presence of tau protein in isolated filaments from detergent-insoluble fractions. Ishihara et al. [13] reported similar tau inclusions in forebrain neurons from aged (18–20 months) Tg mice expressing 3R wild-type human tau. However, the levels of human tau expressed in those mice were 5 times higher than those of endogenous mouse tau. Given the modest (~20%) levels of mutant tau expression in our Tg mice, the development of tau inclusions in similar regions and at similar ages was surprising. This might reflect one of the deleterious properties acquired by the R406W mutation that promotes NFT formation, as in FTDP-17.

The major difference between the tau fibrils in our Tg mice and those in AD or R406W FTDP-17 is that tau filaments recovered from our aged Tg mice were mostly straight. This may be due to altered tau isoform ratios or the combination of human and mouse tau in the filaments. In Tg mice, the 3 endogenous mouse wild-type (4R) and the longest human tau (4R) were expressed and incorporated into the inclusions, whereas in AD or R406W FTDP-17, all 6 tau isoforms (3R and 4R) were incorporated into the NFTs [2]. In contrast, we notice some similarities in tau fibril formation between our Tg mice and AD and R406W FTDP-17. For example, all the tau isoforms expressed were incorporated into the tangles proportionally to their expression levels.

Tg mice showed reduced levels of fear response during the cued but not the contextual testing relatively soon after the initial conditioning (within 48 h). Conditioning to a tone requires at least two different projections: one from the auditory cortex or thalamus to the lateral nucleus of the amygdala, and the other from the lateral nucleus to the central nucleus of the amygdala [12]. Since the latter projection is also used for the contextual memory, the former projection might be disproportionately impaired in aged Tg mice. In addition, we found a significant decrease in the prepulse inhibition of the startle response, suggesting the impairment of inhibitory modulation in the auditory cortex.

Tg mice showed significantly lower levels of fear response during the contextual testing conducted 15 days after conditioning. Recently, a similar pattern of memory loss, i.e., impairment in the establishment of permanent memory, has been reported in mice heterozygous for a null mutation of CaMK-II

(CaMK-II$^{+/-}$) [14]. Since our Tg mice express mutant tau under the control of the CaMK-II promoter, it is quite plausible that the formation of tau inclusions and/or mutant tau expression itself might affect the CaMK-II-dependent establishment of permanent memories in Tg mice.

Our results represent the first step towards elucidating the molecular and cellular mechanisms underlying the R406W tau mutation that cause a tauopathy clinically resembling AD. Since most of the R406W brains lack amyloid-β deposition, the adverse properties exhibited by this mutation might potentially recapitulate some early pathophysiological events in AD. Therefore, an investigation of the molecular mechanisms of memory disturbance in our Tg mice could provide an important insight into the pathogenesis of both R406W FTDP-17 and AD.

Acknowledgements

We thank Dr. M. Mayford for providing the CaMK-II promoter, and Dr. Y. Ihara for providing the AR406 and AW406 antisera. This work was partly supported by CREST (Japan Science and Technology, JST), Grant-in-Aid for Scientific Research on Priority Areas (The Ministry of Education, Culture, Sports, and Technology) and a Grant-in-Aid for Scientific Research (11680746, the Japanese Ministry of Education, Science and Culture).

References

1 Reed LA, Wszolek ZK, Hutton M: Phenotypic correlations in FTDP-17. Neurobiol Aging 2001; 22:89–107.

2 Lee VM, Goedert M, Trojanowski JQ: Neurodegenerative tauopathies. Annu Rev Neurosci 2001; 24:1121–1159.

3 Reed LA, Grabowski TJ, Schmidt ML, Morris JC, Goate A, Solodkin A, Van Hoesen GW, Schelper RL, Talbot CJ, Wragg MA, Trojanowski JQ: Autosomal dominant dementia with widespread neurofibrillary tangles. Ann Neurol 1997;42:564–752.

4 van Swieten JC, Stevens M, Rosso SM, Rizzu P, Joosse M, de Koning I, Kamphorst W, Ravid R, Spillantini MG, Niermeijer, Heutink P: Phenotypic variation in hereditary frontotemporal dementia with tau mutations. Ann Neurol 1999;46:617–626.

5 Miyasaka T, Morishima-Kawashima M, Ravid R, Heutink P, van Swieten JC, Nagashima K, Ihara Y: Molecular analysis of mutant and wild-type tau deposited in the brain affected by the FTDP-17 R406W mutation. Am J Pathol 2001;158:373–379.

6 Mayford M, Wang J, Kandel ER, O'Dell TJ: CaMKII regulates the frequency-response function of hippocampal synapses for the production of both LTD and LTP. Cell 1995;81:891–904.

7 Tanemura K, Akagi T, Murayama M, Kikuchi N, Murayama O, Hashikawa T, Yoshiike Y, Park JM, Matsuda K, Nakao S, Sun X, Sato S, Yamaguchi H, Takashima A: Formation of filamentous tau aggregations in transgenic mice expressing V337M human tau. Neurobiol Dis 2001;8:1036–1045.

8 Tanemura K, Murayama M, Akagi T, Hashikawa T, Tominaga T, Ichikawa M, Yamaguchi H, Takashima A: Neurodegeneration with tau accumulation in a transgenic mouse expressing V337M human tau. J Neurosci 2002;22:133–141.

9 Takashima A, Murayama M, Murayama O, Kohno T, Honda T, Yasutake K, Nihonmatsu N, Mercken M, Yamaguchi H, Sugihara S, Wolozin B: Presenilin 1 associates with glycogen synthase kinase-3beta and its substrate tau. Proc Natl Acad Sci USA 1998;95:9637–9641.

10 Hempen B, Brion JP: Reduction of acetylated alpha-tubulin immunoreactivity in neurofibrillary tangle-bearing neurons in Alzheimer's disease. J Neuropathol Exp Neurol 1996;55:964–972.
11 Greenberg SG, Davies P: A preparation of Alzheimer paired helical filaments that displays distinct tau proteins by polyacrylamide gel electrophoresis. Proc Natl Acad Sci USA 1990;87:5827–5831.
12 LeDoux JE: Emotion circuits in the brain. Annu Rev Neurosci 2000;23:155–184.
13 Ishihara T, Zhang B, Higuchi M, Yoshiyama Y, Trojanowski JQ, Lee VM: Age-dependent induction of congophilic neurofibrillary tau inclusions in tau transgenic mice. Am J Pathol 2001;158:555–562.
14 Frankland PW, O'Brien C, Ohno M, Kirkwood A, Silva AJ: Alpha-CaMKII-dependent plasticity in the cortex is required for permanent memory. Nature 2001;411:309–313.

Akihiko Takashima
Laboratory for Alzheimer's Disease, Brain Science Institute
The Institute of Physical and Chemical Research (RIKEN)
2–1 Hirosawa, Wako-shi, Saitama, 351–0198 (Japan)
Tel. +81 48 467 9704, Fax +81 48 467 5916, E-Mail kenneth@brain.riken.go.jp

Takeda M, Tanaka T, Cacabelos R (eds): Molecular Neurobiology of Alzheimer Disease and Related Disorders. Basel, Karger, 2004, pp 225–235

Activated Protein Kinases and Phosphorylated Tau Protein in Alzheimer Disease

Toshihisa Tanaka[a], *Hidenaga Yamamori*[a], *Kenji Wada-Isoe*[b], *Ichiro Tsujio*[a], *Masatoshi Takeda*[a]

[a] Psychiatry and Behavioral Science, Osaka University, Graduate School of Medicine, Yamadaoka, Suita, Osaka,
[b] Division of Neurology, Faculty of Medicine, Tottori University, Nishimachi, Yonago, Tottori, Japan

Neurofibrillary tangles of paired helical filaments (PHF) are neuropathological hallmarks of Alzheimer disease (AD), and abnormally hyperphosphorylated tau protein is the major protein subunit of PHF [1–4]. Several kinases and phosphatases are thought to be involved in the process [5–11]. However, the mechanisms of phosphorylation of tau protein and neurodegeneration in AD are still unclear.

In this review, phosphorylation of tau protein and the recently found activated protein kinases in AD are overviewed and their involvement in the formation of phosphorylated tau protein and in the neurodegenerative process of AD are discussed.

Phosphorylation of Tau Protein

Tau protein has six isoforms which derive from a single gene by an alternative splicing mechanism, and the longest isoform is composed of 441 amino acids [12] (fig. 1). An alternative splicing mechanism produces tau with two, one and no N-terminal insertion(s); it also produces tau with and without an insertion in microtubule binding domains. Isoforms with an insertion in microtubule binding domains are called 4-repeats tau, and the others are called

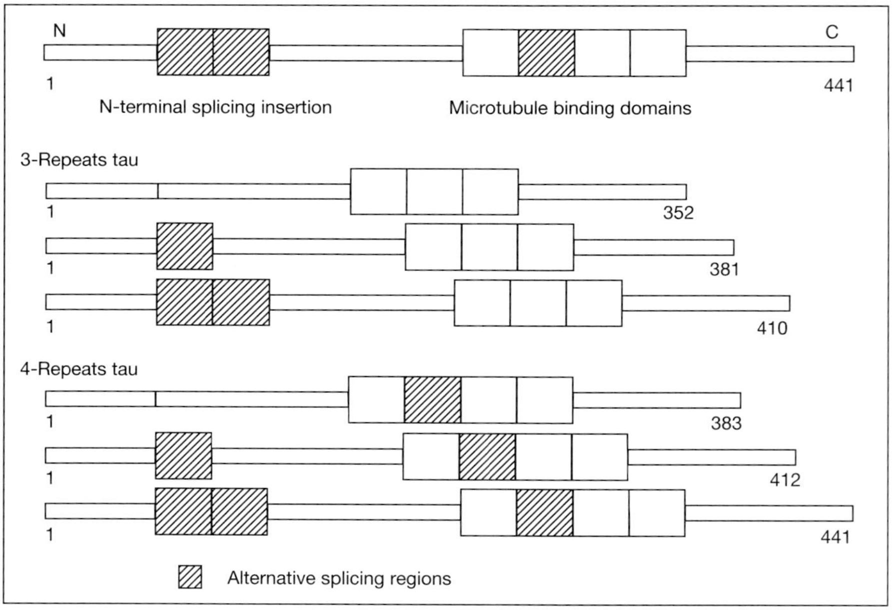

Fig. 1. Structure of tau protein. Tau protein consists of six isoforms which derive from a single gene by an alternative splicing mechanism; the isoforms are composed of 352, 381, 383, 410, 412, and 441 amino acids, respectively. An alternative splicing mechanism produces tau with two, one and no N-terminal insertion(s); it also produces tau with and without an insertion in microtubule binding domains. Isoforms with an insertion in microtubule binding domains are called 4-repeat tau, and others are called 3-repeat tau. Hatched squares indicate alternative splicing regions.

3-repeat tau. The expression of these isoforms is developmentally regulated; fetal tau consists of only 3-repeat tau and adult tau consists of all six isoforms [13]. Tau protein is one of the microtubule-associated proteins and has an ability to promote microtubule assembly. This activity is regulated by phosphorylation, which impairs the ratio of microtubule assembly [14].

Tau protein in normal brain contains 1–2 mol of phosphate(s) per 1 mol of tau protein; however in AD brains, tau protein contains 5–9 mol of phosphates [15]. This abnormal phosphorylated tau protein is present in the affected neurons as amorphous aggregates [16]. Phosphorylation of many Ser/Thr sites on tau protein has already been reported (fig. 2). Among 27 phosphorylated Ser/Thr sites on tau protein in AD, 11 sites are proline-directed (-Ser/Thr-Pro-) sites [17]. Therefore proline-directed protein kinases have been thought to be important in abnormal phosphorylation of tau protein in AD.

Among proline-directed kinases, glycogen synthase kinase-3 (GSK-3) is one of the important kinases in the phosphorylation of tau protein. GSK-3

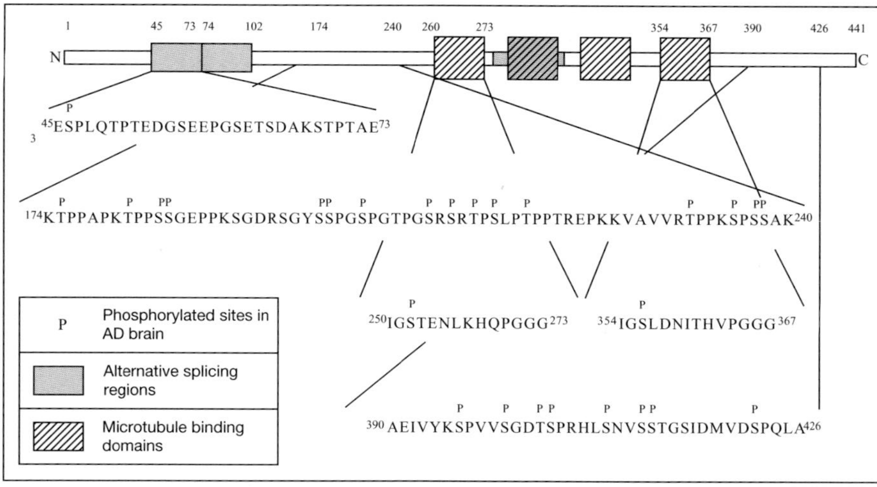

Fig. 2. Phosphorylated sites in an AD brain. The structure of tau is the longest isoform, and P associated with Ser/Thr indicates a phosphorylated site. Gray and hatched boxes indicate alternative splicing regions and microtubule binding domains, respectively.

strongly phosphorylates tau protein in vitro, and is localized in axons [18, 19]. And in fact, previously reported tau protein kinase-I was identified as GSK-3β [20, 21].

Regulation of Activity of GSK-3

GSK-3 is a calcium- and cyclic nucleotide-independent kinase that phosphorylates glycogen synthase, a regulatory enzyme of glycogen. The molecular weights of the α- and β-isoforms are 51 and 46 kD, respectively, and both are known to be able to phosphorylate tau protein [20, 21]. Protein phosphatase-I, G-subunit, the RII subunit of cyclic AMP-dependent protein kinase, phosphatase inhibitor-2, myelin basic protein and neurofilaments are known as substrates for GSK-3 [22–25]. The phosphorylation of GSK-3 regulates its activity and it has been reported that phosphorylation at Tyr 216 and nonphosphorylation at Ser9 are necessary for the activation of GSK-3β [26–28]. The consensus sequence of GSK-3 for phosphorylation is not only proline-directed Ser/Thr, but also Ser-X-X-X-Ser(p) [Ser(p) is prephosphorylated Ser] [29]. Glycogen synthase, β-catenine and tau protein contain Ser/Thr-X-X-X-Ser/Thr sequences and this implies that these proteins have potential sites for phosphorylation by GSK-3 after prime phosphorylation by other kinases [29–31].

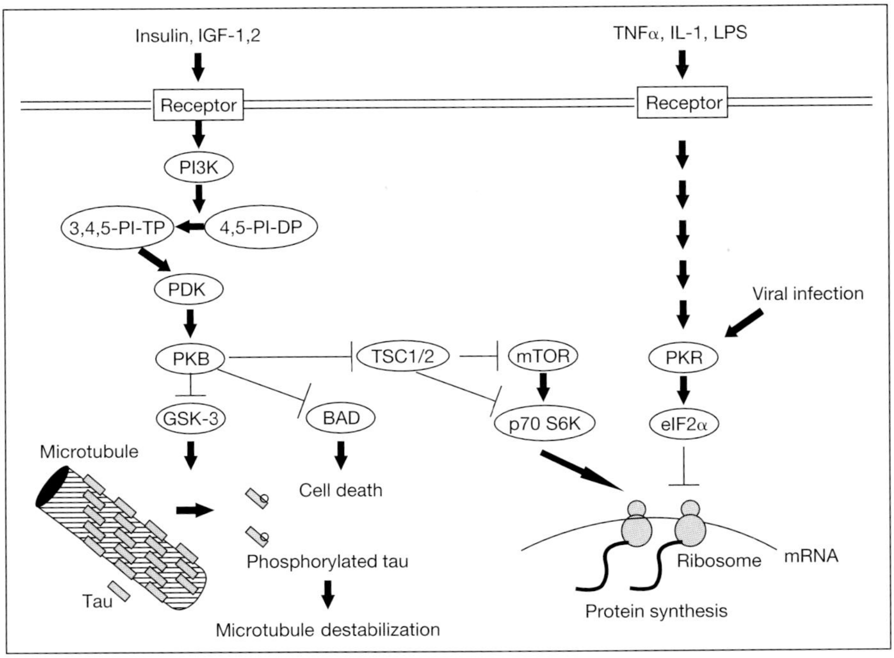

Fig. 3. Intracellular signal transduction on the activated kinases in an AD brain. The PI3K pathway is shown on the left. Binding of insulin, IGF-1 or 2, to the receptor leads to autophosphorylation of the receptor, and PI3K is activated by its binding to the receptor. Activated PI3K converts 4,5-PI-DP to 3,4,5-PI-TP. PDK is activated by its binding with 3,4,5-PI-TP and phosphorylates PKB. Activated PKB (by its phosphorylation) phosphorylates GSK-3 and inhibits the kinase activity of GSK-3. GSK-3 strongly phosphorylates tau protein and phosphorylated tau protein is detached from microtubules, leading to destabilization of microtubules. Then active PKB inhibits tau phosphorylation and microtubule destabilization. In addition, active PKB phosphorylates BAD, a proapoptotic protein, and phosphorylated BAD leads to inhibition of apoptosis. Further active PKB phosphorylates components of TSC, TSC1 and TSC2, and phosphorylation of TSC1/2 inhibits its inhibitory potency to activation of p70 S6K by mammalian target of rapamycin (mTOR). Therefore active PKB accelerates the activation of p70 S6K by mTOR, leading to activation of protein synthesis. On the right, regulation of eIF2α and PKR, and dsRNA-activated protein kinase are shown. TNFα, IL-1 and lipopolysaccharide (LPS) bind their receptors, and induce the expression of PKR; viral infection also activates PKR. Activated PKR phosphorylates eIF2α, and phosphorylated eIF2α inhibits the translational machinery.

The regulation of the signal transduction cascade has been studied, and regulatory mechanisms of GSK-3 have already been identified (fig. 3). The predominant regulatory system of the activity of GSK-3 is the phosphatidylinositol-3 kinase (PI3K) pathway, which is stimulated by insulin, IGF-1 or -2 [32]. The binding of insulin, IGF-1 or -2, to the receptor leads to autophosphorylation of

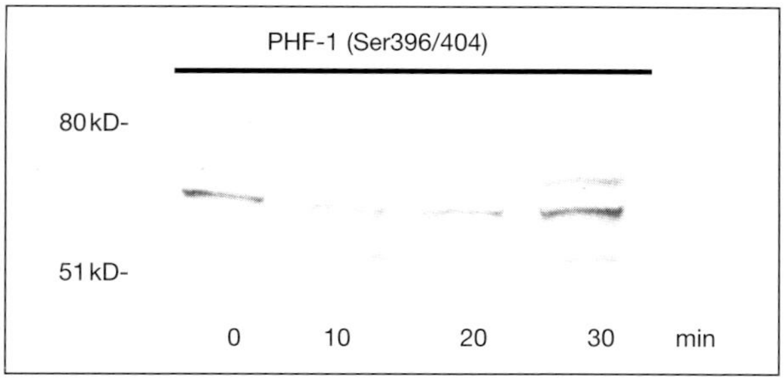

Fig. 4. Phosphorylation levels of tau protein in SH-SY5Y neuroblastoma cells treated with insulin. SH-SY5Y human neuroblastoma cells were treated with 250 μg/ml of insulin and the levels of phosphorylation of tau protein were investigated employing PHF-1 antibody that reacted with tau phosphorylated at Ser396/404. The intensities of bands decreased within 10–20 min, suggesting that dephosphorylation of tau protein was induced. The effect of insulin disappeared after 30 min.

the receptor, and PI3K is activated by its binding to the receptor. Activated PI3K converts 4,5-phosphatidylinositol diphosphate (4,5-PI-DP) to 3,4,5-phosphatidylinositol triphosphate (3,4,5-PI-TP). Phosphatidylinositol-dependent kinase (PDK) is activated by its binding with 3,4,5-PI-TP, and phosphorylates protein kinase B (PKB). Activated PKB (by its phosphorylation) phosphorylates GSK-3, and inhibits the kinase activity of GSK-3 [32]. In addition, active PKB phosphorylates BAD, a pro-apoptotic protein, and phosphorylated BAD is sequestrated from mitochondria, leading to inhibition of apoptosis [33, 34]. Therefore activation of PKB plays an important role in inhibition of apoptosis. Taken together insulin, IGF-1 or -2, is able to reduce the phosphorylation of tau protein through inhibition of GSK-3 and to lead to cell survival.

The level of phosphorylation of tau protein was regulated by this pathway. SH-SY5Y human neuroblastoma cells were treated with 250 μg/ml of insulin and the levels of phosphorylation of tau protein was investigated employing PHF-1 antibody that reacted with tau phosphorylated at Ser396/404 (fig. 4). Insulin is thought to activate the PI3K pathway that reduces the activity of GSK-3. Dephosphorylation of tau protein was induced with 10–20 min, and after 30 min the effect of insulin disappeared.

To observe the role of PI3K in the regulation of tau phosphorylation, wortmannin, an inhibitor of PI3K, was employed in our previous investigation [35]. And increased phosphorylation levels of tau protein at the PHF-1 (Ser 396/404) site in the early phase (1–3 h) were observed, suggesting that PI3K is involved in the regulation of phosphorylation of tau protein [35].

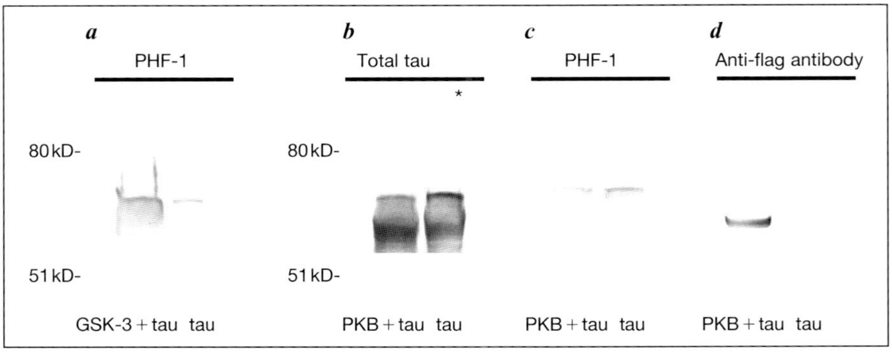

Fig. 5. Phosphorylation of tau protein in 293T cells transfected with tau, GSK-3, and PKB. 293T cells were transfected with tau, GSK-3 and PKB genes. The vectors containing tau (the longest isoform), GSK-3β, and PKBα, were constructed by insertion into pcDNA3.1 (Invitrogen). 293T cells were cotransfected by lipofectamine (Invitrogen) with tau + mock genes, tau + GSK-3 genes, and tau + PKB genes; sufficient expression of gene products was observed 48 h after the transfection. ***a*** The cells were cotransfected with tau + GSK-3 genes and tau + mock genes. The PHF-1 antibody (Ser396/404) revealed more increased phosphorylation of tau protein in cells cotransfected with tau + GSK-3 genes than in cells cotransfected with tau + mock genes (***b–d***) The cells were c-transfected with tau + PKB genes and tau + mock genes. ***b*** Prior to application of the first antibody, membrane was treated with alkaline phosphatase, to observe total tau protein. After treatment, the Tau-1 antibody that reacted with dephosphorylated tau protein at Ser198/199/202 revealed similar expression of tau protein in cells cotransfected with tau + PKB genes and with tau + mock genes. ***c*** The PHF-1 antibody revealed decreased phosphorylation of tau protein in cells co-transfected with tau + PKB genes than in cells co-transfected with tau + mock genes, suggesting that phosphorylation of tau protein by GSK-3 is attenuated by overexpressed PKB. ***d*** Anti-Flag antibody staining was observed only in cells cotransfected with tau + PKB genes, suggesting the PKB gene was appropriately expressed in this experiment.

To confirm the involvement of PKB and GSK-3 in the regulation of phosphorylation of tau protein, 293T cells were transfected with GSK-3 and PKB (fig. 5). The vectors containing tau (the longest isoform), GSK-3β, and PKB-α (generous gifts from M. Goedert, Medical Research Council, UK, A. Takasima, Riken, Japan, and U. Kikkawa, Kobe University, Japan, respectively) were constructed by insertion into pcDNA3.1 (Invitrogen). 293T cells were cotransfected with tau + mock genes, tau + GSK-3 genes, and tau + PKB genes, and sufficient expression of gene products was observed 48 h after the transfection. The PHF-1 antibody (Ser396/404) revealed more increased phosphorylation of tau protein in cells cotransfected with tau + GSK-3 genes than in cells cotransfected with tau + mock genes (fig. 5a). In the membrane treated with alkaline phosphatase, the Tau-1 antibody that reacted with total tau protein revealed a similar expression of tau protein in cells cotransfected with tau + PKB genes, and with tau + mock genes (fig. 5b). Anti-Flag antibody staining was observed only in cells co-transfected with tau + PKB genes, suggesting the PKB gene is appropriately expressed in this experiment (fig. 5d). Then the PHF-1 antibody revealed decreased phosphorylation of tau protein in cells cotransfected with tau + PKB genes than in cells cotransfected with tau + mock genes, suggesting that phosphorylation of tau protein by GSK-3 is attenuated by overexpressed PKB (fig. 5c). Taken together, the PI3K pathway regulates tau phosphorylation in cultured cells.

Activated Kinases Colocalized with Phosphorylated Tau Protein in AD

As described above, GSK-3 might be a major kinase involved in the formation of abnormally phosphorylated tau protein in AD. Biochemical analysis revealed that the levels of GSK-3, as determined by indirect ELISA, are increased by approximately 50% in the postsynaptosomal supernatant from AD brains as compared with the controls [36]. Immunohistlogical analysis revealed that GSK-3 is prominently present in neuronal cell bodies and their processes and colocalizes with neurofibrillary changes in AD brain [36]. Furthermore, active GSK-3β, phosphorylated at Tyr 216, was found to initially accumulate in the cytoplasm of pretangle neurons [37]. And these active GSK-3-positive neurons appear initially in the pre-alpha layer of the entorhinal cortex and extend to other brain regions, coincident with the sequence of the development of neurofibrillary changes [37]. It was also reported that total GSK-3α, total GSK-3β, and active GSK-3β were found to colocalize with the granulovacuolar degeneration and to be associated with the granules of the granulovacuolar bodies [38]. Further, GSK-3β was expressed in neurons containing neurofibrillary tangles, but only a small proportion of intracellular neurofibrillary tangles was observed to be GSK-3β-immunoreactive [38]. It suggests that neurons developing granulovacuolar degeneration sequester an active form of GSK-3 in this compartment or that activation of GSK-3 is involved predominantly in the early stages of neurodegeneration.

As to the intracellular signal transduction pathway, the upstream regulator of GSK-3 was also investigated in AD brains. As mentioned above, PKB is an important intermediate in the PI3K-signaling cascade that acts to phosphorylate GSK-3β at its serine 9 residue, thereby inactivating it. The amount of activated PKB phosphorylated at Thr308 increased in correlation to the progressive sequence of neurofibrillary changes assessed according to Braak's criteria [39]. This activated PKB was found to appear in particular in neurons that are known to later develop NFTs in AD [39]. Western blotting showed that activated PKB was increased by more than 50% in the 16,000 *g* supernatants of AD brains as compared with normal aged controls [39]. This increase in PKB levels corresponded to a several-fold increase in the levels of total tau and abnormally hyperphosphorylated tau. Apparently, it is difficult to explain co-localization of the phosphorylated tau with both active GSK-3 and active PKB, because active PKB is thought to inactivate GSK-3. Presumably, more complex mechanisms are involved in the neurodegenerative process in AD; a decreased activity of protein phosphatases might be able to explain these situations.

In addition to active GSK-3 and PKB, other kinases related to the protein translation system have also been reported to be active in AD brains. Eukaryotic initiation factor-2α (eIF2α) is a regulator of protein translation and a phosphorylated form of eIF2α terminates global protein translation [40]. This eIF2α is phosphorylated by several kinases including protein kinase R (PKR), double-stranded RNA (dsRNA)-activated protein kinase, that is a ubiquitously expressed serine/threonine protein kinase induced by interferon and activated by dsRNA, TNF, IL-1 and lipopolysaccharide, or viral infection [41] (fig. 3). Chang et al. [42] reported that in AD brains both phosphorylated eIF2α and PKR were observed in affected neurons, while they were rarely observed in age-matched control brains.

The ribosomal S6 protein kinase p70 S6 kinase is known for its role in modulating cell size and cell survival [43]. Activated p70 S6 kinase upregulates ribosomal biosynthesis and enhances the translational capacity of the cell. Signal transduction cascade that activates p70 S6 kinase has been investigated, and the mammalian target of rapamycin (mTOR) and tuberous sclerosis complex (TSC) were reported to be important regulators in this translational machinery [44] (fig. 3). Recent reports revealed that phosphorylation of p70 S6 kinase by mTOR induced activation of p70 S6 kinase, leading to activation of translation, and that mTOR and p70 S6 kinase were inhibited by TSC [44]. On the other hand, the inhibitory potential of TSC to mTOR was reported to be downregulated by its phosphorylation by PKB [45, 46]. Therefore, p70 S6 kinase is located down stream of the PI3K pathway. An et al. [47] reported that the levels of phosphorylated p70 S6 kinase (at Thr389 or at Thr421/Ser424) were increased in accordance with the progressive sequence of neurofibrillary changes according to Braak's criteria [47]. Both PKR and p70 S6 kinase Both eIF2α and p70 S6 kinase are working to regulate translation, but the former inhibits translation and, on the other hand, the latter enhances it. Therefore activation of PKB and p70 S6 kinase might be an auto-feedback response to inhibition of translation induced by activated eIF2α and PKR. Until now only little evidence on the involvement of translational machinery in AD pathology has been reported. On the translational system, we have previously reported that peptidyltransferase inhibitors induced the phosphorylation of tau protein and neuronal cell death in SH-SY5Y cells; dysfunction of ribosome in AD was also reported by others [48, 49]. More investigations on this matter will be required to clarify the neurodegenerative mechanisms including AD.

Conclusion

The mechanisms of phosphorylation of tau protein are still unclear; however, the signal transduction pathway of GSK-3, a strong candidate that phosphorylates tau abnormally, was overviewed. The PI3K pathway has an important role in both the regulation of phosphorylation of tau and cell survival. In addition PKB, which participates in the PI3K pathway, also has an important role in the regulation of p70 S6 kinase activity that regulates protein translation. From the reports on the phosphorylation of p70 S6 kinase, eIF2α and PKR in AD brains, both the PI3K pathway and the protein translational machinery might be closely involved in AD pathology, and more investigations will be required to understand tau phosphorylation in neurodegeneration.

References

1 Grundke-Iqbal I, Iqbal K, Tung YC, Quinlan M, Wisniewski HM, Binder LI: Abnormal phosphorylation of the microtubule-associated protein tau (tau) in Alzheimer cytoskeletal pathology. Proc Natl Acad Sci USA 1986;83:4913–4917.

2 Grundke-Iqbal I, Iqbal K, Quinlan M, Tung YC, Zaidi MS, Wisniewski HM: Microtubule-associated protein tau. A component of Alzheimer paired helical filaments. J Biol Chem 1986; 261:6084–6089.

3 Iqbal K, Grundke-Iqbal I, Smith AJ, George L, Tung YC, Zaidi T: Identification and localization of a tau peptide to paired helical filaments of Alzheimer disease. Proc Natl Acad Sci USA 1989; 86:5646–5650.

4 Iqbal K, Smith AJ, Zaidi T, Grundke-Iqbal I: Microtubule-associated protein tau. Identification of a novel peptide from bovine brain. FEBS Lett 1989;248:87–91.

5 Roder HM, Eden PA, Ingram VM: Brain protein kinase PK40erk converts TAU into a PHF-like form as found in Alzheimer's disease. Biochem Biophys Res Commun 1993;193:639–647.

6 Kobayashi S, Ishiguro K, Omori A, et al: A cdc2-related kinase PSSALRE/cdk5 is homologous with the 30 kDa subunit of tau protein kinase II, a proline-directed protein kinase associated with microtubule. FEBS Lett 1993;335:171–175.

7 Ishiguro K, Shiratsuchi A, Sato S, et al: Glycogen synthase kinase 3 beta is identical to tau protein kinase I generating several epitopes of paired helical filaments. FEBS Lett 1993;325: 167–172.

8 Gong CX, Grundke-Iqbal I, Iqbal K: Dephosphorylation of Alzheimer's disease abnormally phosphorylated tau by protein phosphatase-2A. Neuroscience 1994;61:765–772.

9 Gong CX, Grundke-Iqbal I, Damuni Z, Iqbal K: Dephosphorylation of microtubule-associated protein tau by protein phosphatase-1 and -2C and its implication in Alzheimer disease. FEBS Lett 1994;341:94–98.

10 Gong CX, Singh TJ, Grundke-Iqbal I, Iqbal K: Alzheimer's disease abnormally phosphorylated tau is dephosphorylated by protein phosphatase-2B (calcineurin). J Neurochem 1994;62: 803–806.

11 Tanaka T, Zhong J, Iqbal K, Trenkner E, Grundke-Iqbal I: The regulation of phosphorylation of tau in SY5Y neuroblastoma cells: the role of protein phosphatases. FEBS Lett 1998;426: 248–254.
12 Goedert M, Jakes R: Expression of separate isoforms of human tau protein: Correlation with the tau pattern in brain and effects on tubulin polymerization. EMBO J 1990;9:4225–4230.
13 Kosik KS, Orecchio LD, Bakalis S, Neve RL: Developmentally regulated expression of specific tau sequences. Neuron 1989;2:1389–1397.
14 Lindwall G, Cole RD: Phosphorylation affects the ability of tau protein to promote microtubule assembly. J Biol Chem 1984;259:5301–5305.
15 Kopke E, Tung YC, Shaikh S, Alonso AC, Iqbal K, Grundke-Iqbal I: Microtubule-associated protein tau. Abnormal phosphorylation of a non-paired helical filament pool in Alzheimer disease. J Biol Chem 1993;268:24374–24384.
16 Bancher C, Brunner C, Lassmann H, et al: Accumulation of abnormally phosphorylated tau precedes the formation of neurofibrillary tangles in Alzheimer's disease. Brain Res 1989;477: 90–99.
17 Morishima-Kawashima M, Hasegawa M, Takio K, et al: Proline-directed and non-proline-directed phosphorylation of PHF-tau. J Biol Chem 1995;270:823–829.
18 Hanger DP, Hughes K, Woodgett JR, Brion JP, Anderton BH: Glycogen synthase kinase-3 induces Alzheimer's disease-like phosphorylation of tau: Generation of paired helical filament epitopes and neuronal localisation of the kinase. Neurosci Lett 1992;147:58–62.
19 Takahashi M, Tomizawa K, Kato R, et al: Localization and developmental changes of tau protein kinase I/glycogen synthase kinase-3 beta in rat brain. J Neurochem 1994;63:245–255.
20 Ishiguro K, Omori A, Takamatsu M, et al: Phosphorylation sites on tau by tau protein kinase I, a bovine derived kinase generating an epitope of paired helical filaments. Neurosci Lett 1992;148: 202–206.
21 Ishiguro K, Takamatsu M, Tomizawa K, et al: Tau protein kinase I converts normal tau protein into A68-like component of paired helical filaments. J Biol Chem 1992;267:10897–10901.
22 Guan RJ, Khatra BS, Cohlberg JA: Phosphorylation of bovine neurofilament proteins by protein kinase FA (glycogen synthase kinase 3). J Biol Chem 1991;266:8262–8267.
23 Hemmings BA, Resink TJ, Cohen P: Reconstitution of a Mg-ATP-dependent protein phosphatase and its activation through a phosphorylation mechanism. FEBS Lett 1982;150: 319–324.
24 Hemmings BA, Aitken A, Cohen P, Rymond M, Hofmann F: Phosphorylation of the type-II regulatory subunit of cyclic-AMP-dependent protein kinase by glycogen synthase kinase 3 and glycogen synthase kinase 5. Eur J Biochem 1982;127:473–481.
25 Yang SD: Identification of the ATP.Mg-dependent protein phosphatase activator (FA) as a myelin basic protein kinase in the brain. J Biol Chem 1986;261:11786–11791.
26 Hughes K, Nikolakaki E, Plyte SE, Totty NF, Woodgett JR: Modulation of the glycogen synthase kinase-3 family by tyrosine phosphorylation. EMBO J 1993;12:803–808.
27 Sutherland C, Leighton IA, Cohen P: Inactivation of glycogen synthase kinase-3 beta by phosphorylation: New kinase connections in insulin and growth-factor signalling. Biochem J 1993; 296(pt 1):15–19.
28 Wang QM, Fiol CJ, DePaoli-Roach AA, Roach PJ: Glycogen synthase kinase-3 beta is a dual specificity kinase differentially regulated by tyrosine and serine/threonine phosphorylation. J Biol Chem 1994;269:14566–14574.
29 Yu JS, Yang SD: Protein kinase FA/glycogen synthase kinase-3 predominantly phosphorylates the in vivo site Thr97-Pro in brain myelin basic protein: Evidence for Thr-Pro and Ser-Arg-X-X-Ser as consensus sequence motifs. J Neurochem 1994;62:1596–1603.
30 Fiol CJ, Wang A, Roeske RW, Roach PJ: Ordered multisite protein phosphorylation. Analysis of glycogen synthase kinase 3 action using model peptide substrates. J Biol Chem 1990;265: 6061–6065.
31 Aberle H, Bauer A, Stappert J, Kispert A, Kemler R: Beta-catenin is a target for the ubiquitin-proteasome pathway. EMBO J 1997;16:3797–3804.
32 Cross DA, Alessi DR, Cohen P, Andjelkovich M, Hemmings BA: Inhibition of glycogen synthase kinase-3 by insulin mediated by protein kinase B. Nature 1995;378:785–789.

33 del Peso L, Gonzalez-Garcia M, Page C, Herrera R, Nunez G: Interleukin-3-induced phosphorylation of BAD through the protein kinase Akt. Science 1997;278:687–689.
34 Datta SR, Dudek H, Tao X, et al: Akt phosphorylation of BAD couples survival signals to the cell-intrinsic death machinery. Cell 1997;91:231–241.
35 Tsujio I, Tanaka T, Kudo T, et al: Inactivation of glycogen synthase kinase-3 by protein kinase C delta: Implications for regulation of tau phosphorylation. FEBS Lett 2000;469: 111–117.
36 Pei JJ, Tanaka T, Tung YC, Braak E, Iqbal K, Grundke-Iqbal I: Distribution, levels, and activity of glycogen synthase kinase-3 in the Alzheimer disease brain. J Neuropathol Exp Neurol 1997;56: 70–78.
37 Pei JJ, Braak E, Braak H, et al: Distribution of active glycogen synthase kinase 3beta (GSK-3beta) in brains staged for Alzheimer disease neurofibrillary changes. J Neuropathol Exp Neurol 1999; 58:1010–1019.
38 Leroy K, Boutajangout A, Authelet M, Woodgett JR, Anderton BH, Brion JP: The active form of glycogen synthase kinase-3beta is associated with granulovacuolar degeneration in neurons in Alzheimer's disease. Acta Neuropathol (Berl) 2002;103:91–99.
39 Pei JJ, Khatoon S, An WL, et al: Role of protein kinase B in Alzheimer's neurofibrillary pathology. Acta Neuropathol (Berl) 2003;105:381–392.
40 Kimball SR: Eukaryotic initiation factor eIF2. Int J Biochem Cell Biol 1999;31:25–29.
41 Gil J, Esteban M: Induction of apoptosis by the dsRNA-dependent protein kinase (PKR): Mechanism of action. Apoptosis 2000;5:107–114.
42 Chang RC, Wong AK, Ng HK, Hugon J: Phosphorylation of eukaryotic initiation factor-2alpha (eIF2alpha) is associated with neuronal degeneration in Alzheimer's disease. Neuroreport 2002; 13:2429–2432.
43 Shima H, Pende M, Chen Y, Fumagalli S, Thomas G, Kozma SC: Disruption of the p70(s6k)/p85(s6k) gene reveals a small mouse phenotype and a new functional S6 kinase. EMBO J 1998; 17:6649–6659.
44 Kwiatkowski DJ: Tuberous sclerosis: From tubers to mTOR. Ann Hum Genet 2003;67:87–96.
45 Gao X, Zhang Y, Arrazola P, et al: Tsc tumour suppressor proteins antagonize amino-acid-TOR signalling. Nat Cell Biol 2002;4:699–704.
46 Inoki K, Li Y, Zhu T, Wu J, Guan KL: TSC2 is phosphorylated and inhibited by Akt and suppresses mTOR signalling. Nat Cell Biol 2002;4:648–657.
47 An WL, Cowburn RF, Li L, Braak H, et al: Up-regulation of phosphorylated/activated p70 S6 kinase and its relationship to neurofibrillary pathology in Alzheimer's disease. Am J Pathol. 2003; 163:591–607.
48 Wada K, Tanaka T, Wakutani Y, et al: Phosphorylation of tau protein and neuronal cell death induced by peptidyltransferase inhibitors apoptosis (abstract) 8th International Conference on Alzheimer Disease and Related Disorders (Jul 20–25,2002, Stockholm, Sweden).
49 Payao SL, Smith MA, Winter LM, et al: Ribosomal RNA in Alzheimer's disease and aging. Mech Ageing Dev 1998;105:265–272.

Toshihisa Tanaka, MD, PhD
Department of Clinical Neuroscience, Psychiatry
Osaka University, Graduate School of Medicine
D-3, 2–2 Yamadaoka, Suita, Osaka 565–0871 (Japan)
Tel. +81 6 6879 3051, Fax +81 6 6879 3059, E-Mail tanaka@psy.med.osaka-u.ac.jp

Takeda M, Tanaka T, Cacabelos R (eds): Molecular Neurobiology of Alzheimer Disease and Related Disorders. Basel, Karger, 2004, pp 236–285

A Functional Genomics Approach to the Analysis of Biological Markers in Alzheimer Disease

Ramón Cacabelos[a,d], Valter Lombardi[b], Lucía Fernández-Novoa[c], Yasuhiko Kubota[a], Lola Corzo[a], Victor Pichel[a], Masatoshi Takeda[e]

[a]Department of Clinical Neuroscience, Institute for CNS Disorders, EuroEspes Biomedical Research Center, Departments of [b]Biotechnology, and [c]Molecular and Clinical Genetics, Ebiotec, Coruña, [d]EuroEspes Chair of Biotechnology and Genomics, Camilo José Cela University, Madrid, Spain, [e]Department of Psychiatry and Behavioral Proteomics, Osaka University Medical School, Osaka, Japan

The growing knowledge on structural and functional genomics is helping to elucidate many black holes in the genetics of Alzheimer disease (AD) and other complex disorders in which mendelian genetics do not fully explain the pathogenic mechanisms underlying these multifactorial/multicomponent processes [1–3]. More than 50 genes distributed across the human genome are potentially involved in the etiopathogenesis of AD, and it is very likely that many other genes will be associated with neurodegeneration in the coming future considering that approximately 80% of the genes present in the human genome are expressed in the brain [4–6]. Functional interactions among some of the AD-related genes, especially APP, APOE and PSs, can lead to amyloid deposition in senile plaques and premature neuronal death in transgenic animals and humans [7–10]. However, many other factors have been identified as potential risk modifiers in AD, including education and early life experience, mental and leisure activity, smoking, Down syndrome, depression, traumatic head injury, cardiovascular disease and related vascular risk factors, anti-inflammatory agents, antioxidants, hormone replacement, lipid-lowering agents, and gene-environment interactions [11]. Notwithstanding, aging and genetic factors still remain the most relevant risk factors for AD together with cerebrovascular dysfunction and stroke as major determinants of vascular dementia [1, 12]. Furthermore, primary cerebrometabolic abnormalities, cerebrovascular pathology, ischemic brain damage and autonomic dysregulation inducing brain hypoperfusion seem to be fundamental in the pathogenesis of dementia [13, 14].

Functional genomics strategies can help to improve our understanding of the interactions between genomic and environmental factors in the activation of pathogenic mechanisms responsible for neurodegeneration [6]. For instance, the analysis of genotype-phenotype correlations has revealed that the presence of the APOE-4 allele in AD, in conjunction with other genetic loci, can influence disease onset, brain atrophy, cerebrovascular perfusion, blood pressure, β-amyloid deposition, serum ApoE secretion, lipid metabolism, brain bioelectrical activity, cognition, apoptosis and treatment outcome [4–6, 12, 15–17]. In addition, pharmacogenomics studies also indicate that the therapeutic response in AD is genotype specific and that approximately 15–20% of the cases with efficacy and/or safety problems are associated with a defective CYP2D6 gene linked to the phenotype of poor metabolizers and ultrarapid metabolizers in whom the pharmacological treatment currently administered induces toxicity or is merely useless [6]. The understanding of functional genomics in AD is particularly important for therapy, and the future development of pharmacogenomics as the AD population exhibits a higher genetic variation rate than the healthy population without a family history of dementia [18].

The aim of the present study was to investigate the potential influence of (a) age, (b) sex, (c) mental deterioration, (d) APOE genotypes, and (e) genetic haplotypes (APOE+PS1 clusters) on the following biological parameters: total cholesterol (CHO), HDL, LDL, VLDL, triglycerides, serum apolipoprotein (ApoE) levels, serum nitric oxide (NO) levels, serum β-amyloid (1–42) (BAP) levels, and serum histamine (HA) levels in patients with AD.

Patients and Methods

One hundred AD patients (DSM-IV, NINCDS-ADRDA criteria) of both sexes were randomly selected from the EuroEspes Medical Center Database, including 54 females (age: 74.42 ± 7.84 years; range: 54–88 years) and 46 males (age: 72.10 ± 8.86 years; range: 51–88.6 years) (table 1). All patients underwent the EuroEspes Neuroscience Diagnostic Protocol including: (a) clinical examination, (b) neuropsychological assessment (MMSE, ADAS, GDS, FAST, BCRS, Hamilton-A/D, Hachinski scales), (c) clinical laboratory analysis (blood, urine), (d) CT scan, (e) chest and neck X-ray, (f) brain mapping (qEEG), (g) cerebrovascular evaluation (transcranial Doppler ultrasonography), and (h) genetic testing (APOE, PS1) (with written informed consent) [19, 20].

Biochemical analysis in fresh blood samples included the following parameters: CHO, HDL, LDL, VLDL, and triglycerides (TG). BAP, NO, and HA were measured in serum with chemiluminometric, Griess reagent, and HPLC methods, respectively.

APOE and PS1 genotyping was performed as previously reported [21, 22].

Table 1. Sex-related differences in biological parameters of patients with AD

Parameter	Females	Males
Number of patients	54	46
Age, years	74.42 ± 7.84	72.10 ± 8.86
Age range, years	54–88	51–93
MMSE score	12.88 ± 8.68	15.86 ± 8.16[1]
ADAS-Cog score	23.04 ± 14.04	20.05 ± 12.84
Serum ApoE, mg/dl	5.01 ± 1.05	4.65 ± 1.06[2]
Serum CHO, mg/dl	230.40 ± 39.29	204.18 ± 41.16[3]
Serum HDL, mg/dl	55.83 ± 15.99	44.82 ± 9.48[4]
Serum LDL, mg/dl	148.85 ± 35.29	138.12 ± 36.63
Serum VLDL, mg/dl	25.71 ± 14.82	21.16 ± 8.29[5]
Serum TG, mg/dl	128.59 ± 74.05	105.76 ± 41.65[6]
Serum NO, μM	46.68 ± 20.30	43.90 ± 23.68
Serum BAP, pg/ml	17.27 ± 5.87	21.11 ± 16.94
Blood HA, ng/ml	93.11 ± 72.41	89.65 ± 71.35

All values are expressed as mean ± standard deviation.
[1] $p < 0.05$. [3] $p < 0.001$. [5] $p < 0.05$.
[2] $p < 0.05$. [4] $p < 0.00008$. [6] $p < 0.05$.

In the present study, the following analyses were carried out: (a) age-related differences in biological parameters to evaluate the influence of age on biochemical factors and cognition; (b) cognition-related differences in biological parameters; (c) sex-related differences in biological parameters; (d) APOE-4-related differences in biological parameters; (e) APOE genotype-related differences in biological parameters; and (f) haplotype (APOE+PS1 bigenic clusters)-related differences in biological parameters.

The statistical analysis of the data was performed using the SPSS, Sigma-Stat and Sigma-Plot computer programs. Linear and nonlinear correlation analysis, Student's t test, ANOVA, Durbin-Watson statistics, normality test, constant variance test, and regression diagnostics parameters were evaluated. All data are expressed as mean ± standard deviation in tables and figures. Differences are considered to be significant when $p < 0.05$.

Results

Age-Related Differences in Biological Parameters

From 50 to 90 years of age, the influence of age per se on cognition is minimal in terms of either MMSE score (<3 points) or ADAS-Cog score (<5 points) in patients with AD (fig. 1). Serum ApoE levels are not affected by age (fig. 2). Serum CHO ($r = 0.02$, $p < 0.04$), LDL ($r = 0.27$, $p < 0.006$),

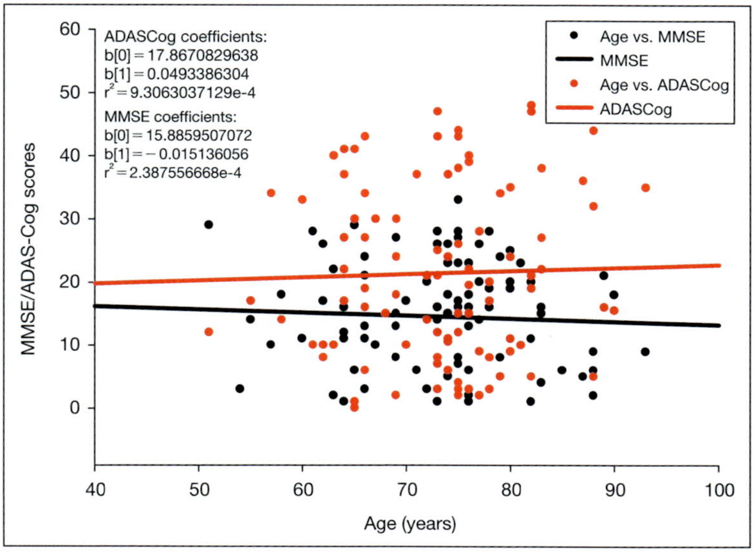

Fig. 1. Age-related cognitive decline in AD.

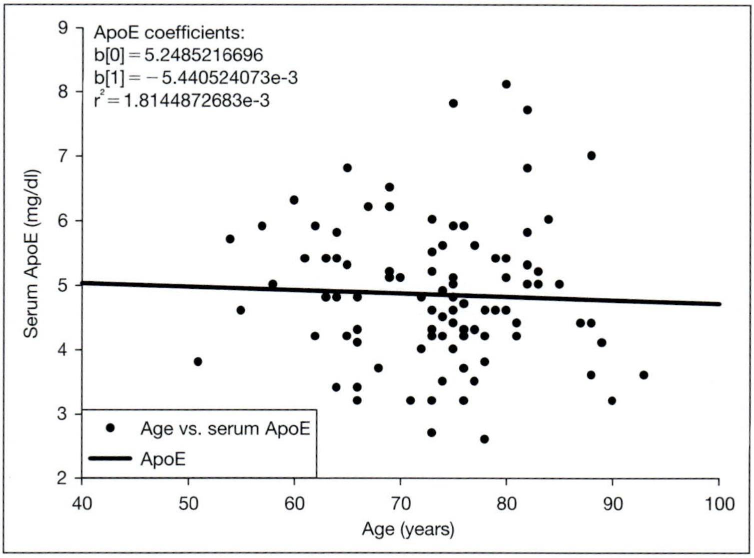

Fig. 2. Age-related serum ApoE levels in AD.

VLDL ($r = 0.19$, $p < 0.05$), and TG levels ($r = 0.20$, $p < 0.04$) tend to increase with age (fig. 3). Neither serum NO (fig. 4) nor BAP levels (fig. 5) show a significant age-dependent change in AD, though both parameters tend to slightly increase with age (fig. 4, 5). In contrast, blood HA levels clearly exhibit an age-related surge in AD ($r = 0.025$, $p < 0.01$) (fig. 6).

Cognition-Related Differences in Biological Parameters

Cognitive deterioration (MMSE scores) does not seem to influence serum ApoE levels (fig. 7), blood lipid levels (fig. 8), serum BAP levels (fig. 10) or blood HA levels (fig. 11) (or vice versa), but tends to increase in parallel with serum NO levels in AD ($p < 0.05$) (fig. 9).

Sex-Related Differences in Biological Parameters

In our samples, females showed a lower cognitive profile as compared with males ($p < 0.05$) (table 1). Serum ApoE ($p < 0.05$), CHO ($p < 0.001$), HDL ($p < 0.00008$), VLDL ($p < 0.05$), and TG levels ($p < 0.05$) were significantly lower in males than in females (table 1).

When we compared females and males correlating age with the parameters evaluated, no significant differences were found between sexes in cognitive decline (fig. 12, 13), serum ApoE levels (fig. 14), serum BAP levels (fig. 21) and blood HA levels (fig. 22); however, CHO ($p < 0.0004$) (fig. 15), LDL ($p < 0.05$) (fig. 17), and NO ($p < 0.01$) (fig. 20) showed and age-dependent decrease in males, whereas HDL ($p < 0.05$) (fig. 16), VLDL ($p < 0.01$) (fig. 18) and TG levels ($p < 0.01$) (fig. 19) decreased in females with age, indicating that age per se can influence both lipid metabolism and NO levels in AD.

APOE-4-Related Differences in Biological Parameters

The differentiation of AD patients in APOE-4 carriers (APOE-4+) and APOE-4 noncarriers (APOE-4−) does not affect any of the parameters studied except blood HA levels, which show a marked decrease in APOE-4+ ($p < 0.02$) (table 2).

When APOE-4+ and APOE-4− patients are compared as a function of age, both groups do not show any significant difference in cognition (fig. 23 and 24), serum ApoE levels (fig. 25), HDL (fig. 27), VLDL (fig. 29), TG (fig. 30), NO (fig. 31) or BAP levels (fig. 32); however, CHO levels decrease with age in APOE-4+ ($p < 0.05$) (fig. 26), LDL levels decrease in both APOE-4+ ($p < 0.03$) and APOE-4− ($p < 0.05$) (fig. 28), and blood HA levels increase in APOE-4− ($p < 0.01$) (fig. 33).

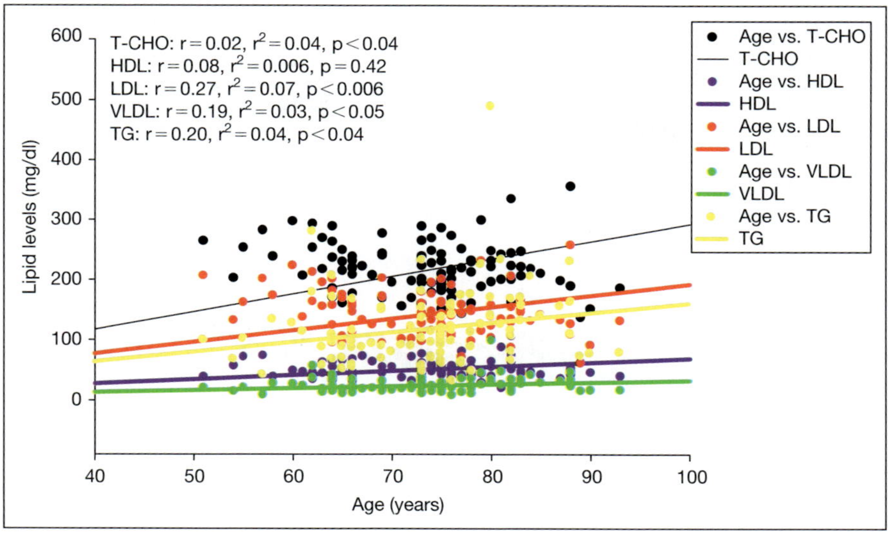

Fig. 3. Age-related serum lipid levels in AD.

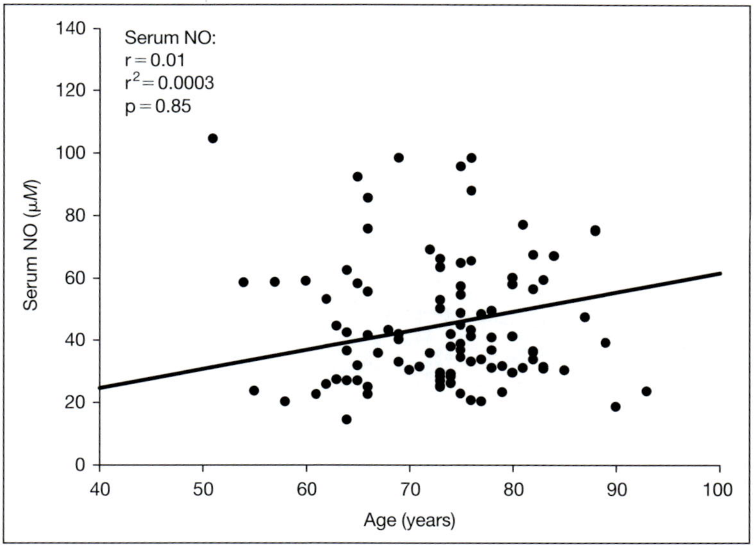

Fig. 4. Age-related serum NO levels in AD.

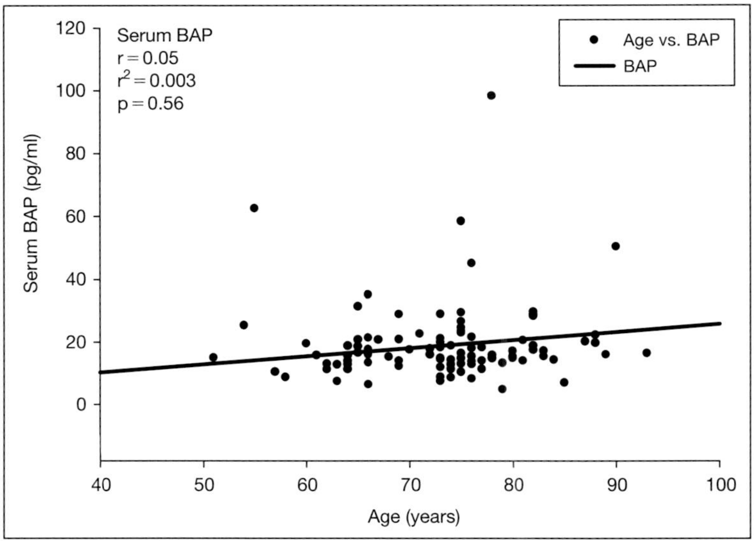

Fig. 5. Age-related serum BAP levels in AD.

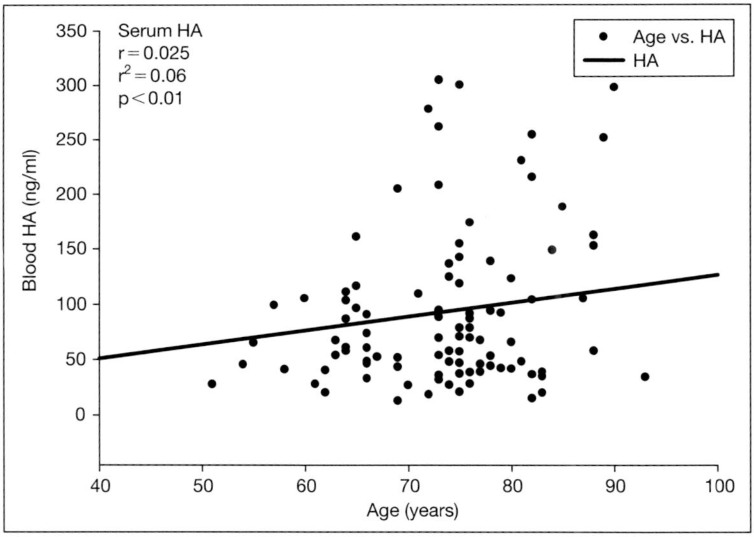

Fig. 6. Age-related blood HA levels in AD.

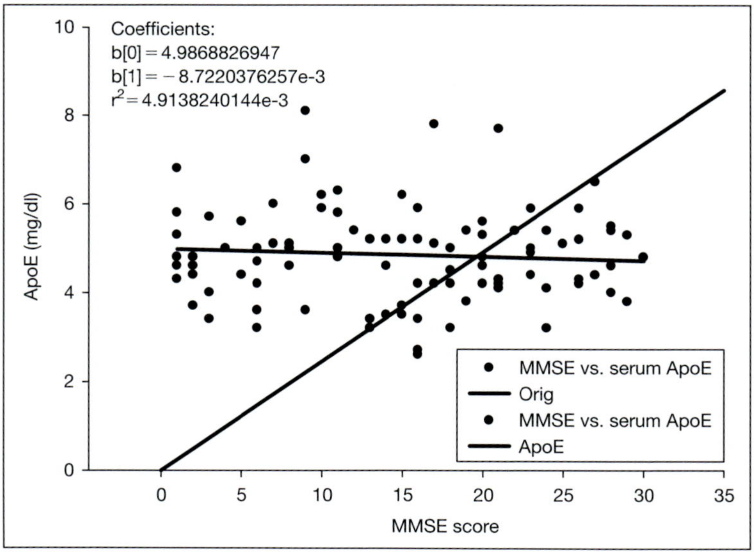

Fig. 7. Cognitive performance vs. serum ApoE levels in AD.

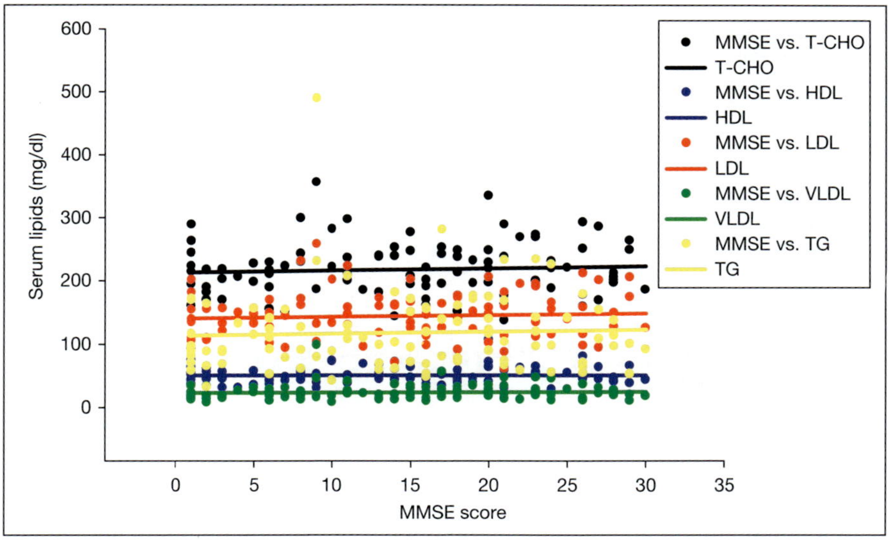

Fig. 8. Cognitive performance vs. blood lipid levels in AD.

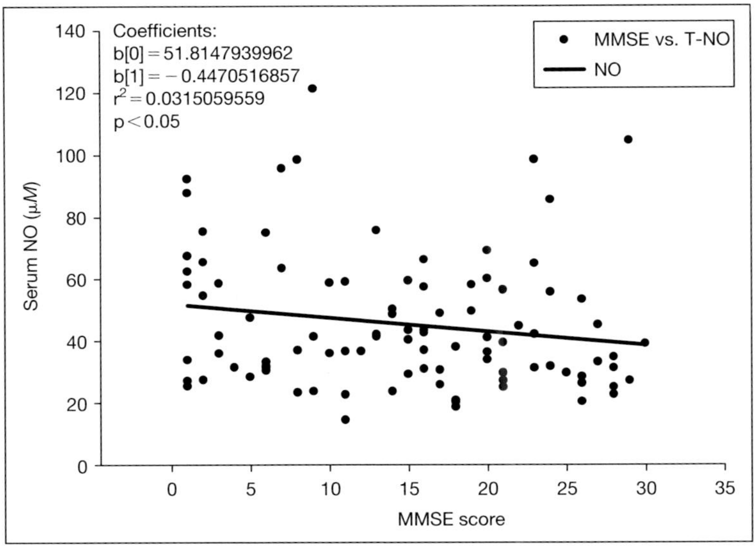

Fig. 9. Cognitive performance vs. serum NO levels in AD.

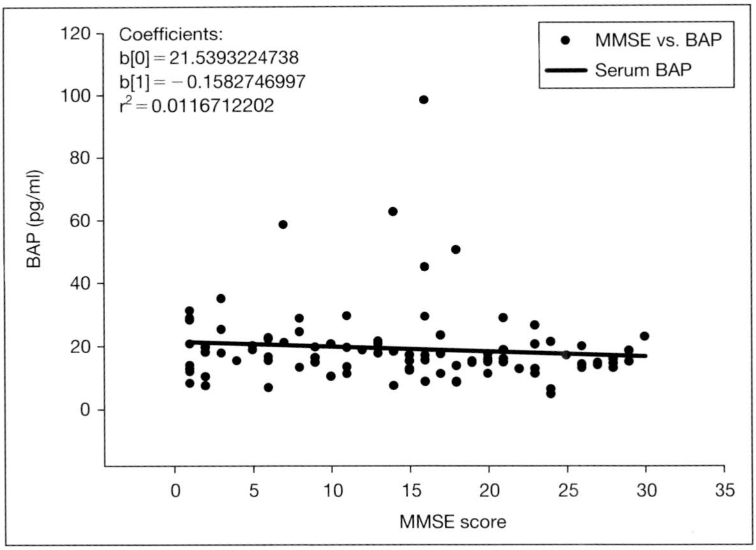

Fig. 10. Cognitive performance vs. serum BAP levels in AD.

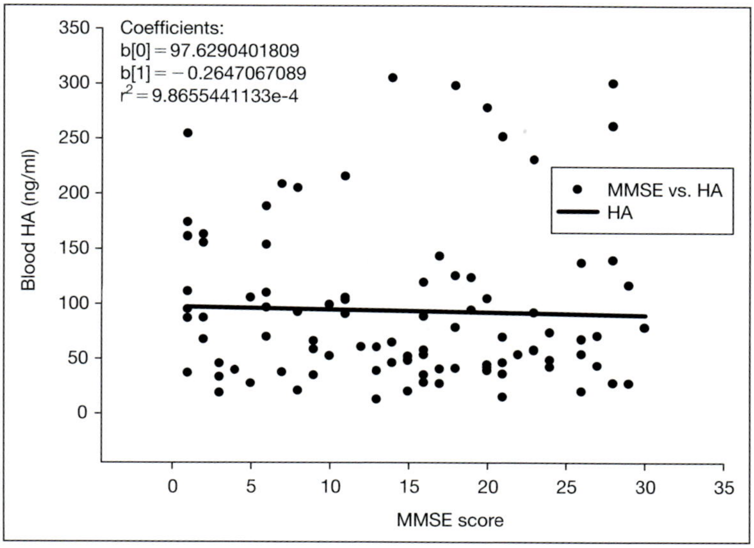

Fig. 11. Cognitive performance vs. blood HA levels in AD.

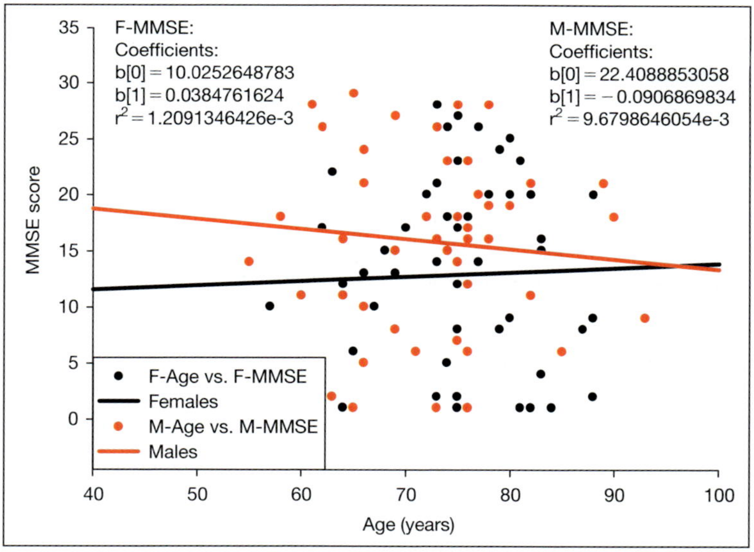

Fig. 12. Age- and sex-related cognitive deterioration in AD (MMSE score).

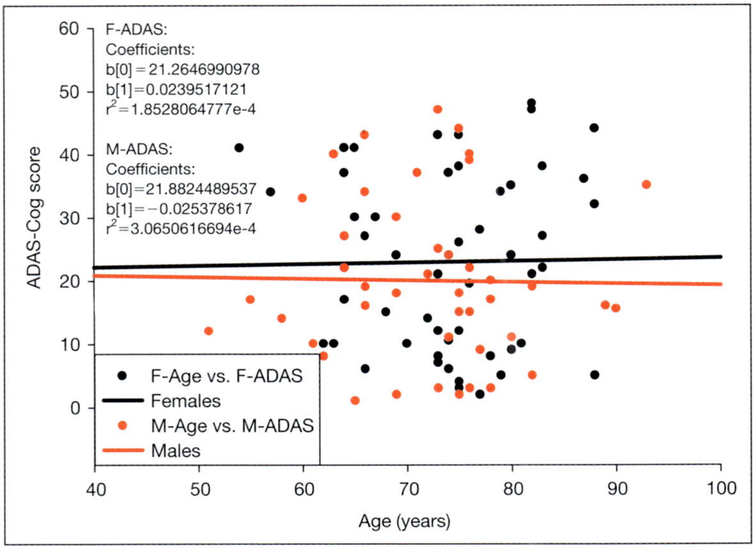

Fig. 13. Age- and sex-related cognitive deterioration in AD (ADAS-Cog score).

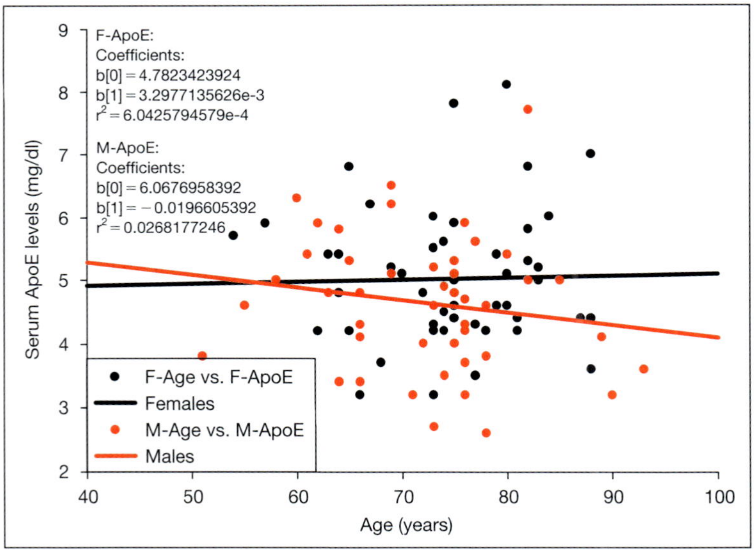

Fig. 14. Age- and sex-related serum ApoE levels in AD.

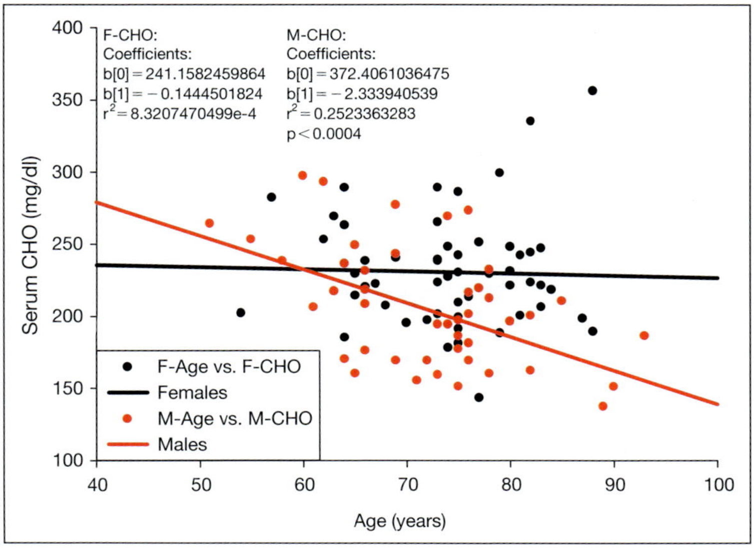

Fig. 15. Age- and sex-related serum CHO levels in AD.

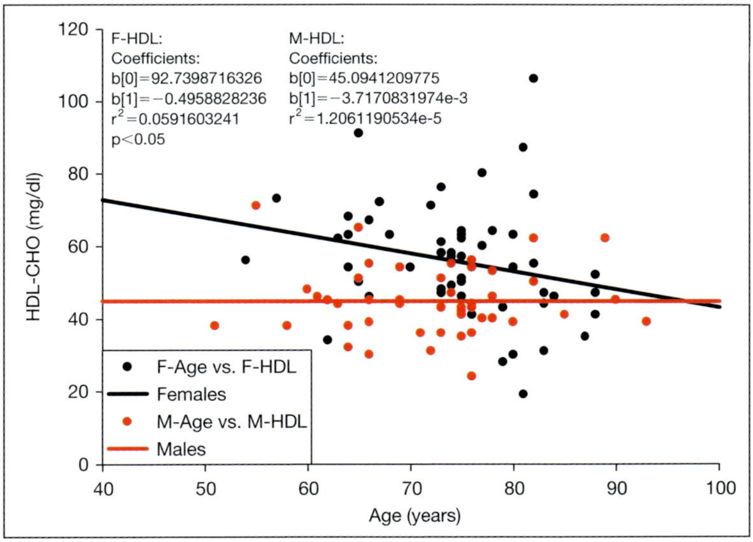

Fig. 16. Age- and sex-related serum HDL levels in AD.

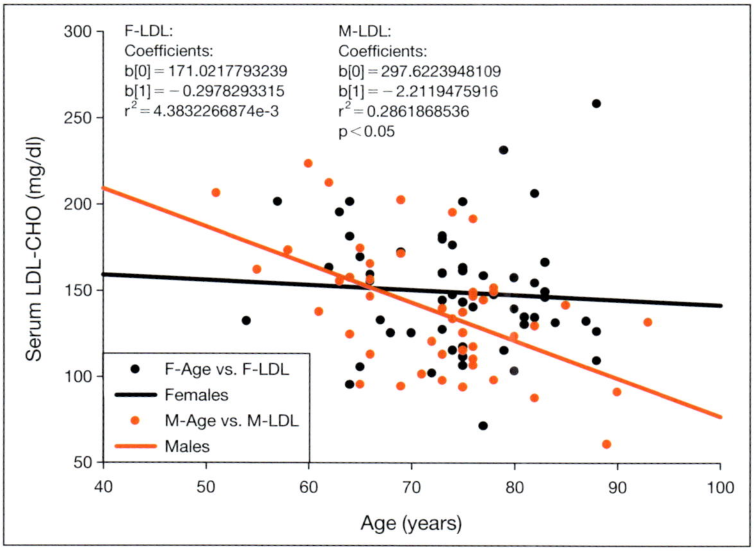

Fig. 17. Age- and sex-related serum LDL levels in AD.

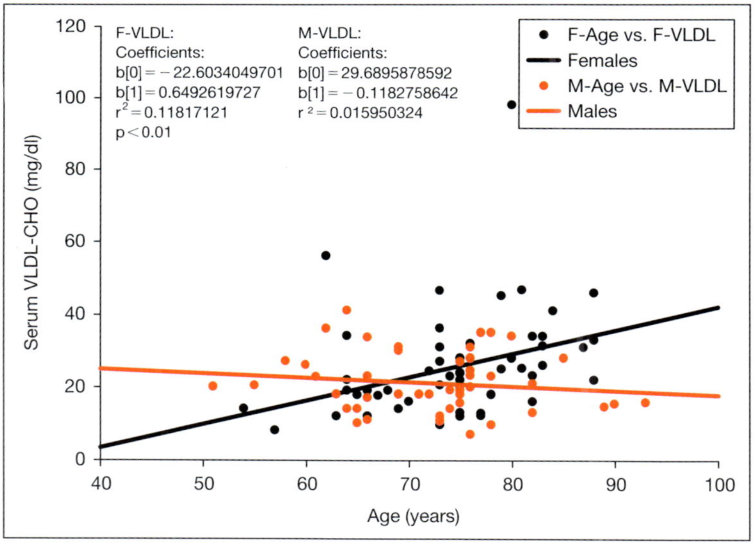

Fig. 18. Age- and sex-related serum VLDL levels in AD.

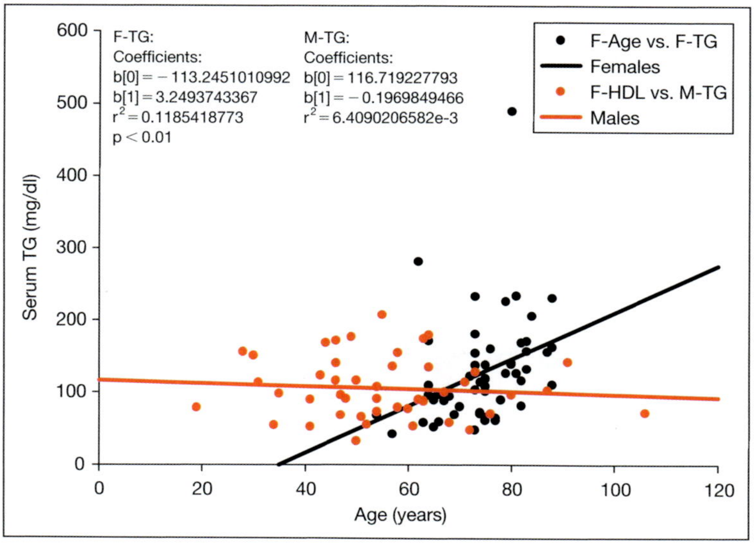

Fig. 19. Age- and sex-related serum TG levels in AD.

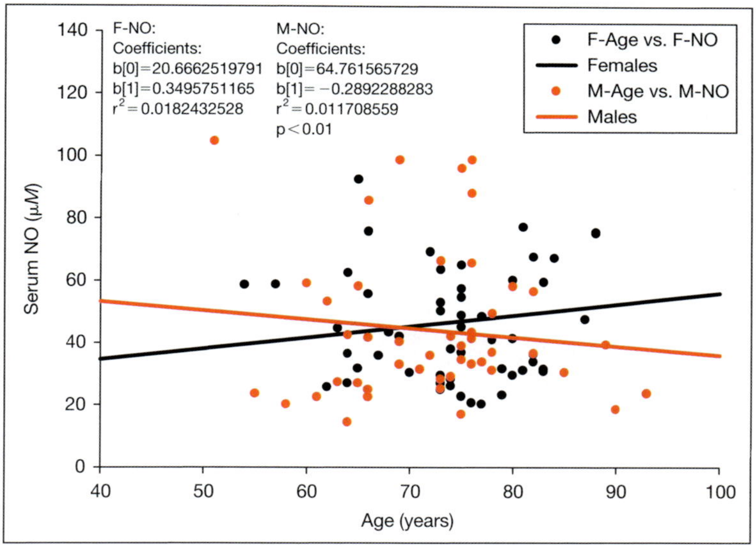

Fig. 20. Age- and sex-related serum NO levels in AD.

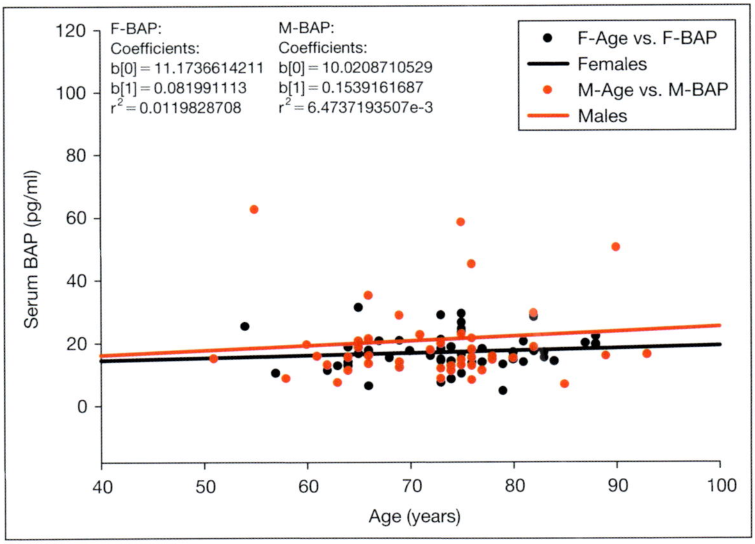

Fig. 21. Age- and sex-related serum BAP levels in AD.

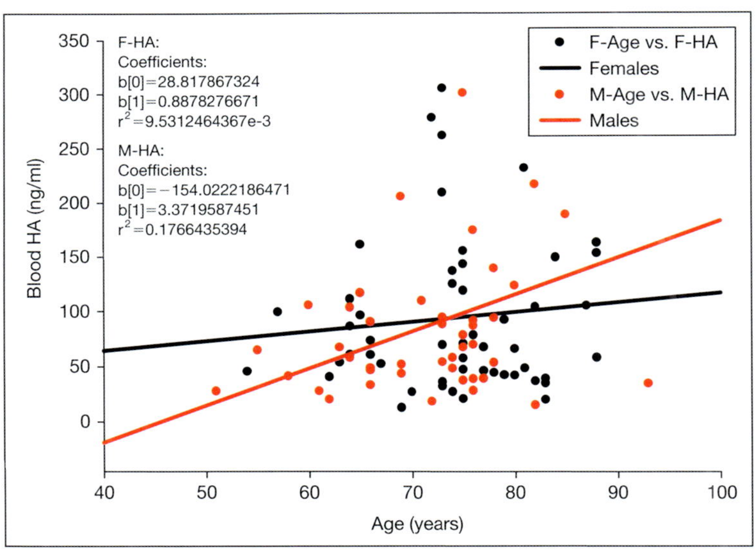

Fig. 22. Age- and sex-related blood HA levels in AD.

Table 2. APOE-4-related differences in biological parameters of patients with AD

Parameter	APOE-4−	APOE-4+/carriers
Number of patients	54	46
Age, years	73.83 ± 9.80	72.73 ± 6.48
Age range, years	51–93	57–83
MMSE score	14.71 ± 9.20	13.95 ± 7.49
ADAS-Cog score	21.07 ± 14.38	22.83 ± 12.29
Serum ApoE, mg/dl	5.01 ± 0.87	4.68 ± 1.25
Serum CHO, mg/dl	212.43 ± 40.75	224.36 ± 43.28
Serum HDL, mg/dl	48.56 ± 10.93	52.67 ± 17.08
Serum LDL, mg/dl	139.90 ± 36.61	148.22 ± 35.79
Serum VLDL, mg/dl	23.95 ± 9.05	23.46 ± 15.32
Serum TG, mg/dl	119.64 ± 46.93	117.47 ± 76.60
Serum NO, μM	45.76 ± 24.46	45.53 ± 18.56
Serum BAP, pg/ml	19.04 ± 10.66	19.15 ± 14.28
Blood HA, ng/ml	106.92 ± 81.29	74.98 ± 50.04[1]

All values are expressed as mean ± standard deviation.
[1]$p < 0.02$.

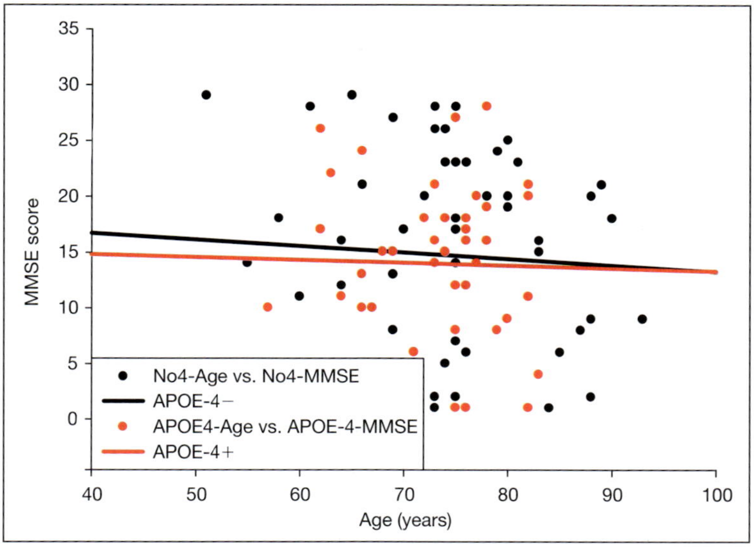

Fig. 23. Age- and APOE-4-related cognitive deterioration in AD (MMSE score).

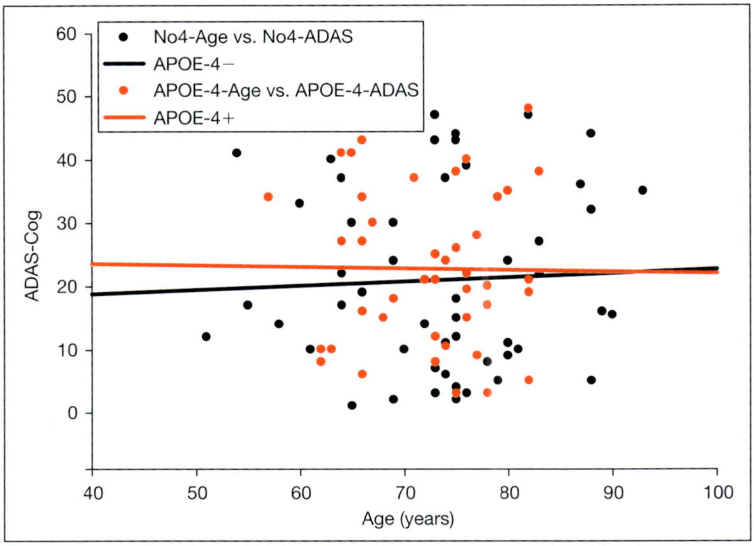

Fig. 24. Age- and APOE-4-related cognitive deterioration in AD (ADAS-Cog score).

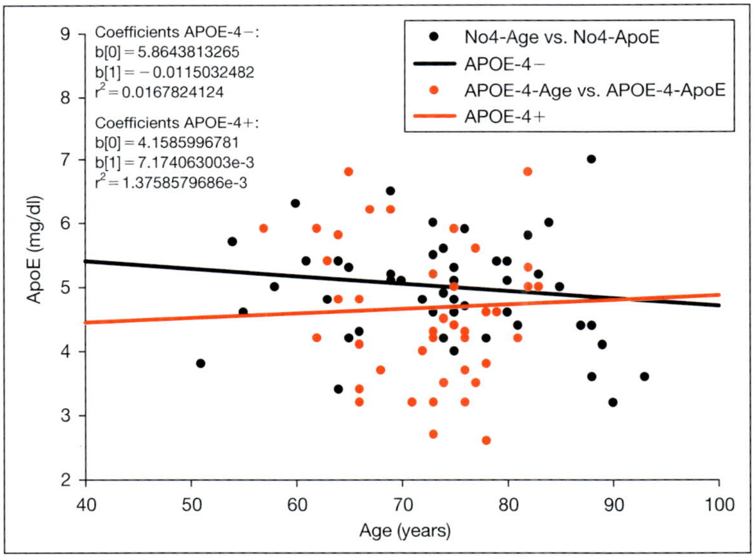

Fig. 25. Age- and APOE-4-related serum ApoE levels in AD.

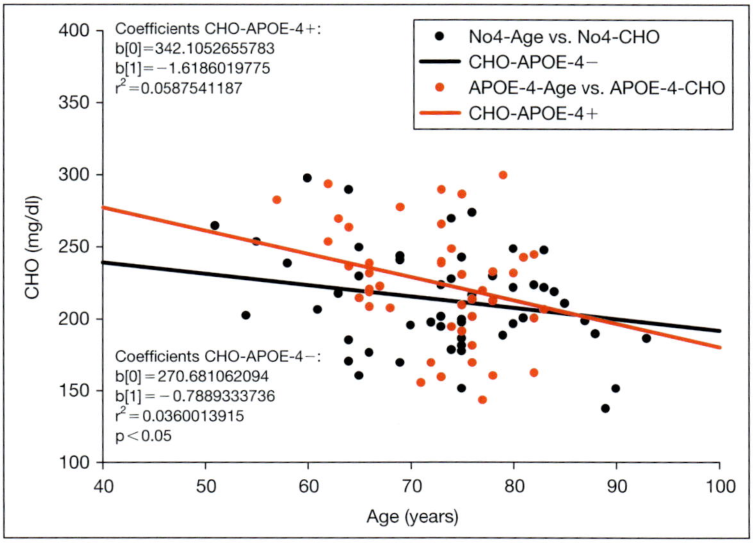

Fig. 26. Age- and APOE-4-related serum CHO levels in AD.

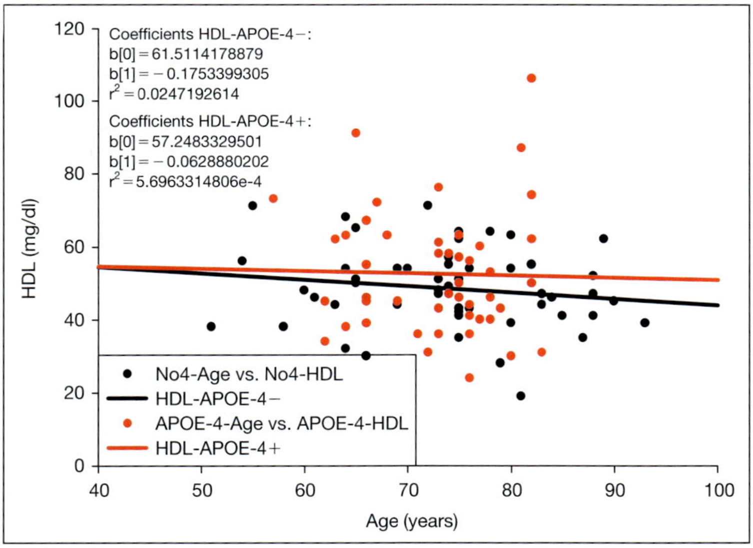

Fig. 27. Age- and APOE-4-related serum HDL levels in AD.

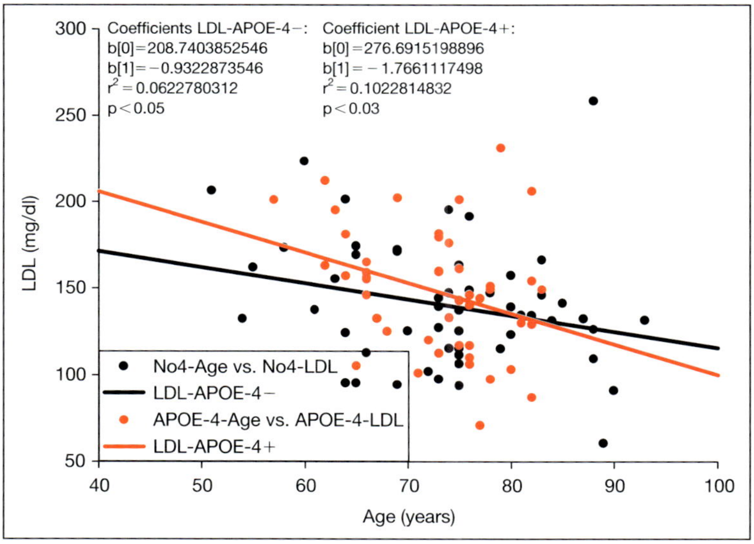

Fig. 28. Age- and APOE-4-related serum LDL levels in AD.

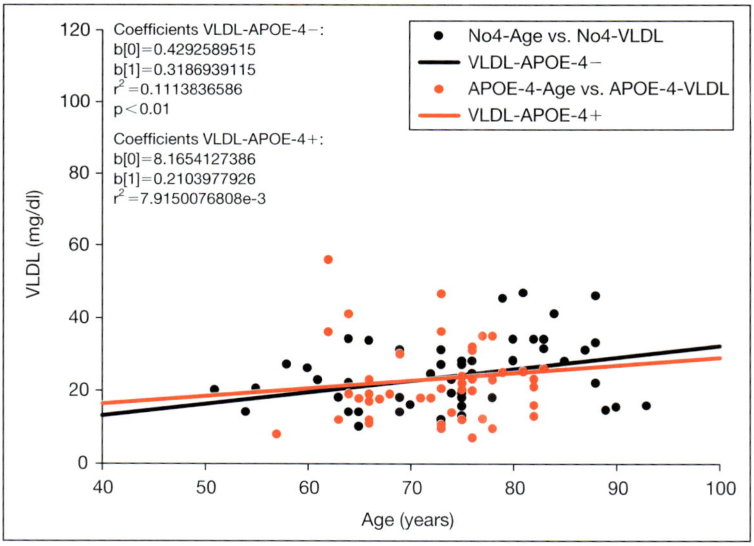

Fig. 29. Age- and APOE-4-related serum VLDL levels in AD.

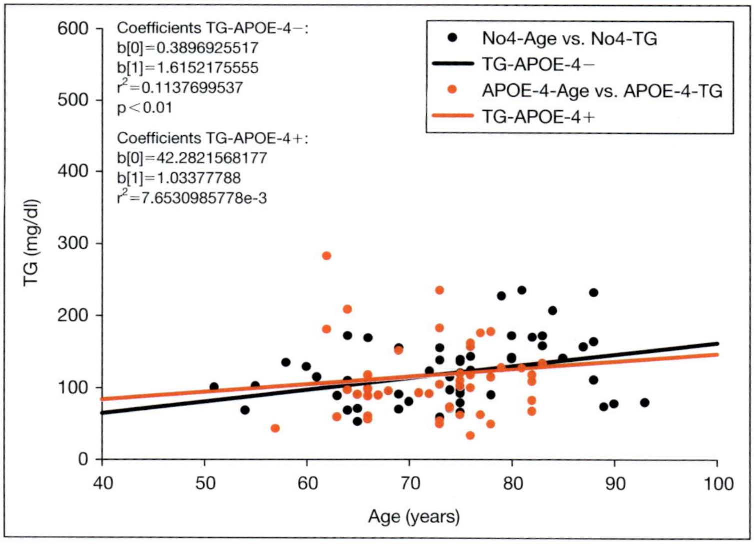

Fig. 30. Age- and APOE-4-related serum TG levels in AD.

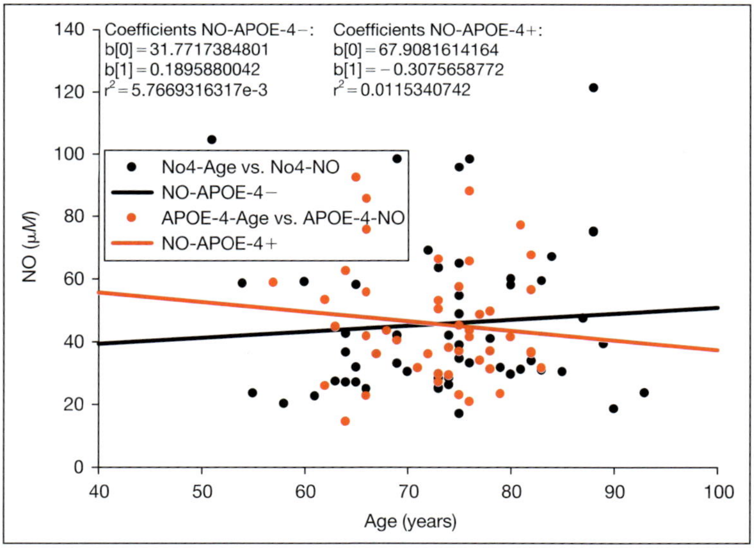

Fig. 31. Age- and APOE-4-related serum NO levels in AD.

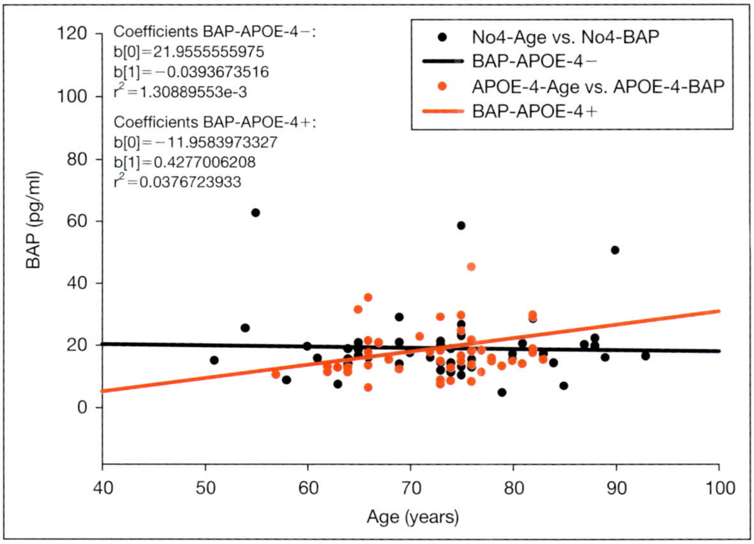

Fig. 32. Age- and APOE-4-related serum BAP levels in AD.

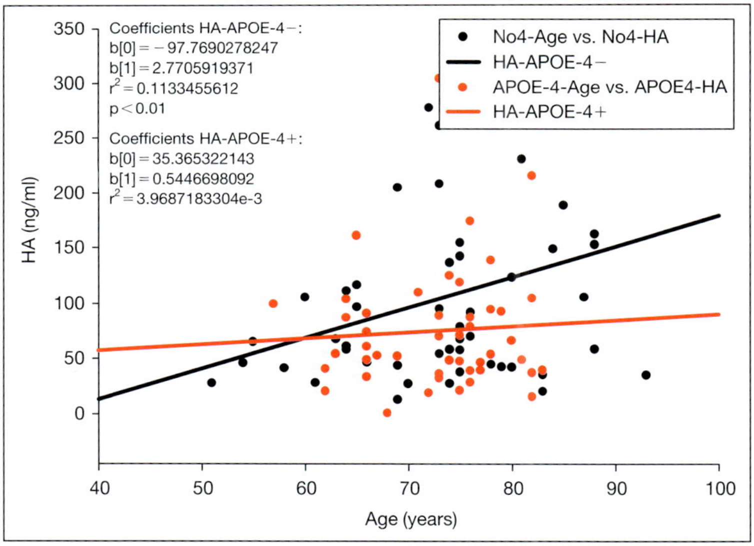

Fig. 33. Age- and APOE-4-related blood HA levels in AD.

APOE Genotype-Related Differences in Biological Parameters

The distribution of APOE genotypes in our sample was the following: APOE-2/2: 0%; APOE-2/3: 4%; APOE-2/4: 2%; APOE-3/3: 49%; APOE-3/4: 35%; and APOE-4/4: 9% (table 3). Significant differences were found in most parameters according to APOE genotype including age, ApoE, CHO, HDL, LDL, NO, BAP, and HA. Only VLDL and TG did not show any difference among APOE genotype clusters (table 3). No differences were detected in APOE-related cognitive performance (fig. 34). APOE-4/4 carriers are the youngest patients of the cohort ($p < 0.02$) (table 3). Serum ApoE levels are significantly lower in APOE-3/4 than in APOE-2/3 ($p < 0.03$) and APOE-2/4 ($p < 0.006$), and in APOE-4/4 than in APOE-2/3 ($p < 0.001$), APOE-2/4 ($p < 0.00003$), APOE-3/3 ($p < 0.001$), and APOE-3/4 ($p < 0.05$) (table 3; fig. 35). CHO levels are higher in APOE-4/4 than in APOE-2/3 ($p < 0.04$), APOE-2/4 ($p < 0.05$) and APOE-3/3 ($p < 0.03$) (table 3; fig. 36). HDL levels tend to be higher in APOE-2/4 than in the other groups (table 3; fig. 37); however, any parametric difference between APOE-2/4 carriers and other APOE genotypes should be interpreted with caution due to the small number of APOE-2/4 cases in this sample. LDL levels tend to be higher in APOE-4/4 than in APOE-2/3, APOE-2/4, and APOE-3/3 (table 3; fig. 38). NO levels tend to be lower in APOE-3/4 and APOE-4/4 with respect to APOE-2/4 ($p < 0.05$) (table 3; fig. 39). BAP levels are also lower in APOE-3/3, APOE-3/4 and APOE-4/4 than in APOE-2/4, but no differences are apparent between APOE-3/3 and APOE-3/4 or APOE-4/4 (table 3; fig. 40). In contrast, HA levels in APOE-4/4 ($p < 0.03$) and APOE-3/4 ($p < 0.05$) are significantly lower than in the other groups reflecting a surprising HA hypoactivity in APOE-4 carriers (table 3; fig. 41).

When all the biochemical parameters are analyzed as dependent variables of age in the most frequent APOE genotypes (APOE-3/3, APOE-3/4, APOE-4/4), no significant differences are detected in ApoE (fig. 42), HDL (fig. 44), LDL (fig. 45), VLDL (fig. 46), TG (fig. 47), NO (fig. 48), BAP (fig. 49), and HA levels (fig. 50); however, CHO levels were found to be significantly increased in APOE-4/4 carriers in an age-dependent fashion ($p < 0.05$) (fig. 43). A similar tendency was detected in HDL and LDL levels, and the opposite (APOE-4/4- and age-related decline) was seen in VLDL, TG, NO and HA (fig. 46–48, 50, respectively).

Bigenic Haplotypes (APOE+PS1)-Related Differences in Biological Parameters

The distribution of the most frequent bigenic (APOE+PS1) haplotypes in AD patients was the following: (a) 3311: 16%; (b) 3312: 27%; (c) 3322: 10%; (d) 3411: 12%; (e) 3412: 20%; (f) 3422: 5.5%; (g) 4411: 5.5%; and (h) 4412: 4% (table 4).

Table 3. APOE genotype-related differences in biological parameters of patients with AD

Parameter	APOE-2/3	APOE-2/4	APOE-3/3	APOE-3/4	APOE-4/4
Number	4	2	49	35	9
Age, years	73.50 ± 3.00	75.50 ± 12.02	73.85 ± 10.17	73.57 ± 6.49	69.33 ± 4.79[1]
Age range, years	69–75	65–82	51–93	57–83	62–75
MMSE score	18.50 ± 8.60	11.00 ± 14.14	14.44 ± 9.20	13.70 ± 7.09	16.00 ± 7.98
ADAS-Cog score	15.50 ± 19.40	23.00 ± 25.45	21.55 ± 14.03	23.65 ± 11.57	19.50 ± 13.53
Serum ApoE, mg/dl	6.10 ± 1.30	7.25 ± 0.63	4.92 ± 0.77[2,3]	4.71 ± 1.21[4,5]	3.98 ± 0.53[6–9]
Serum CHO, mg/dl	190.25 ± 39.77	188.00 ± 36.76	214.24 ± 40.69	220.74 ± 43.53	246.55 ± 37.05[10–12]
Serum HDL, mg/dl	50.25 ± 9.94	76.50 ± 20.50[13]	48.42 ± 11.09[14]	50.62 ± 16.70[15]	55.33 ± 15.30
Serum LDL, mg/dl	115.40 ± 32.78	96.10 ± 12.86	141.91 ± 36.48[16]	146.45 ± 36.26[17]	166.68 ± 22.85[18–20]
Serum VLDL, mg/dl	24.60 ± 6.68	15.40 ± 3.39	23.90 ± 9.59	23.65 ± 15.73	24.53 ± 15.84
Serum TG, mg/dl	122.25 ± 32.78	77.50 ± 16.26	119.42 ± 48.15	118.40 ± 78.54	122.77 ± 79.46
Serum NO, μM	60.42 ± 26.76	74.15 ± 25.44	44.57 ± 24.17[21]	44.34 ± 17.84[22]	43.81 ± 17.00[23]
Serum BAP, pg/ml	30.45 ± 19.32	24.95 ± 8.83	18.11 ± 9.38[24]	18.88 ± 15.84[25]	18.91 ± 7.87[26]
Blood HA, ng/ml	69.97 ± 49.08	87.61 ± 103.04	110.07 ± 83.05	80.81 ± 59.21[27]	49.48 ± 22.81[28]

All values are expressed as mean ± standard deviation.

[1] $p < 0.02$ vs. APOE-3/4.
[2] $p < 0.008$ vs. APOE-2/3.
[3] $p < 0.001$ vs. APOE-2/4.
[4] $p < 0.03$ vs. APOE-2/3.
[5] $p < 0.006$ vs. APOE-2/4.
[6] $p < 0.001$ vs. APOE-2/3.
[7] $p < 0.00003$ vs. APOE-2/4.
[8] $p < 0.001$ vs. APOE-3/3.
[9] $p < 0.05$ vs. APOE-3/4.
[10] $p < 0.04$ vs. APOE-2/3.
[11] $p < 0.05$ vs. APOE-2/4.
[12] $p < 0.03$ vs. APOE-3/3.
[13] $p < 0.05$ vs. APOE-2/3.
[14] $p < 0.001$ vs. APOE-2/4.
[15] $p < 0.04$ vs. APOE-2/4.
[16] $p < 0.05$ vs. APOE-2/4.
[17] $p < 0.05$ vs. APOE-2/4.
[18] $p < 0.007$ vs. APOE-2/3.
[19] $p < 0.002$ vs. APOE-2/4.
[20] $p < 0.05$ vs. APOE-3/3.
[21] $p < 0.05$ vs. APOE-2/4.
[22] $p < 0.02$ vs. APOE-2/4.
[23] $p < 0.05$ vs. APOE-2/4.
[24] $p < 0.02$ vs. APOE-2/3.
[25] $p < 0.05$ vs. APOE-2/3.
[26] $p < 0.05$ vs. APOE-2/3.
[27] $p < 0.05$ vs. APOE-3/3.
[28] $p < 0.03$ vs. APOE-3/3.

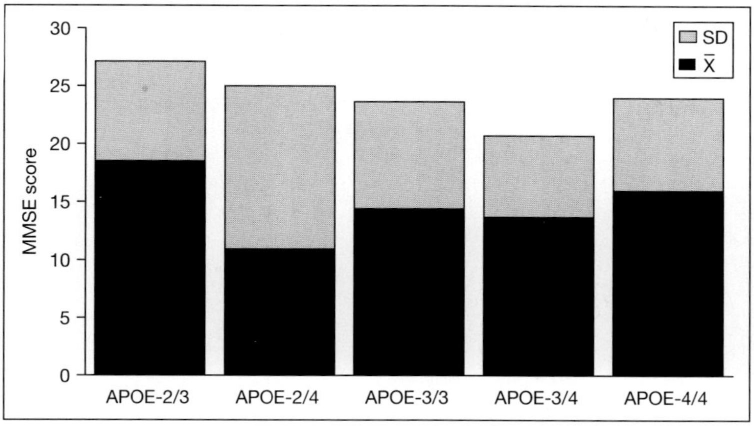

Fig. 34. APOE genotype-related cognitive performance in AD.

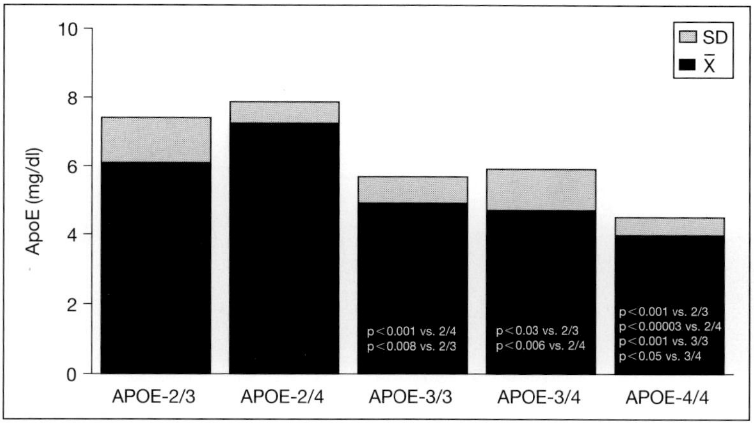

Fig. 35. APOE genotype-related serum ApoE levels in AD.

The differentiation of AD by bigenic haplotypes yields highly heterogeneous AD phenotypes in which practically all biological parameters exhibit a clear haplotype-related pattern (table 4). The youngest patients of the cohort are those with the 4411 and 4412 haplotypes (APOE-4/4-related) (table 4). Cognitive performance also shows great variability with the lowest score in 3422. Serum ApoE levels were found lower in 4412 ($p < 0.02$) and 4411 ($p < 0.05$) than in the other clusters (table 4; fig. 51). CHO levels tend to be higher in 3411, 3422, and 4411 than in 3311, 3312, and 3322 (table 4; fig. 52).

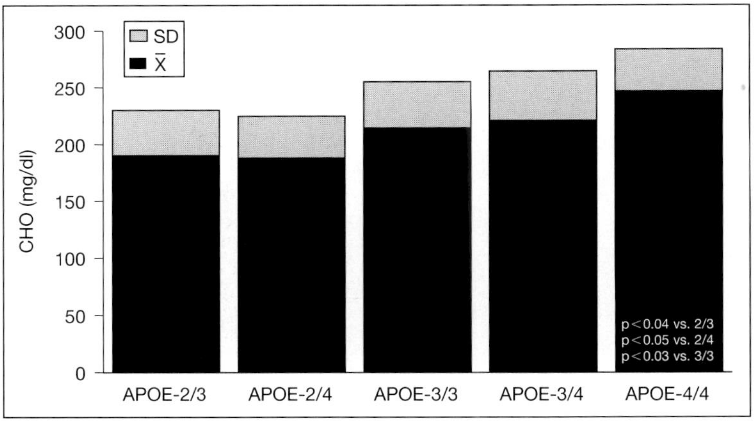

Fig. 36. APOE genotype-related serum CHO levels in AD.

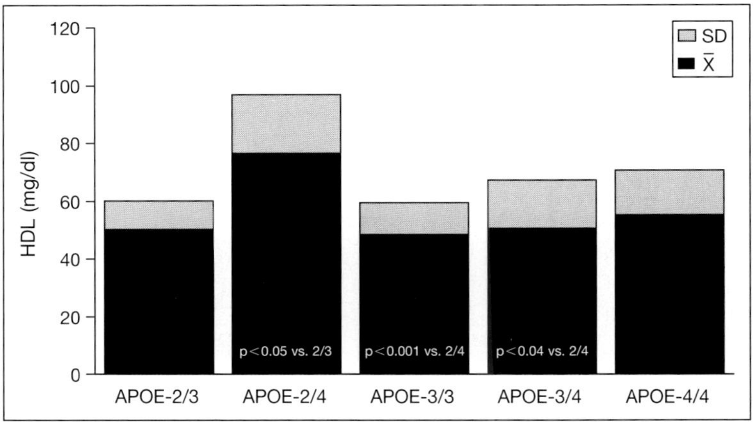

Fig. 37. APOE genotype-related serum HDL levels in AD.

HDL levels are elevated in 3322, 3411 and 4411 (table 4; fig. 53), and LDL levels mimic the phenotypic pattern of cluster-related CHO levels (table 4; fig. 54). VLDL (fig. 55) and TG levels (fig. 56) are significantly reduced in 4412 (table 4; fig. 55, 56). NO levels are increased in 3312 and 3412 (table 4; fig. 57). BAP levels are increased in 3322, 3422, and 4412 with respect to the other clusters (table 4; fig. 58); and HA levels are dramatically reduced in 4412 > 3422 > 4411 > 3412 as compared with HA levels in 3322 > 3311 > 3411 > 3312 (table 4; fig. 59).

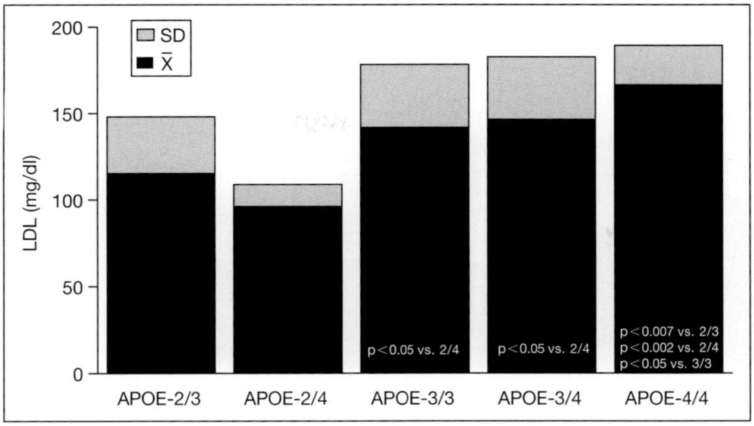

Fig. 38. APOE genotype-related serum LDL levels in AD.

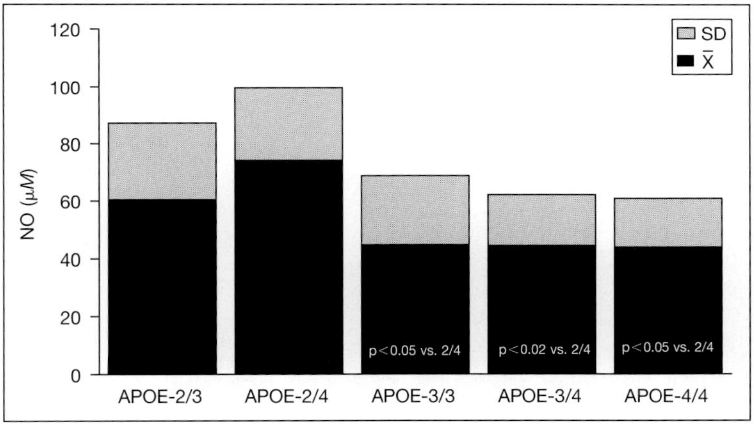

Fig. 39. APOE genotype-related serum NO levels in AD.

The analysis of biological parameters in bigenic haplotypes as dependent variables of age revealed cluster-specific nonsignificant differences following an age-related pattern. For instance, serum ApoE levels tended to increase with age in 3312, 3422 and 4412, whereas in the other clusters the regular tendency was to decrease with age (fig. 60). CHO levels follow an age-dependent increase in 4412, 3422, and 4411, no change in 3312 and 3411, and decrease in the other clusters (fig. 61). HDL levels increase with age in 3411, 3422, 4411 and 4412, with minor age-related changes in the remaining clusters (fig. 62). LDL levels increase in 4411 and decrease in 3412 without apparent differences

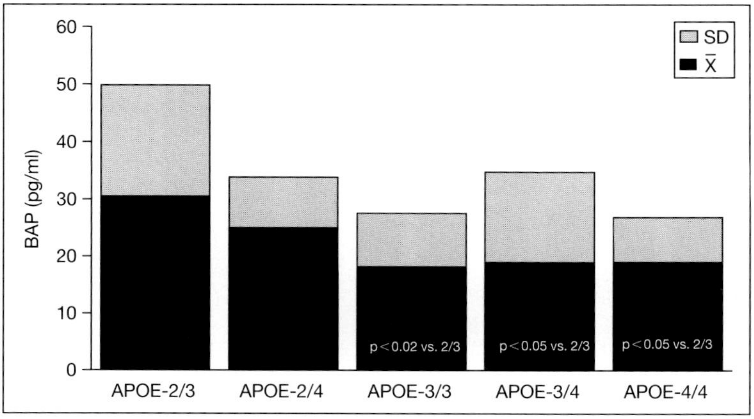

Fig. 40. APOE genotype-related serum BAP levels in AD.

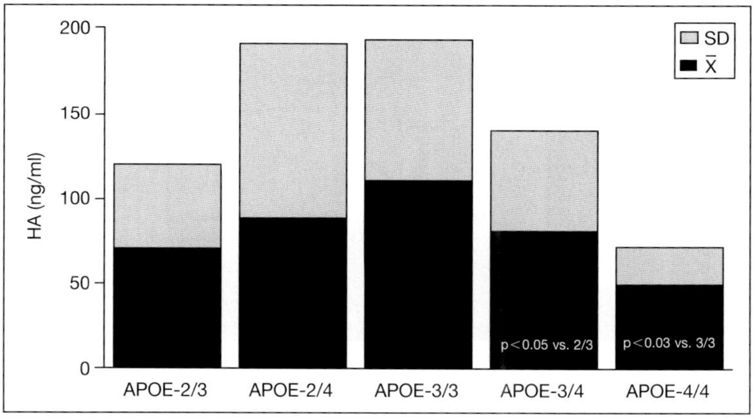

Fig. 41. APOE genotype-related blood HA levels in AD.

among the other clusters (fig. 63). No major changes are seen in VLDL (fig. 64) and TG levels (fig. 65). NO levels clearly show an age-related increase in all clusters except in 4411 (fig. 66). BAP levels decrease with age in 4412, 3422 and 3322, increase in 4411 and 3412, and remain unchanged with age in the other clusters (fig. 67). Finally, HA levels only show a mild age-dependent increase in 3322, 3412, and 4412, while the general tendency is to keep stable levels with age or to decrease minimally (fig. 68).

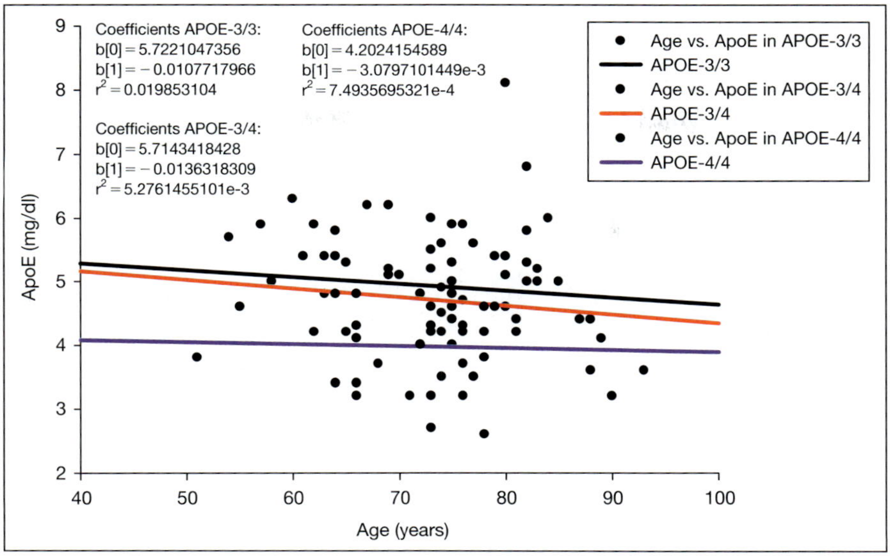

Fig. 42. Age- and APOE genotype-related serum ApoE levels in AD.

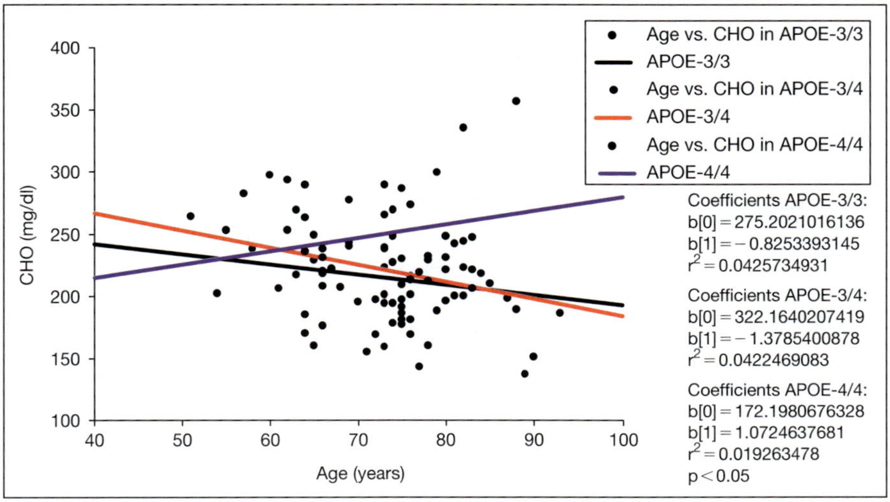

Fig. 43. Age- and APOE genotype-related serum CHO levels in AD.

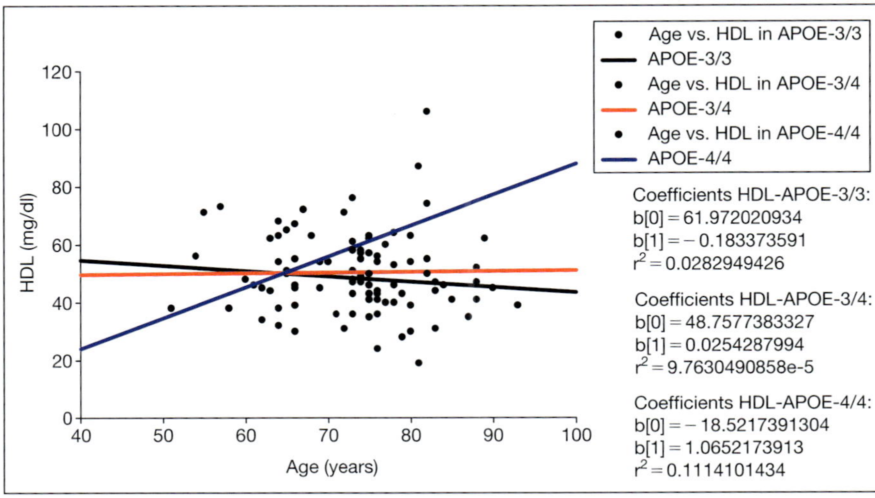

Fig. 44. Age- and APOE genotype-related serum HDL levels in AD.

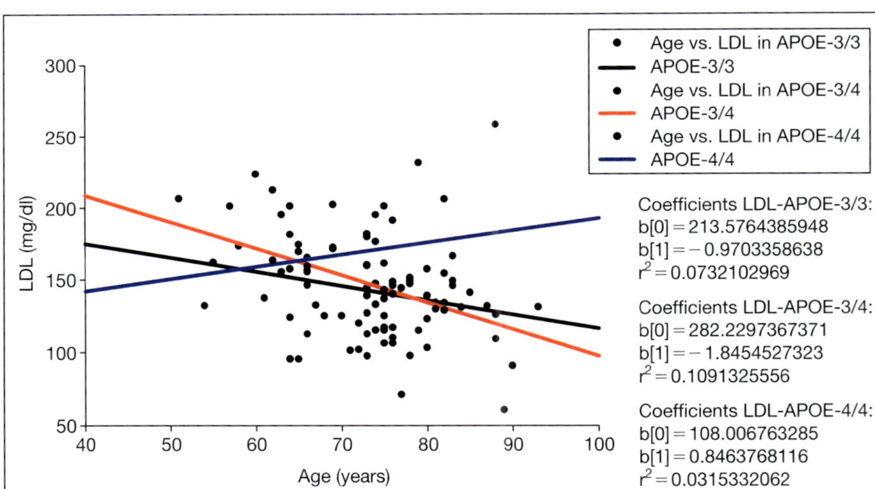

Fig. 45. Age- and APOE genotype-related serum LDL levels in AD.

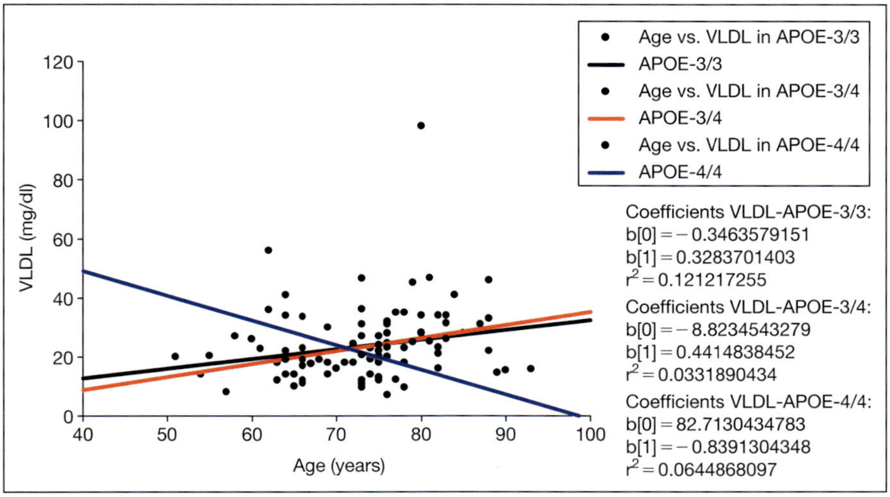

Fig. 46. Age- and APOE genotype-related serum VLDL levels in AD.

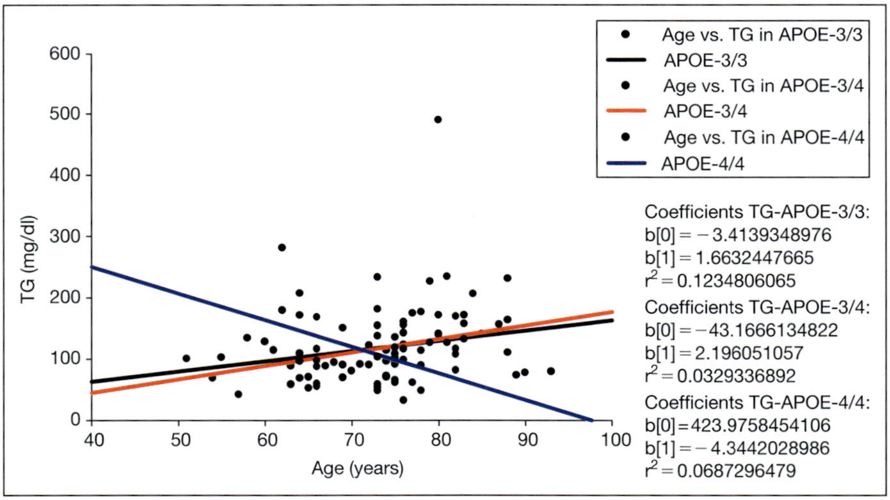

Fig. 47. Age- and APOE genotype-related serum TG levels in AD.

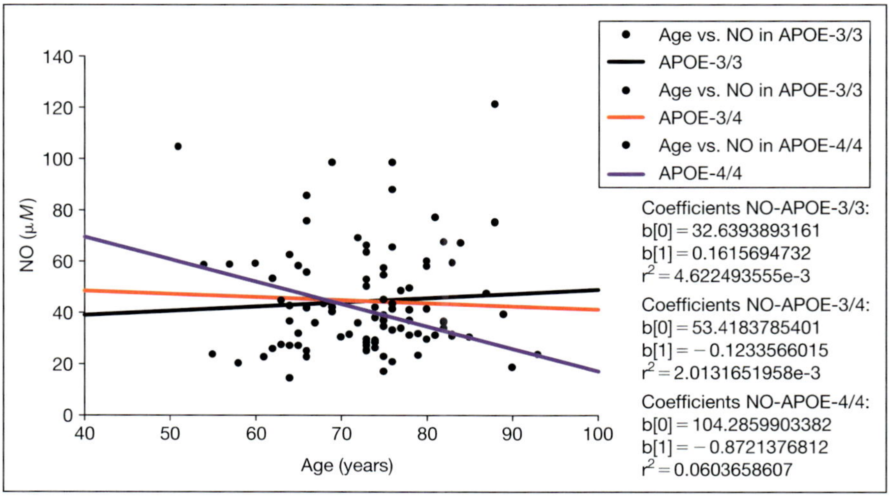

Fig. 48. Age- and APOE genotype-related serum NO levels in AD.

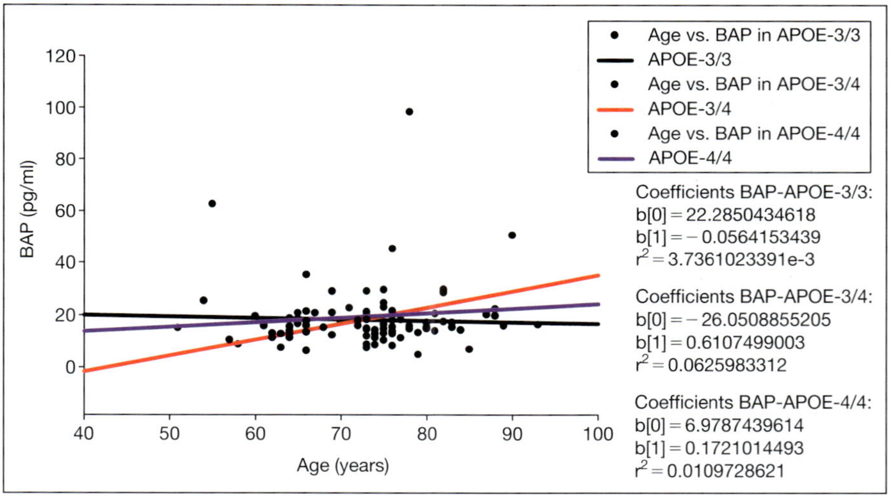

Fig. 49. Age- and APOE genotype-related serum BAP levels in AD.

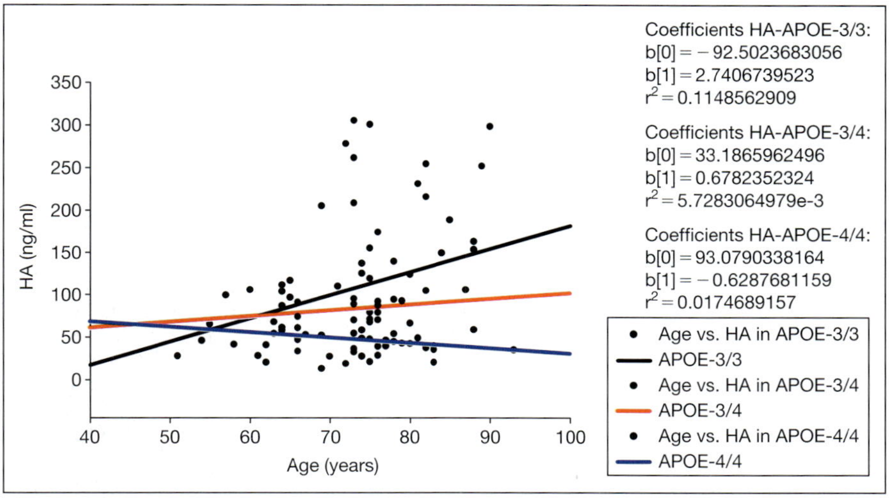

Fig. 50. Age- and APOE genotype-related blood HA levels in AD.

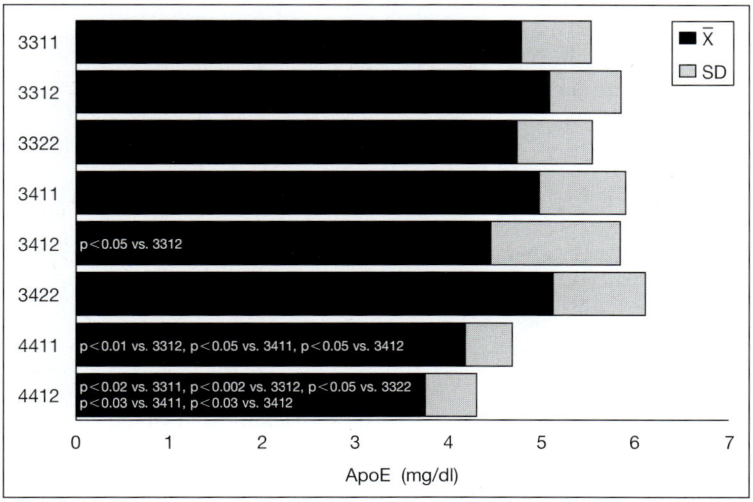

Fig. 51. Bigenic haplotype-related serum ApoE levels in AD.

Table 4. Bigenic (APOE+PS1) Genotype-related differences in biological parameters of patients with AD

Parameter	3311	3312	3322	3411	3412	3422	4411	4412
Number	15	25	9	11	19	5	5	4
Age, years	73.53 ± 9.60	73.28 ± 10.32	76.00 ± 11.55	70.45 ± 7.42	73.94 ± 5.65	79.00 ± 3.53[1,2]	67.60 ± 5.12[3,4]	71.50 ± 3.870[5]
Age range, years	58–93	51–88	55–90	57–82	62–82	75–83	62–73	66–75
MMSE score	17.93 ± 9.75	12.92 ± 9.01	12.88 ± 8.06	16.72 ± 4.64	14.00 ± 7.22	6.00 ± 6.28[6–8]	14.60 ± 8.29	18.33 ± 8.50[9]
ADAS-Cog score	13.78 ± 12.15	25.87 ± 13.85[10]	22.18 ± 13.29	19.18 ± 10.46	24.17 ± 10.76[11]	33.75 ± 13.81[12,13]	19.60 ± 14.11	19.33 ± 15.56
Serum ApoE, mg/dl	4.78 ± 0.75	5.08 ± 0.77	4.73 ± 0.81	4.97 ± 0.93	4.45 ± 1.39[14]	5.12 ± 0.99	4.18 ± 0.50[15–17]	3.75 ± 0.55[18–22]
Serum CHO, mg/dl	207.40 ± 33.63	223.00 ± 42.05	201.33 ± 46.21	241.72 ± 40.91[23,24]	205.68 ± 41.59[25]	231.80 ± 41.22	256.20 ± 21.18[26–29]	234.50 ± 52.10
Serum HDL, mg/dl	46.80 ± 8.54	47.28 ± 11.67	54.33 ± 12.35[30]	56.36 ± 19.96[31]	49.15 ± 13.70	43.60 ± 19.19	56.60 ± 13.04[32]	53.75 ± 19.78
Serum LDL, mg/dl	138.82 ± 28.01	149.74 ± 38.27	125.28 ± 41.19	161.41 ± 30.00[33,34]	133.03 ± 35.42[35]	164.56 ± 37.36[36]	168.76 ± 11.46[37–39]	164.10 ± 34.67
Serum VLDL, mg/dl	21.77 ± 9.64	25.97 ± 10.41	21.71 ± 6.06	23.94 ± 10.83	23.49 ± 19.86	23.64 ± 5.74	30.84 ± 19.24	16.65 ± 5.17[40]
Serum TG, mg/dl	109.40 ± 48.13	129.28 ± 52.56	108.77 ± 30.77	119.90 ± 54.09	117.36 ± 99.21	119.00 ± 28.64	154.40 ± 96.64	83.25 ± 25.26[41]
Serum NO, μM	33.81 ± 15.77	50.49 ± 26.40[42]	46.05 ± 25.49	41.01 ± 14.45	48.07 ± 19.34[43]	37.50 ± 18.60	43.92 ± 23.27	43.69 ± 7.10
Serum BAP, pg/ml	16.24 ± 3.63	16.50 ± 6.03	25.73 ± 17.79[44]	12.93 ± 6.35[45]	21.06 ± 19.46	23.68 ± 13.40[46–48]	16.96 ± 7.03	21.35 ± 9.23[49,50]
Blood HA, ng/ml	114.49 ± 82.56	95.47 ± 72.28	147.95 ± 108.67	109.87 ± 73.75	78.30 ± 51.28[51]	48.44 ± 25.34[52–54]	58.39 ± 20.95[55,56]	38.34 ± 22.47[57–59]

All values are expressed as mean ± standard deviation

[1] $p < 0.02$ vs. 3411.
[2] $p < 0.05$ vs. 3412.
[3] $p < 0.03$ vs. 3412.
[4] $p < 0.003$ vs. 3422.
[5] $p < 0.01$ vs. 3422.
[6] $p < 0.001$ vs. 3311.
[7] $p < 0.001$ vs. 3411.
[8] $p < 0.02$ vs. 3412.
[9] $p < 0.05$ vs. 3422.
[10] $p < 0.01$ vs. 3311.
[11] $p < 0.01$ vs. 3311.
[12] $p < 0.01$ vs. 3311.
[13] $p < 0.02$ vs. 3411.
[14] $p < 0.05$ vs. 3312.
[15] $p < 0.01$ vs. 3312.
[16] $p < 0.05$ vs. 3411.
[17] $p < 0.05$ vs. 3422.
[18] $p < 0.02$ vs. 3311.
[19] $p < 0.002$ vs. 3312.
[20] $p < 0.05$ vs. 3322.
[21] $p < 0.03$ vs. 3411.
[22] $p < 0.03$ vs. 3422.
[23] $p < 0.02$ vs. 3311.
[24] $p < 0.05$ vs. 3322.
[25] $p < 0.02$ vs. 3411.
[26] $p < 0.007$ vs. 3311.
[27] $p < 0.05$ vs. 3312.
[28] $p < 0.02$ vs. 3322.
[29] $p < 0.01$ vs. 3412.
[30] $p < 0.05$ vs. 3311.
[31] $p < 0.05$ vs. 3312.
[32] $p < 0.05$ vs. 3311.
[33] $p < 0.05$ vs. 3311.
[34] $p < 0.03$ vs. 3222.
[35] $p < 0.03$ vs. 3411.
[36] $p < 0.05$ vs. 3412.
[37] $p < 0.03$ vs. 3311.
[38] $p < 0.04$ vs. 3322.
[39] $p < 0.03$ vs. 3412.
[40] $p < 0.05$ vs. 3312.
[41] $p < 0.05$ vs. 3422.
[42] $p < 0.03$ vs. 3311.
[43] $p < 0.02$ vs. 3311.
[44] $p < 0.05$ vs. 3311.
[45] $p < 0.03$ vs. 3322.
[46] $p < 0.05$ vs. 3311.
[47] $p < 0.05$ vs. 3312.
[48] $p < 0.02$ vs. 3411.
[49] $p < 0.05$ vs. 3311.
[50] $p < 0.05$ vs. 3411.
[51] $p < 0.03$ vs. 3322.
[52] $p < 0.05$ vs. 3311.
[53] $p < 0.05$ vs. 3322.
[54] $p < 0.05$ vs. 3411.
[55] $p < 0.05$ vs. 3311.
[56] $p < 0.05$ vs. 3322.
[57] $p < 0.05$ vs. 3311.
[58] $p < 0.05$ vs. 3322.
[59] $p < 0.05$ vs. 3411.

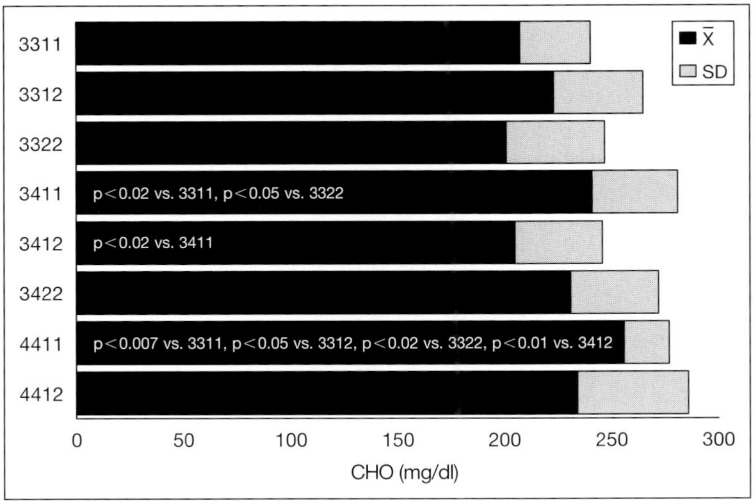

Fig. 52. Bigenic haplotype-related serum CHO levels in AD.

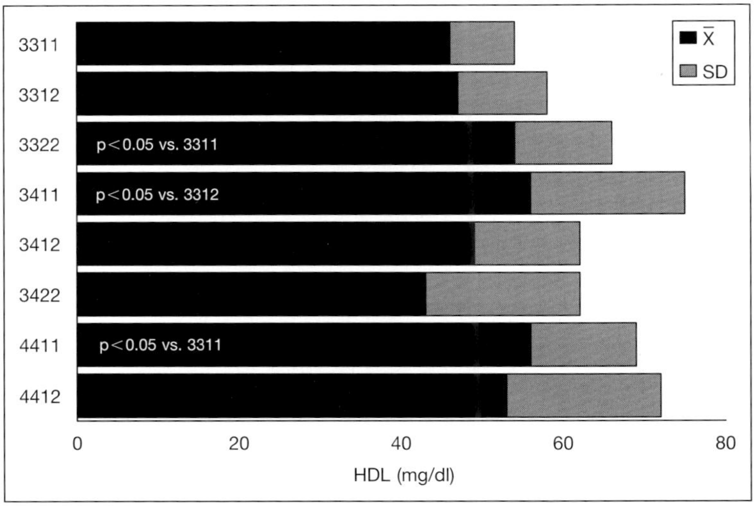

Fig. 53. Bigenic haplotype-related serum HDL levels in AD.

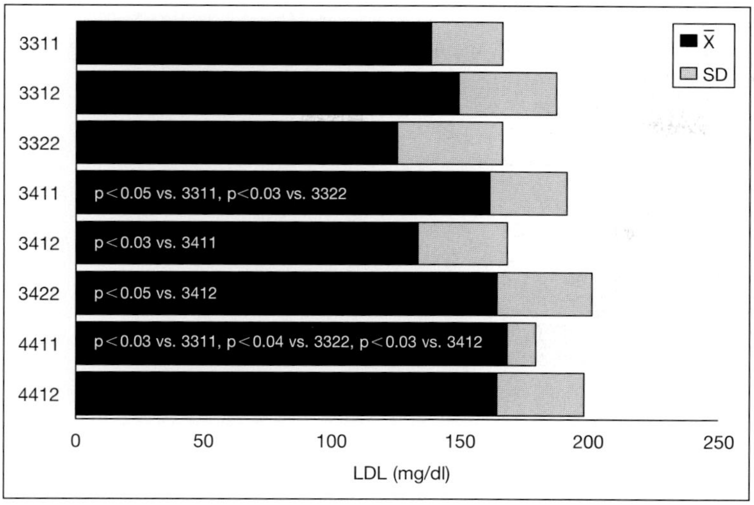

Fig. 54. Bigenic haplotype-related serum LDL levels in AD.

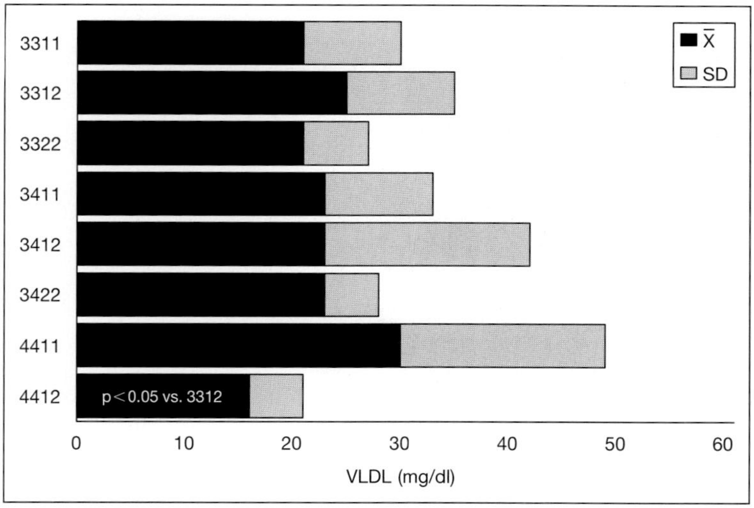

Fig. 55. Bigenic haplotype-related serum VLDL levels in AD.

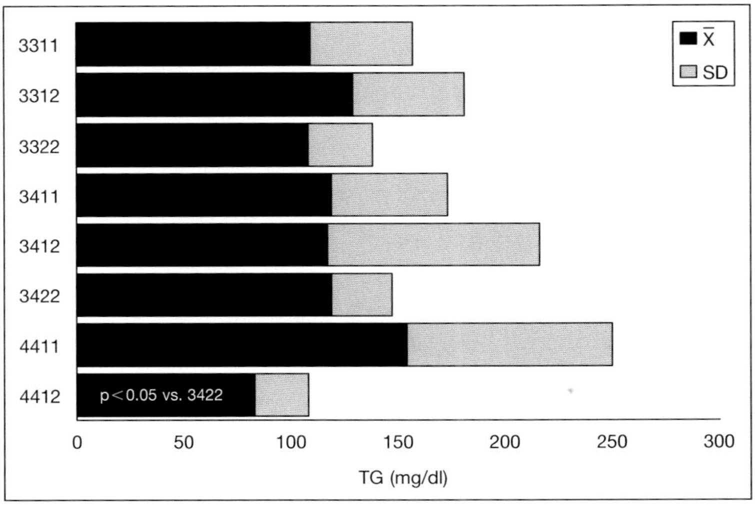

Fig. 56. Bigenic haplotype-related serum TG levels in AD.

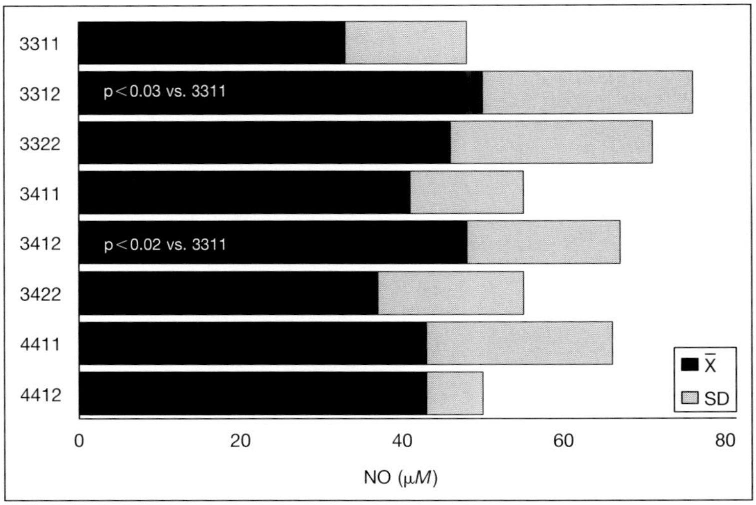

Fig. 57. Bigenic haplotype-related serum NO levels in AD.

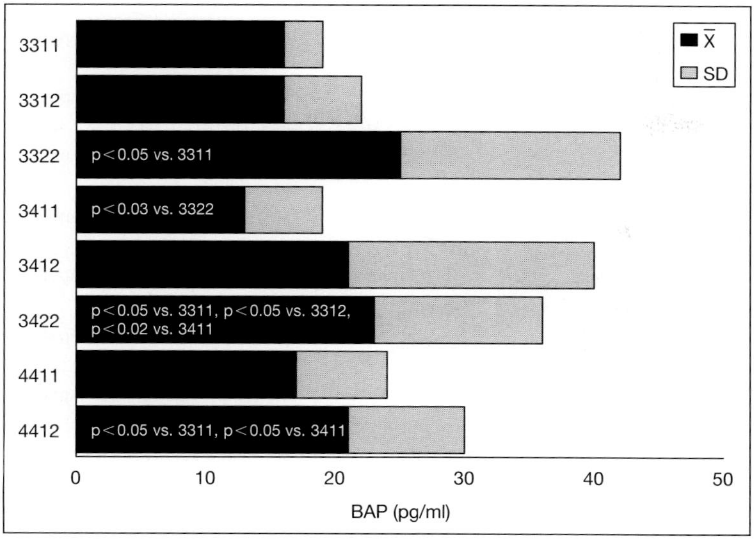

Fig. 58. Bigenic haplotype-related serum BAP levels in AD.

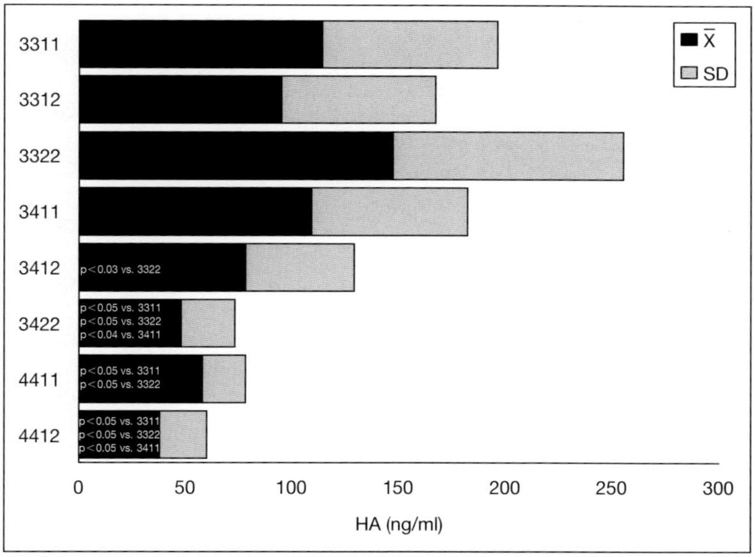

Fig. 59. Bigenic genotype-related blood HA levels in AD.

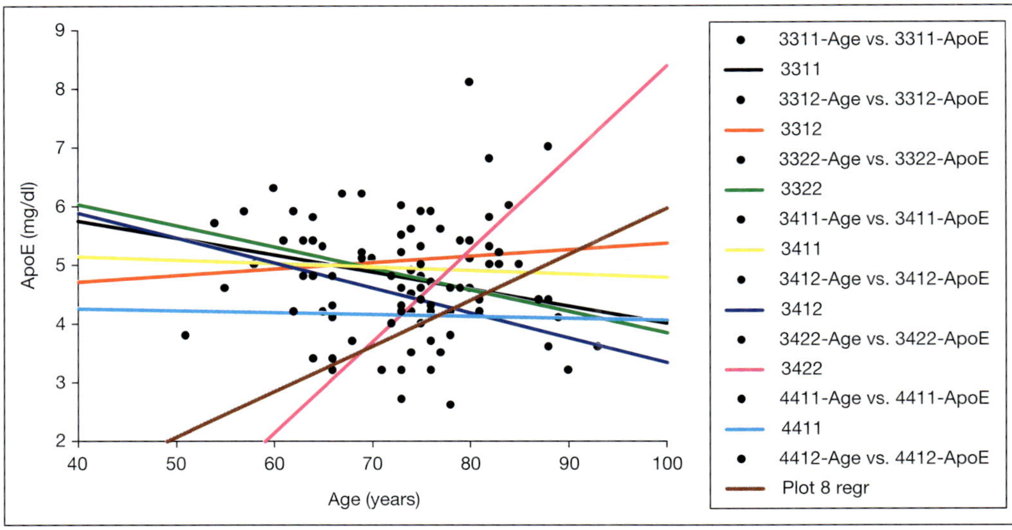

Fig. 60. Age- and bigenic haplotype-related serum ApoE levels in AD.

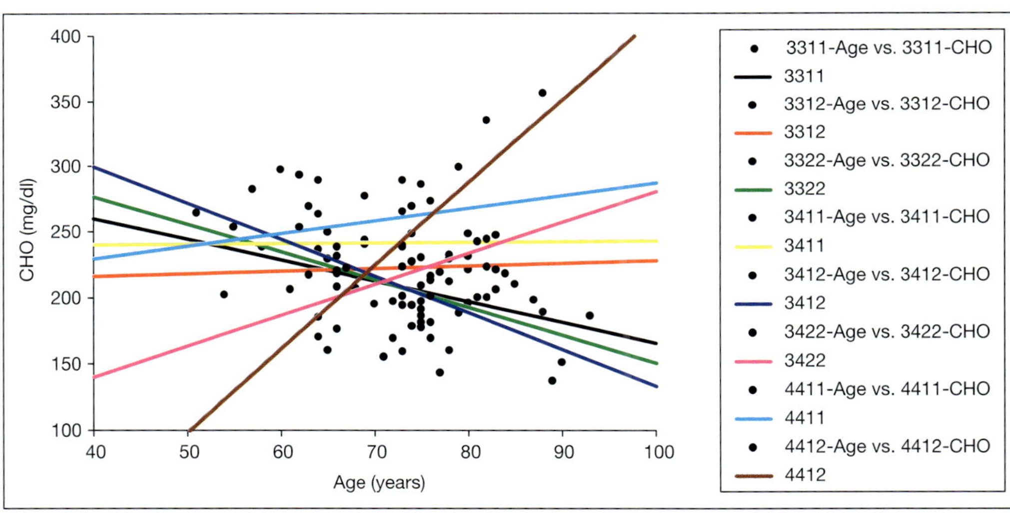

Fig. 61. Age- and bigenic haplotype-related serum CHO levels in AD.

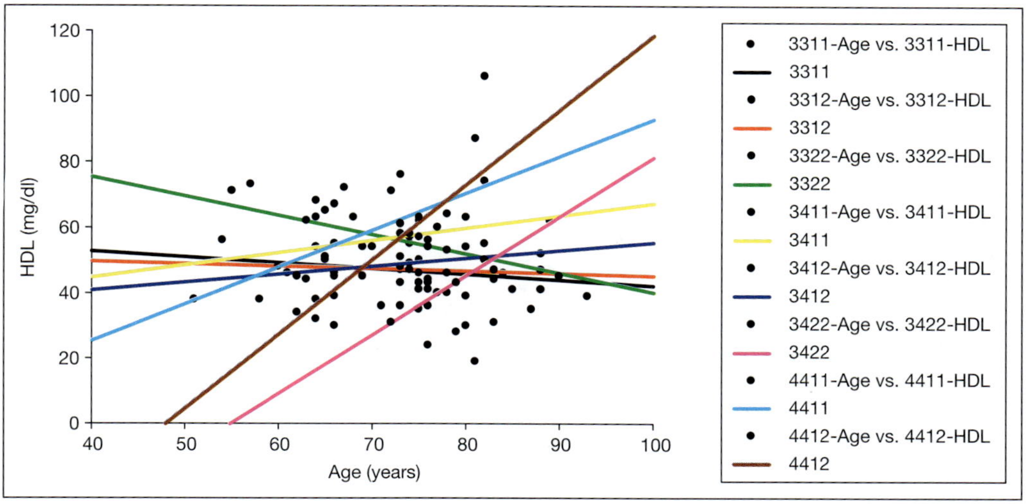

Fig. 62. Age- and bigenic haplotype-related serum HDL levels in AD.

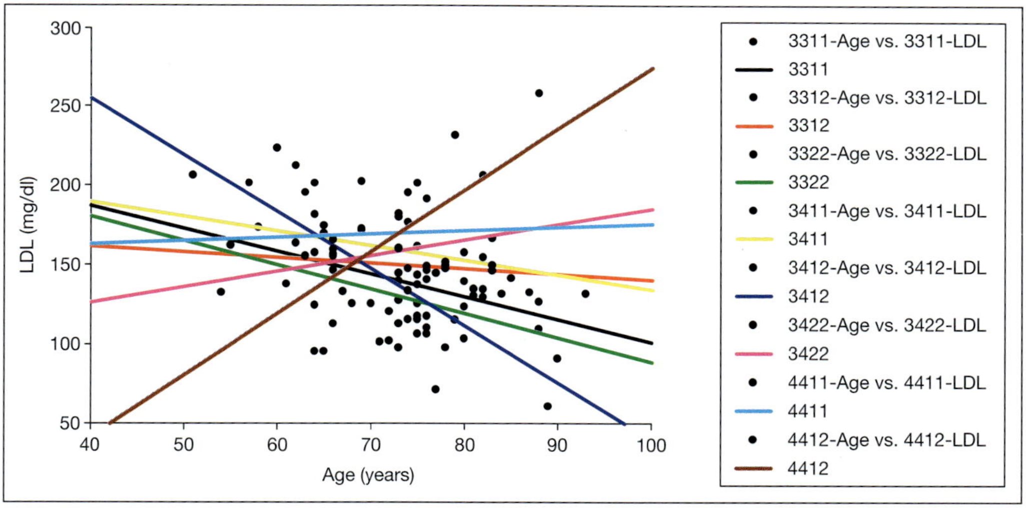

Fig. 63. Age- and bigenic haplotype-related serum LDL levels in AD.

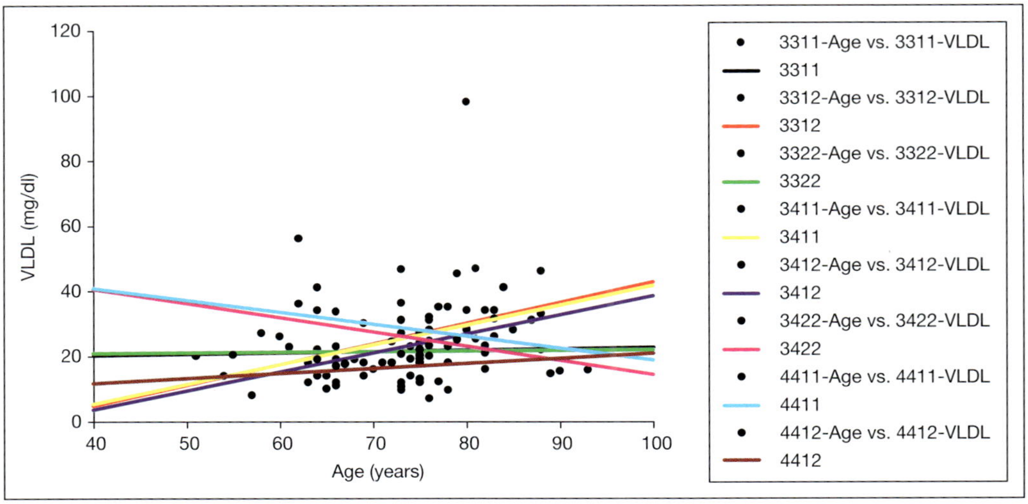

Fig. 64. Age- and bigenic haplotype-related serum VLDL levels in AD.

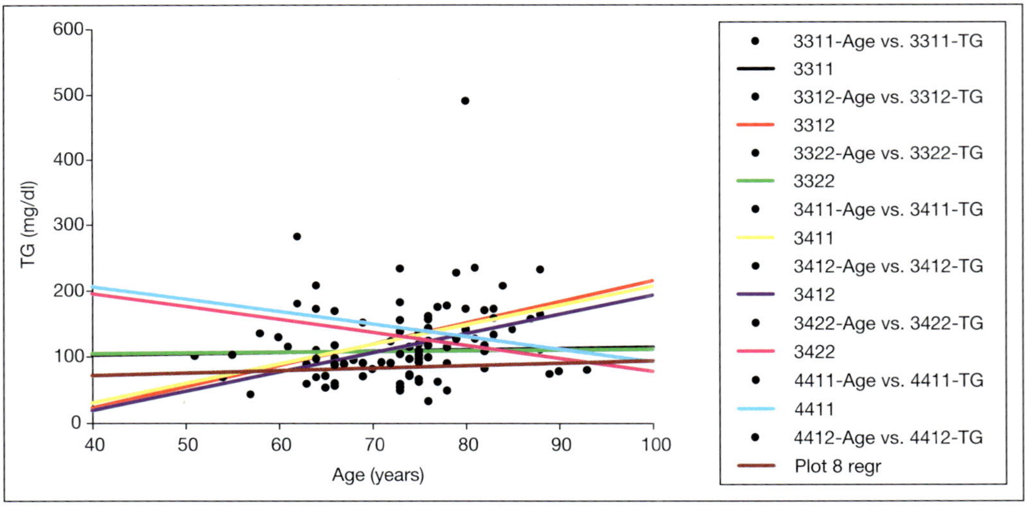

Fig. 65. Age- and bigenic haplotype-related serum TG in AD.

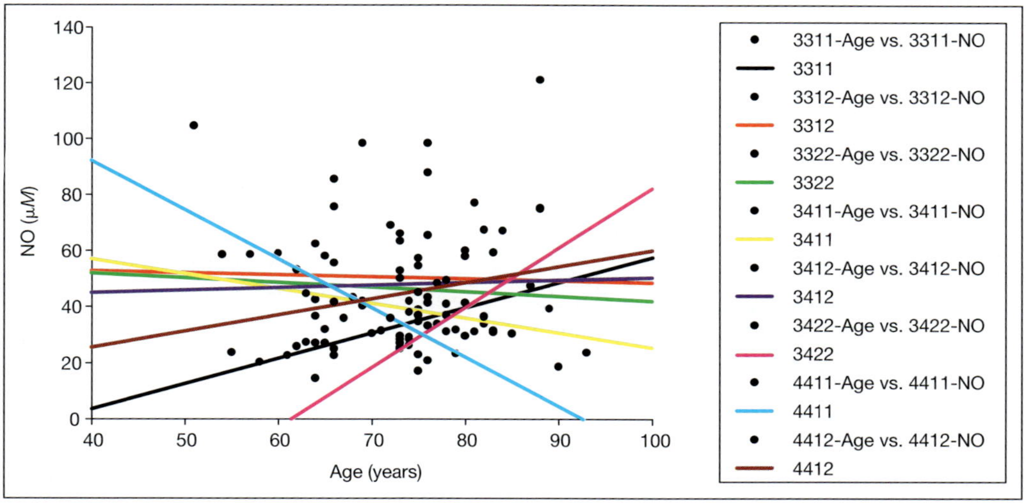

Fig. 66. Age- and bigenic haplotype-related serum NO levels in AD.

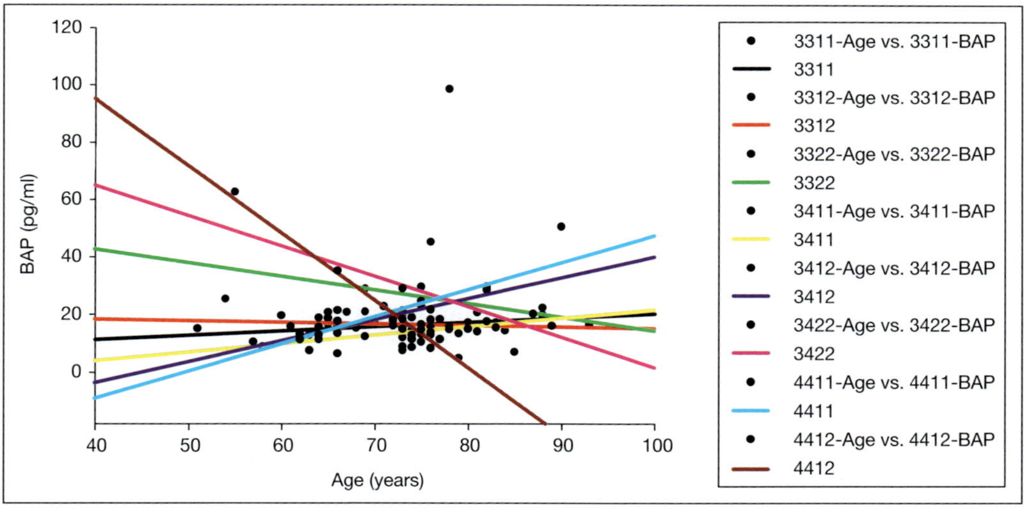

Fig. 67. Age- and bigenic haplotype-related serum BAP levels in AD.

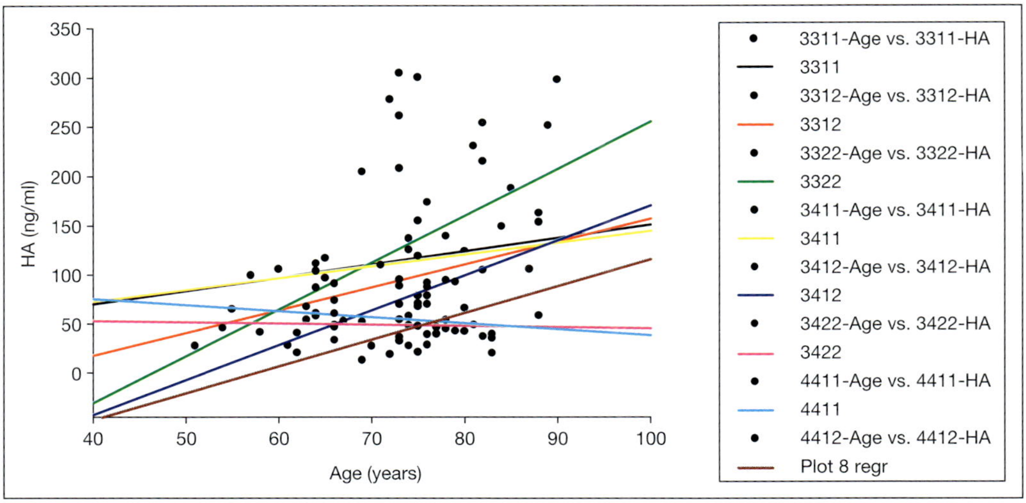

Fig. 68. Age- and bigenic haplotype-related blood HA levels in AD.

Discussion

In this study, we attempted to identify the influence of age, gender, cognition, and genetic factors on several biological parameters which might be altered in AD or that are recognized to be involved in the pathogenesis of AD. The strategy we have followed was to quantify basic parameters in randomly selected patients with AD and then to evaluate the parametric values as dependent variables of age, sex, cognitive performance, APOE-4+ vs. APOE-4−, APOE genotypes, and bigenic haplotypes integrating the allelic variants of the APOE and PS1 genes in single clusters. This strategy allows to correlate biological parameters with determinants of aging or neurodegeneration in a straightforward manner as well as to simplify a functional genomics approach to AD by using bigenic clusters of genes associated with AD as single haplotypes to genetically homogenize the AD sample [4–6]. The biological parameters we decided to evaluate were serum ApoE, CHO, HDL, LDL, VLDL, TG, NO, BAP and HA, since most of these biochemical factors are associated with either primary pathogenic events in AD (i.e., BAP, ApoE) [7–9], risk factors for cerebrovascular disorders (i.e., lipids) [12], regulators of immune function and inflammation (i.e., HA) [23] or regulators of endothelial function and vascular reactivity (i.e., NO) [12, 23].

Age-Related Variation

Although age is currently implicated as a risk factor for dementia from an epidemiological perspective, our data clearly indicate that age does not influence the cognitive decline at all in AD (fig. 1, 2). At the same time, both the MMSE and the ADAS-Cog scales show age-related superimposed values, indicating a similar accuracy for this purpose (fig. 1, 2). It is also clear that age does not influence serum ApoE, NO, BAP, and HA levels, although a small tendency to increase with age is seen in NO, BAP, and HA levels (fig. 4–6). In contrast, most blood lipids increase with age in AD probably contributing to atheromatosis, artery obliteration and cerebrovascular dysfunction [12–14].

Cognitive Performance-Related Variation

Changes in cognitive function reflected by fluctuation and/or progressive decline in AD do not correlate with serum ApoE, CHO, HDL, LDL, VLDL, TG, BAP and HA levels and vice versa. In contrast, serum NO levels increase in parallel with worsening in cognitive performance independently from age or other factors (fig. 9). Serum NO levels differ in AD and controls [12]. Both populations show a divergent age-dependent profile, with increasing levels in controls, and no change or progressive decrease in AD [12]. Dysregulation in NO synthesis and release has been postulated as a pathogenic factor in AD-related cerebrovascular dysfunction [12, 24, 25].

Sex-Related Variation

It appears that gender can influence cognitive performance and biological parameters in AD (table 4). Our results indicate that cognition seems to be poorer in females than in males (table 2) and that this cognitive pattern is unrelated to age (fig. 12, 13). The biological profile of males also differs from that of females in age-related lower levels of ApoE, CHO, LDL, and NO, without any difference in BAP. Hormonal differences and/or physical exercise might account for this gender-related variation in AD. However, the higher levels of lipids in females might also be responsible for increased cerebrovascular dysfunction and poorer mental performance [12].

APOE-4-Related Variation

The differentiation between APOE-4+ and APOE-4− patients does not yield any informative advantage, indicating that the APOE-4 allele load isolated does not determine a phenotypic profile reflected in peripheral biological markers. The only exception to this tentative rule is serum HA levels which appear to be markedly diminished in APOE-4 carriers (table 2; fig. 33). This is a very striking finding since HA levels are currently augmented in serum [26], CSF [27], and brain tissue [23, 28–31] of AD patients as compared to controls [23, 27, 29, 30]. A dysregulation in brain cytokine-HA networks has been invoked as a potential mechanism for neuroinflammation in AD [31].

CHO and LDL levels tend to show a sharper age-related decline in APOE-4+ (fig. 26) with no apparent age-related differences in HDL, VLDL, and TG levels. Although this phenotypic pattern seems to be paradoxical due to the CHO carrier function of ApoE, it is interesting to note that ApoE levels are not modified by age in APOE-4+ (fig. 25) whereas both CHO and LDL levels decrease with age in this genetic cluster (fig. 26, 28). If this is true, the administration of statins or other cholesterol-lowering compounds to AD patients may not be a reasonable decision at the present time [32]. Furthermore, serum NO levels follow a declining age-related pattern similar to CHO and LDL, probably indicating mechanistic interactions at the endothelial level between NO and circulating lipids. Moreover, the age-related lowering profile of CHO in APOE-4+ contrasts with the progressive increase in serum BAP in APOE-4 carriers (fig. 32). In conclusion, with the exception of HA, which shows a clear differential profile either in total serum levels or in APOE-4 carriers, most APOE-4-related changes in biological parameters seem to be age-dependent rather than APOE-4-dependent, leaving the APOE-4+ vs. APOE-4− strategy as of poor utility in defining phenotypic profiles.

APOE Genotype-Related Variation

Clear differences have been identified in most biological markers according to the APOE genotype in AD patients (table 3). It is interesting to reconfirm, as previously reported [33, 34], that APOE-4/4 carriers develop AD earlier than other genotype carriers (table 3). It has also been indicated that the accumulation of genetic defects in AD patients anticipates the onset of the disease by several years and that the earlier onset is more frequent in patients with the APOE-4/4 genotype [1, 6, 33, 34]. In our sample, no major differences in cognitive performance have been found (table 3; fig. 34). However, in previous studies we have demonstrated that patients with the APOE-4/4 genotype deteriorate faster than patients with other APOE genotypes [4, 5, 15, 16]. Our results clearly show marked differences in ApoE (fig. 35), CHO (fig. 36), LDL (fig. 38), NO (fig. 39), BAP (fig. 40), and HA levels (fig. 41) in APOE-4/4 carriers (table 3). Low levels of serum ApoE, NO, BAP, and HA parallel increased levels of CHO and LDL in APOE-4/4, suggesting a potential interplay of all these biochemical factors in atheromatosis, cerebrovascular alterations, lipid dysfunction, and immune dysregulation potentially leading to neuronal damage [12, 35]. As individual factors or as average values in AD cohorts vs. controls, serum ApoE, NO and BAP levels are noninformative parameters for AD. Similarly, serum BAP levels do not change with age, cognitive performance or gender (tables 1, 2). However, when we differentiate AD patients according to their APOE genotype, a clear APOE-dependent phenotypic profile emerges, indicating that ApoE, CHO, LDL, NO, BAP, and HA are potential players in the pathogenesis of AD associated with APOE and other genetic factors and partially related to age and environmental factors. ApoE levels are lower in APOE-4/4 and APOE-3/4 than in the other genotypes and do not change with age (fig. 42). CHO, HDL and LDL levels are higher in APOE-4/4 and progressively increase with age in APOE-4/4, decreasing with age in APOE-3/4 and APOE-3/3 (fig. 43–45). In contrast, VLDL and TG levels, which show similar values among the different APOE genotypes, tend to decrease with age in APOE-4/4 and increase with age in APOE-3/4 and APOE-3/3 (fig. 46, 47). NO levels do not vary with age in APOE-3/3 and APOE-3/4 and progressively decrease with age in the APOE-4/4 homozygous (fig. 48). BAP levels do not change with age and are significantly lower in APOE-4 carriers (table 3). HA levels in APOE-4/4 and APOE-3/4 remain stable with age and only APOE-3/3 carriers show an age-related increase (fig. 50). Since APOE-3/3 carriers roughly account for 50% of the AD population [1], it is very likely that the high levels of serum HA present in this APOE cluster are responsible for the increased levels of HA found in AD patients when all the data are pooled together [23, 29, 30]. In conclusion, although APOE genotyping alone is not enough for diagnostic purposes, in the light of these findings it seems plausible that the differentiation of AD patients in APOE clusters is very helpful as a functional genomics strategy and as a discriminating aid in pharmacogenomics. In fact, it has been convincingly demonstrated that the presence of the APOE-4 allele in AD influences disease onset, brain atrophy, cerebrovascular hemodynamics, blood pressure, amyloid deposition, ApoE secretion, lipid metabolism, brain bioelectrical activity, cognition, apoptosis and the response to conventional drugs [4–6, 15–18].

Bigenic Haplotype-Related Variation

The construction of bigenic, trigenic, tetragenic, pentagenic and polygenic haplotypes integrating the allelic variants of major AD-related genes has been used to perform studies of structural genomics [18], functional genomics [6, 12] and pharmacogenomics in AD [4–6, 15]. In this study we tried to identify the potential influence of both APOE- and PS1-related products on several biological parameters and also to differentiate the effect of PS1 from APOE on serum ApoE, lipids, NO, BAP, and HA levels. Of the possible 18 haplotypes constructed as the result of integrating the allelic variants of the APOE and PS1 genes, we have selected only 8 clusters with more than 4–5 patients per cluster (table 4). Basically, the analyzed clusters include bigenic haplotypes resulting from the integration of the APOE-3/3, APOE-3/4, APOE-4/4 genotype variants with the PS1–1/1, PS1–1/2, and PS1–2/2 variants (table 4; fig. 51–59).

The 4411 and 4412 haplotypes show a significant reduction in ApoE levels as compared with the other haplotypes (fig. 51), indicating that in this case APOE-4/4 directly influences ApoE production because the phenotypic profile of ApoE in figure 35 does not differ from that in figure 51. A very similar profile is seen in the levels of lipids (fig. 52–56) when we compare APOE-related phenotypes (fig. 36–38) with bigenic haplotype-related phenotypes (fig. 52–56). According to these results, it appears that lipid metabolism is directly influenced by APOE-related factors with a minor involvement of PS1-related factors. There is also a very striking parallelism between CHO and LDL levels in the bigenic haplotypes (fig. 52, 54, respectively) and in the APOE-related genotypes (fig. 36, 38, respectively). Furthermore, despite phenotypic similarities, bigenic haplotyping increases the accuracy in differentiating lipid levels in a genotype-dependent manner.

NO levels are significantly lower in APOE-3/3, APOE-3/4 and APOE-4/4 with no major differences among these genotypes (fig. 39); however, bigenic haplotypes show that patients with the 3312 and 3412 clusters exhibit a higher level of peripheral NO, probably indicating that PS1–1/2-related factors might influence NO secretion (fig. 57). The results of both BAP and HA are clear-cut and very striking since both substances display an antagonistic phenotype in which it is very likely that both APOE- and PS1-related factors are involved. BAP levels are very similar in APOE-3/3, APOE-3/4 and APOE-4/4 carriers (fig. 40); however, in the bigenic haplotype-associated phenotype, the highest levels of BAP are present in those patients who carry a PS1–2/2 or PS1–1/2 genotype completely unrelated to the presence of heterozygous or homozygous APOE-4 (fig. 58), indicating that peripheral BAP levels are influenced by PS1-related factors associated with byproducts of PS1 genes in which the 2/2 or 1/2 allelic variants are present, but not the PS1–1/1 genotype (no major PS1 mutations have been identified in these clusters). On the contrary, it appears that blood HA levels are mainly associated with the presence of the APOE-4 allele irrespective of the PS1-related variants (tables 2–4; fig. 41, 59). It is also important to mention that most of the biological markers associated with bigenic haplotypes do not show major age-related changes with the exception of lipids whose serum levels can also depend upon nutritional variables and physical exercise. These results invite to conclude that the biological phenotypic variants of AD are primarily associated with complex genetic factors determining neuronal susceptibility to premature functional derailment and neurodegeneration, and that age-related cerebrovascular risk factors may also contribute to accelerate neuronal death and dementia as the result of genomic-environmental interactions.

Conclusions

1. Age does not influence cognition in AD (fig. 1, 2).
2. Serum ApoE levels do not show any change with age in AD (fig. 2).

3. Serum NO, BAP, and HA levels tend to show an age-dependent nonsignificant increase in AD (fig. 4–6).
4. CHO, LDL, VLDL, and TG levels tend to increase with age in AD (fig. 3).
5. Changes in cognitive function do not influence any variation in serum biochemical parameters (ApoE, CHO, HDL, LDL, VLDL, TG, BAP, HA), and only NO levels increase in parallel with cognitive deterioration in AD (fig. 9).
6. Sex-related differences may influence cognitive performance and lipid levels in AD (table 1; fig. 15–19).
7. No major differences in biological parameters and cognition are seen between APOE-4+ and APOE-4− AD patients, except in serum HA levels which are markedly diminished in APOE-4 carriers (table 2; fig. 33).
8. There is a parallel age-related decline in CHO, LDL, and NO levels in APOE-4 carriers which is antagonistic to the progressive increase in BAP levels with age in this genetic cluster.
9. Different APOE genotypes clearly distinguish biological phenotypes which are age independent (table 3).
10. APOE-4/4 carriers show a marked phenotypic profile of risk reflected by significant changes in most biological parameters, including ApoE (fig. 35), lipids (fig. 36–38), NO (fig. 39), BAP (fig. 40) and HA (fig. 41). APOE-related alterations in these biochemical factors in association with other genetic and environmental factors may accumulate with age, inducing a progressive deterioration in cerebrovascular function and brain perfusion this leading to neuronal damage and premature neurodegeneration in patients carrying a haplotype of risk.
11. Bigenic haplotyping integrating APOE+PS1 allelic variants allows the identification of genetic clusters whose biological phenotype is potentially influenced by APOE- and/or PS1-related byproducts independently or in association (table 4).
12. Patients with the 4411 and 4412 haplotypes show the lowest ApoE levels (table 4; fig. 51).
13. Patients with the 3411, 3412, 3422, 4411, and 4412 haplotypes show the highest levels of CHO and LDL in serum with a striking parallelism in both phenotypes (table 4; fig. 52, 54).
14. Only the 4412 haplotype exhibits a marked decrease in both VLDL and TG (table 4; fig. 55, 56).
15. The 3312 and 3412 haplotypes differentially show higher levels of serum NO probably indicating the involvement of PS1-1/2-related factors in peripheral NO secretion (table 4; fig. 57).
16. PS1-2/2- and PS1-1/2-related factors, irrespective of the APOE-4 allele, might influence peripheral BAP secretion since the higher levels of serum

BAP are present in patients with the 3322 and 3422 haplotypes (table 4; fig. 58).

17. Serum HA levels seem to be directly influenced by the presence of the APOE-4 allele since only the 3412, 3422, 4411, and 4412 haplotypes exhibit a marked decrease in the concentration of serum HA (table 4; fig. 59).

Our main conclusions from the present results are that: (a) genetic factors are more relevant than age to AD; (b) most biological markers reflecting a phenotypic profile of AD are primarily genotype dependent and some of them vary with age, showing an age-related progression; (c) cognitive decline is not determinant in inducing major changes in biological markers; (d) changes in most biological markers are the phenotypic expression of genetic variation rather than age or disease progression; (e) APOE-4/4 carriers and, to a lesser extent, APOE-3/4 carriers are by large a population at risk with a well-defined phenotypic profile reflected by striking changes in biological markers, and (f) the construction of bigenic haplotypes integrating APOE+PS1 allelic variants can help to further differentiate AD-related genetic clusters for functional genomics and pharmacogenomics studies.

References

1 Cacabelos R: Handbook of Neurogeriatrics: Alzheimer Disease and Other Dementias. Epidemiology and Genetics. Masson, Barcelona, 1999.
2 Cacabelos R: Psychogeriatric research. A conceptual introduction to geriatric neuroscience. Psychogeriatrics 2001;1:158–188.
3 Saunders AM: Gene identification in Alzheimer's disease. Pharmacogenomics 2001;2:239–249.
4 Cacabelos R: Pharmacogenomics for the treatment of dementia. Ann Med 2002;34:357–379.
5 Cacabelos R: Pharmacogenomics in Alzheimer's disease. Min Rev Med Chem 2002;2:59–84.
6 Cacabelos R: The application of functional genomics to Alzheimer's disease. Pharmacogenomics 2003;4:597–621.
7 Selkoe DJ: Alzheimer's disease: Genes, proteins, and therapy. Physiol Rev 2001;81:741–766.
8 Selkoe D, Kopan R: Notch and presenilin: Regulated intrammebrane proteolysis links development and degeneration. Annu Rev Neurosci 2003;26:565–597.
9 Bonini NM, Fortini ME: Human neurodegenerative disease modeling using Drosophila. Annu Rev Neurosci 2003;26:627–656.
10 Phinney A, Horne P, Yang J, Janus C, Bergeron C, Westaway D: Mouse models of Alzheimer's disease: The long and filamentous road. Neurol Res 2003;25:590–600.
11 Mayeux R: Epidemiology of neurodegeneration. Annu Rev Neurosci 2003;26:81–104.
12 Cacabelos R, Fernández-Novoa L, Lombardi V, Corzo L, Pichel V, Kubota Y: Cerebrovascular risk factors in Alzheimer's disease: Brain hemodynamics and pharmacogenomic implications. Neurol Res 2003;25:567–580.
13 Kalaria RN: Comparison between Alzheimer's disease and vascular dementia: Implications for treatment. Neurol Res 2003;25:661–664.
14 Blass JP: Cerebrometabolic abnormalities in Alzheimer's disease. Neurol Res 2003;25:556–566.
15 Cacabelos R, Alvarez A, Fernández-Novoa L, Lombardi V: A pharmacogenomic approach to Alzheimer's disease. Acta Neurol Scand Suppl 2000;176:12–19.

16 Cacabelos R: Pharmacogenomics in Alzheimer's disease. Drug News Perspect 2000;13:252–254.
17 Cacabelos R, Alvarez XA, Lombardi V, Fernández-Novoa L, Corzo L, Pérez P, et al: Pharmacological treatment of Alzheimer disease: From psychotropic drugs and cholinesterase inhibitors to pharmacogenomics. Drugs Today 2000;36:415–499.
18 Cacabelos R, Fernández-Novoa L, Lombardi V, Takeda M: Genetic variation and pharmacogenomics in Alzheimer disease. Psychiat Neurol Jap 2003;105:47–67.
19 Cacabelos R: Dementia; in Jobe TH, Gaviria M, Kovilparambil A (eds): Clinical Psychiatry. Oxford, Blackwell, 1997, pp 73–122.
20 Cacabelos R: Diagnosis of Alzheimer's disease: Defining genetic profiles (genotype vs. phenotype). Acta Neurol Scand Suppl 1996;165:72–84.
21 Beyer K, Lao JI, Alvarez XA, Cacabelos R: A general method for DNA polymorphism identification in genetic assessment and molecular diagnosis. Methods Find Exp Clin Pharmacol 1997; 19:87–91.
22 Beyer K, Lao JI, Alvarez XA, Cacabelos R: Different implications of APOE-4 in Alzheimer's disease and vascular dementia in the Spanish population. Alzheimer's Res 1996;2:215–220.
23 Fernández-Novoa L, Cacabelos R: Histamine function in brain disorders. Behav Brain Res 2001; 124:213–233.
24 De la Torre JC: Alzheimer disease as a vascular disorder. Stroke 2002;33:1152–1162.
25 De la Torre JC, Stefano GB: Evidence that Alzheimer's disease is a microvascular disorder: The role of constitutive nitric oxide. Brain Res Rev 2000;34:119–136.
26 Cacabelos R, Fernández-Novoa L, Pérez-Trullén JM, Franco-Maside A, Alvarez XA: Serum histamine in Alzheimer's disease and multi-infarct dementia. Methods Find Exp Clin Pharmacol 1992;14:711–715.
27 Cacabelos R: Histaminergic system: Neuroendocrine function of brain histamine. Methods Find Exp Clin Pharmacol 1990;12:341–376.
28 Cacabelos R, Yamatodani A, Niigawa H, Hariguchi S, Tada K, Nishimura T, Wada H, Brandeis L, Pearson J: Brain histamine in Alzheimer's disease. Methods Find Exp Clin Pharmacol 1989; 11:353–360.
29 Fernández-Novoa L, Cacabelos R: The histaminergic system in Alzheimer's disease. Ann Psychiat 1995;5:127–158.
30 Cacabelos R, Fernández-Novoa L, Franco-Maside A, Alvarez XA: Neuroimmune function of brain histamine: Implications for neurotrophic activity and neurotoxicity. Ann Psychiat 1992;3: 147–200.
31 Cacabelos R: The histamine-citokine network in Alzheimer disease: Etiopathogenic and pharmacogenomic implications. Mapping the Progress of Alzheimer's and Parkinson's Disease. Adv Behav Biol 2002;51:59–64.
32 Rockwood K, Darvesh S: The risk of dementia in relation to statins and other lipid-lowering agents. Neurol Res 2003;25:601–604.
33 Cacabelos R, Beyer K, Lao JI, Mesa MD, Fernández-Novoa L: Association of genetic risk factors in Alzheimer's disease and a novel mutation in the predicted TM2 domain of presenilin-2 in late-onset AD; in Iqbal K, Swaab DF, Winblad B, Wisniewski HM (eds): Alzheimer's Disease and Related Disorders. Chichester, Wiley, 1999, pp 93–102.
34 Mesa MD, Fernández-Novoa L, Cacabelos R: Association of genetic risk factors in Alzheimer's disease. Ann Psychiat 1999;7:35–45.
35 Channey MO, Baundy J, Esh C, Childress J, Luehrs D, Kokjohn TA, Roher AE: Aβ, aging, and Alzheimer's disease: A tale, models, and hypotheses. Neurol Res 2003;25:581–589.

Prof. Dr. Ramón Cacabelos
EuroEspes Biomedical Research Center
Institute for CNS Disorders
ES–15166 Bergondo, Coruña (Spain)
Tel. +34 981 780505, Fax +34 981 780511, E-Mail cacabelos@euroespes.com

Takeda M, Tanaka T, Cacabelos R (eds): Molecular Neurobiology of Alzheimer Disease and Related Disorders. Basel, Karger, 2004, pp 286–288

Epilogue

It is an honor for me to write this epilogue to the book prepared by Professor Masatoshi Takeda and his group after the excellent Meeting held in Osaka at the Sun Palace Hotel in October 5–6, 2002, under the title of 'Molecular Neurobiology of Alzheimer Disease and Related Disorders'. Many colleagues from different countries responded to the invitation to attend the meeting, one more among the many scientific events that are organized every year all over the world, with the difference that Japan is always an attractive country to be visited and that the organizer is a leading scientist in the field of dementia and chairman of one of the Japanese Departments of Psychiatry with an extensive tradition in research on Alzheimer disease (AD).

From the early times of Professor Kaneko to the productive leadership of Professor Nishimura in the 1970s and 1980s, the Department of Psychiatry at Osaka University Medical School underwent a progressive transformation, adapting educational programs and research projects to the specific demands of the Japanese society and the rapidly evolving challenges of scientific progress. I had the privilege of joining the Department of Psychiatry in 1982, which was then chaired by Professor Nishimura, when Professor Takeda – then a PhD student – was initiating his successful career as a bright scientist and psychiatrist. I spent a beautiful time of my life (almost a decade) in the Department of Psychiatry at Osaka University Hospital, when the University Hospital and the Faculty of Medicine were separated by the calmed waters of the Yodogawa river as the physical frontier between Fukushima-ku and Nakanoshima-ku. During those days, I eagerly absorbed the flavor of science transmitted through the walls of the old buildings, the historical reservoirs of the scientific knowledge and scholarly mind of the Japanese colleagues with whom I shared years of fruitful studies and in whom I have discovered the deepest values of the Japanese mentality. This is my psychological debt to Professor Takeda and other Japanese colleagues who influenced my Western personality, transforming me into a quasi-Japanese *gaijin* brother, of which I feel very proud.

Besides the fruitful human experience in Japan during the 1980s, when the Japanese economy was in an almost perfect shape, and my baldness was still to appear, I could also realize how education and correct political decisions can distinguish a developed country. I remember that in the late 1970s, *Koseisho* (the Japanese Ministry of Health and Welfare) had launched a national program to investigate untreatable diseases, including AD and Creutzfeldt-Jakob disease, almost 20 years before prion disorders became a popular topic in Western countries and mad cow's disease burst into our lives as an international threat with unpredictable consequences. During the past 50 years, Japanese neuropathologists have importantly contributed to our understanding of the spongiform encephalopathy, but many reports from Japanese scientists were never translated into other languages. The Department of Psychiatry at Osaka University also participated in this research, and at that time a modest brain bank multidisciplinary was already available. Group cooperation was always cultivated; regular seminars and lectures were organized with the participation of staff members and invited speakers from other Japanese and foreign institutions, including neuropsychologists, pathologists, pharmacologists, biologists and scientists from many other disciplines. At the same time, the progression of the Department in research on AD was always outstanding, with important contributions in different areas ranging from aluminum studies, to β-amyloid, tau protein, neuropeptides, biogenic amines, behavior, molecular biology, oxidative stress, genetics and proteomics for the past 20 years. The Department was also involved in the development of different therapeutic strategies for AD, including synthetic neuropeptides and cholinesterase inhibitors as well as novel strategies with newer compounds. In addition, the Department has always been a reference institution for specialized training in psychiatry and clinical care of patients with mental disorders.

Basic and clinical neurosciences, in which psychiatry and neurology are immersed, are constantly faced with new challenges to cope with the socioeconomic and health problems posed by an aging population. The remodeling of universities, hospitals, and departments is under way worldwide, and new models of funding for both research and clinical care have to be developed due to the economic crisis of the old national security systems that are still operative in many countries reluctant to change under political pressure. Japan is no exception to this phenomenon, and the remodeling process will affect structural, functional, and personal aspects of the academic life. Research on AD is pursued in many different areas of knowledge: epidemiology, social science, neuropsychology, molecular biology, genetics, neuroimaging, neuropharmacology, functional genomics, proteomics, pharmacogenomics, and health care in nursing homes, specialized centers, day care facilities and home care programs. The 'alzheimerization' of our society is like a Damocles' sword, with conflicts of interest in many different sectors and limited resources to satisfy the needs for the care of patients and the implementation of competitive research programs with immediate benefit on patient's health and welfare. The pharmacological treatment of AD is still far from successful, and the delay (and resistance) in introducing predictive markers in AD does not help to postulate preventive strategies aimed at delaying the onset of the disease or to stop neuronal death in the population at risk prior to the development of the first clinical symptoms. Therefore, it appears that there still is a long way ahead to overcome AD as a medical, social, familial and economic condition.

As the Spanish writer and Jesuit Baltasar Gracián (1601–1658) pointed out long ago: 'self-reflection is the school of wisdom'. In this regard, I wish the best to Professor Takeda and his team in their daily effort to understand the mechanisms responsible for neurodegeneration in dementia. I would also like to thank the friend behind the scientist to whom life capriciously brings sad moments to test his wisdom. Hermann Hesse (1877–1962) had written in *Siddharta* that 'knowledge can be communicated but not wisdom'. That was probably one of the reasons why Oliver Wender Holmes (1809–1894) said that 'it is the province of knowledge to speak and it is the privilege of wisdom to listen'.

Prof. Ramón Cacabelos, MD, PhD, DMSc
Madrid

Author Index

Subject Index